The Management of Enclosed and Domesticated Deer

John Fletcher
Editor

The Management of Enclosed and Domesticated Deer

International Husbandry Systems and Diseases

 Springer

Editor
John Fletcher
Reediehill Deer Farm
Auchtermuchty, UK

ISBN 978-3-031-05388-7   ISBN 978-3-031-05386-3   (eBook)
https://doi.org/10.1007/978-3-031-05386-3

This Springer imprint is published by the registered company Springer Nature Switzerland AG
The registered company address is: Gewerbestrasse 11, 6330 Cham, Switzerland

Introduction and General Biology of Deer

This book is devoted to contributions from those who work with enclosed deer. The exclusion of the far more numerous unconfined wild deer will strike many as a rather artificial distinction. Yet that distinction is fundamental. Once humans have enclosed a species, they can exert some degree of control over its breeding. For thousands of years, people have kept deer within enclosures (Fletcher 2011). Solely by the exercise of selective culling within parks, it may sometimes have been possible to effect genetic change but this is as nothing compared to the rapid changes in genotype which can be effected on farms. In English and other European enclosures, selection of male red deer from as early as the seventeenth, and especially during the nineteenth century, created animals with larger antlers which were more highly valued for their trophies. However, apart from that, very little genetic change has occurred during the centuries in which deer have been enclosed.

It was only recently by developing techniques for handling deer, recording their performance, and through the ability to manipulate the structure of mating groups that significant selective breeding and genetic change could be achieved. Once the techniques of artificial breeding using semen collection, freezing, and insemination, as well as manipulation of oocytes and ova and in vitro fertilization, become available then extremely rapid genetic change can be achieved and the resulting animals can be described as domesticated.

Most accepted definitions of a domestic animal require that it has been subjected to selective breeding by people causing genetic changes distinct from those caused by natural selection. This was understood by Charles Darwin and it is this distinction which is generally accepted as defining domestication (Darwin 1868). These deer are therefore domesticated and can be said to be farmed.

The first animal to be domesticated was probably the wolf, resulting in domestic dogs about 15,000 years ago, whilst our livestock, the sheep, goats, cattle, and pigs, appeared around ten or eleven thousand years ago. Since then, few new livestock have appeared. However, the large-scale farming of red deer in New Zealand and elsewhere during the second half of the twentieth century has created a new domestic animal. Arguably, therefore, farmed red deer are the first new large mammal domesticate to have been produced during the last ten thousand years.

It is commonly suggested that it was through the domestication of animals and crops that human populations were able to make massive changes from a mobile hunter-gatherer existence to a more sedentary farmer lifestyle. This transition over many centuries is commonly considered to be key to the development of people from Mesolithic, predominantly hunter-gathering communities, to Neolithic farmers, with their more settled way of life. It is generally believed that it was within these farming communities that, among other fundamental developments, writing appeared and the seeds were sown for our modern cultures.

Some species of deer, the Père David and the fallow, were introduced into enclosures many centuries ago and have been contained and managed ever since. Whilst so protected, the free-living wild populations whence they came have dwindled and, in the case of the Père David, become extinct so that this species survives only as a result of having been imparked although it has now been returned to China and exists again in a free-living state. Neither of those two species could ever be defined as domesticated and the change in their appearance, their phenotype, is very slight implying virtually no effects of selective breeding. The occupants of these deer parks do not appear to be significantly different from wild deer.

Reindeer have been herded from time immemorial and although only enclosed rarely and for a few critical moments in the year they have been tamed and their behaviour is very different from their wild counterpart, the caribou. So, with the possible exception of reindeer, no species of deer could, at least until the late twentieth century, be categorised as domestic livestock. Now, increasingly, several species of deer around the world are being manipulated for economic ends in ways which constitute domestication. The application of husbandry methods depends on an ability to physically handle the animal, and the design of handling systems to permit regular handling of the different species of deer is discussed by Ian Thorleifson in this volume. It has been this development of sophisticated handling systems which has permitted deer to be farmed.

If we can argue that farmed deer are indeed domesticated, then the very recent and rapid establishment of deer farms around the world is a highly significant historical event. It is to publicise the ways in which deer are farmed, and to let the reader decide for themselves whether farmed deer are truly domesticated, that has provided much of the motivation behind this book.

In most parts of the world, the wild progenitors of our livestock have disappeared leaving farmed red deer, and perhaps domestic pigs, as the only livestock sharing an environment with a wild counterpart. That creates interesting cultural, and consequently legal and regulatory problems peculiar to farmed red deer. This has probably become most visible in European countries where hunting wild red deer has retained a cultural importance not always readily appreciated by those from a different background. For many Europeans, fallow deer are less wild and have very different 'cultural baggage' compared to the red deer. They only arrived in many European countries in the last few hundred years and their dispersion from south west Asia is due to human agency. It was, for instance, almost certainly the high esteem and near reverence in which red deer were held that encouraged Germany to experiment in the twentieth century with the farming of fallow deer and to discourage an exploration of the feasibility of farming red deer. The same is probably also true of other European countries. Friction between deer farmers and hunters has perhaps lessened but it still exists. It is exacerbated by the fact that deer can be managed and bred on some farms to produce antler trophies which are greater than those carried by wild deer. There have been widely publicised cases where it has transpired that remarkable trophies from deer shot in the wild were in fact from deer that had been bred on farms or in parks and subsequently released into the wild. In addition, a practice, which is widespread in North America, of breeding deer with very large trophies on farms or in parks specifically for enthusiastic trophy hunters to shoot within the enclosure is becoming more common in other parts of the world. It has attracted the soubriquet: 'the canned hunt'.

Hunters in some parts of the world remain unconcerned about the concept of the canned hunt whilst others are deeply upset. Associated with this is the breeding of extremely large trophies which are often the culmination of careful breeding programmes using every modern technique including artificial insemination and embryo transfer.

It seems to me very important that in discussing enclosed deer there should be a line drawn between, on the one hand, those that are kept within an enclosure but over which no sustained selective breeding is imposed, and, on the other hand, those which are subject to selection by the manipulation of their breeding. The former are wild and the latter are farmed. In Britain, a convention has been accepted amongst those working with deer that enclosures in which deer are kept more or less without significant selective breeding are called 'deer parks' whilst those in which the deer are handled and identified, where their performance is to a greater or lesser extent monitored and recorded, and where individual animals are selected for breeding, should be called 'deer farms'. British regulations in relation to meat hygiene and the control of notifiable diseases have to take account of these distinctions between 'farmed' and 'park' deer.

Britain, and especially England, has more long established deer parks, that is enclosures containing wild deer, than any other nation (Fletcher 2011). It has been convincingly argued that it is because of these parks that Britain has more ancient trees than other countries. Such trees are of great ecological value—Oliver Rackham pointed out that *'a single 400 year old oak can generate a whole ecosystem for which ten thousand 200 year old oaks are no use at all'* (Rackham 2006). And yet Britain also has a vibrant deer farming industry so that the rationale for the distinctions outlined above is especially pertinent. Peter Green in this volume discusses the background to English deer parks in detail while Alan Sneddon explains the quite complex husbandry required to achieve a productive deer farm in Britain.

There is another relevant difference between parks and farms because it has only been through the development of deer farms and the ability to handle deer that we can easily learn about the biology of the different species. For example, it was only through the opportunity to gain 'hands on' experience of farmed red deer in Scotland, and at about the same time in New Zealand, during the 1960s and 1970s, that the oestrous cycle, gestation length, and the control of the antler cycle became well understood (Lincoln 1971; Guinness et al. 1971). No such reliable data ever emerged from deer parks.

During a long life with deer, I have been fortunate to have had the opportunity of working with enclosed deer in many places around the world. Often these have a long history and are laden with complex cultural backgrounds but by contrast there are now many fast growing new industries associated with the husbandry of deer.

Different nationalities keep deer in enclosures in order to satisfy a number of specific needs. Simply put, the products fall into the following categories: meat, velvet antler, and other products for the traditional Chinese medicine trade, hard antler trophies, amenity or enhancement of the landscape, provision of hunting experiences, and inevitably a few other niche products, such as the recent development of farms in New Zealand based around demand for the milk of red deer. Each production system demands a different management strategy. What to one nationality seems a very obvious and comprehensible reason for enclosing deer and managing them in a specific way will be perceived by another as bizarre or even unacceptable.

The large industry which revolves around the harvesting of growing velvet antlers has reached the highest levels of sophistication in New Zealand as Tony Pearse describes in his account of the New Zealand deer farming industry. The amputation of the antlers is carried out under anaesthesia in highly regulated conditions. By contrast, the traditional procedures used in Asia seem very old-fashioned to western eyes. Antlers are removed from Russian maral, without anaesthesia, but very fast using electric saws on deer contained in drop floor crushes. Such a procedure is abhorrent to us but we have to remember that this is being carried out by farmers in keeping with long traditions and with limited resources. The absence of broken antlers makes it very clear that these people handle their deer with consummate ease and great rapidity. Before we make value judgments about the ethics of people in distant environments, we should also consider the removal of horns from cattle and other practices adopted in modern 'western' agriculture. Horns are permanent organs which grow slowly year on year whilst antlers are deciduous. Watching wild deer over time, one is bound to see stags breaking their velvet antlers and showing little sign of pain. The removal of a horn is a more radical procedure than the removal of an antler.

Welfare and ethical issues around deer and their antlers are complex. If it is accepted that the velvet antler is a valuable medicine, then their removal is surely justifiable provided that it is carried out painlessly. In New Zealand, the procedure is highly regulated to ensure analgesia and optimum welfare but this is not always the case in other parts of the world.

The removal of the horns of cattle as a part of the meat industry is perhaps comparable to the de-antlering of stags. Dehorning is a once in a lifetime procedure whereas removing the growing antlers is annual. Dehorning is anatomically a more radical procedure than amputating growing antlers. Is meat eating more or less deserving than the supply of velvet to the traditional Chinese medicine trade? In any case, clearly the ultimate objective of a procedure is entirely irrelevant to the individual animal whose welfare is at stake.

In the same way, much of the debate about hunting deer takes no account of the welfare of the animal but rather tends to hinge on whether the hunter is acting in a way which is considered to be 'sporting'. Thus, hunters are strongly divided as to whether bucks shot in 'canned hunts', that is while the deer are confined within a fenced enclosure, are fair sport. The general public, if ever they consider these issues, will also heavily oppose the shooting of deer within enclosures for trophies. Yet from the welfare point of view, the buck is unlikely to be concerned whether it is shot in a fenced enclosure or free range provided the marksman takes an accurate and humane shot. Bowhunting is relevant here. Those who use bows and arrows to kill deer argue that their ability to kill humanely is at a high level; nevertheless, the use of bows to kill deer is illegal in many countries such as the UK. Insofar as the 'hunter' shooting deer within an enclosure is likely to be able to get a much closer shot than if he is shooting deer on free range then it could even be argued that canned hunts are more humane.

In discussing the ethics of deer management, the trophy industry probably attracts the most attention. This is not only amongst the hunters polarised into different camps arguing as to whether canned hunts are fair play. There are also concerns amongst vets and others about the relentless increase in the size of the trophies. The growing interest in shooting big antlered deer whose antlers will be measured, entered into competition, and displayed on the wall has created rapid growth especially in the American whitetail industry. This is because of its scale. It is a multibillion dollar business in the USA alone, where the most valuable whitetail bucks can now be valued at up to a million dollars each, justified by their lifetime semen production and their eventual trophy value.

The rapid increase in whitetail deer farms directed entirely towards the trophy industry provides a growing pool of deer which together with the evolution of successful techniques for artificial breeding permits very rapid increases in antler size and weight.

For some years, there have been growing rumours within the veterinary profession that this has created deer with antlers that are so heavy that the animal struggles to lift his head. I was sceptical of this knowing the massive neck muscles which in a red deer would easily permit a mature stag to turn a small car over. However, my scepticism is misplaced. I read 'the super buck—a buck with antlers so big he cannot hold his head up' (Knox 2011). And again '. . .deer are so bred for antlers that they become unhealthy. They literally have to cut the antlers off of them because they cannot lift their heads' (McConnel et al. 2018). Most compelling of all was the film by 'Keith Warren Hunting' on YouTube dated 14/11/19 (ref V0fCNWYG7US) entitled 'Amputating Giant Antlers off a Whitetail Buck'. Here we see a buck with growing antlers clearly unable to hold his head up for more than a few moments and the commentary makes it clear that the reason the antlers are being removed is because the buck cannot hold his head up.

It is a common practice on deer farms to remove the antlers of the males each year to prevent injury to themselves, the fences, or the people working with them. That in itself need not be a significant welfare problem but since each year the deer are unable to develop neck muscles associated with carrying increasing weights of antler so they become annually more prone to being overwhelmed by the weight of the antlers. This may reach a tipping point when they are expected to carry their largest antlers for a period of time prior to them being shot as a trophy. Normally deer allowed to carry their hardened antlers before, during, and after the breeding season spend many hours sparring with other males as well as participating in the more serious actual fights during the mating season. In all these interactions, they are exercising and developing their muscles. If the antlers are removed each year as they finish growing in the period leading up to the rutting season, then they cannot develop that musculature in the same way. Undoubtedly stags and bucks without antlers will still test each other and fight but not to anything like the same extent. The plight of these animals which struggle to carry such huge antlers is therefore no surprise.

Other worrying accounts exist of mature red deer which have had their growing antlers removed each year in velvet but which are then allowed to harden a final set of antlers as a trophy. In the same way as the whitetail bucks which cannot carry the weight of the antlers because they have never had the opportunity to develop appropriate musculature, so stags which have been repeatedly velveted may also have insufficient bone in their skulls to carry such colossal weight. It is common knowledge that bone responds to loads. It hypertrophies, and becomes stronger depending on the application of mechanical stresses. In those stags which have had their antlers removed each year and thus may never have borne a full set of hard antlers, the bone has not had the opportunity to develop strength each year. The sudden weight in a stag which has finished its life as an annual supplier of velvet and which is now selected to provide a trophy and which may be turned out onto free range or into a large enclosure to provide a hunter with sport can then be too much causing the skull to break.

I have no direct experience of this horrific phenomenon but physiologically it makes sense and if it is occurring then it deserves to be debated.

Within this book, the trophy industry is not discussed in detail. It is the elephant in the room. Efforts to find anyone from within that sector who would be willing to contribute a section to this book were unsuccessful. Yet the obtaining of trophies from enclosed deer impacts every other element of the deer industry. Briefly, for those who are unfamiliar with the collection of trophies, there now follows a short explanation.

People from a variety of backgrounds exhibit the antlers of deer which they claim to have shot on the walls of their houses. These may be mounted as simple antlers, usually on a wooden shield, through varying degrees of taxidermy right up to full mounts of adult deer in dioramas in private museums. The size and weight of the antlers is important and is closely related to the price paid by the trophy hunter to shoot the deer. To rationalise the measurement, several different measuring systems are used: the SCI (Safari Club International) system is perhaps the most widely used internationally but there exists also the CIC (Conseil International de la Chasse) which originated in Europe, and several other measurement systems (Whitehead 1993). These are not restricted to deer but extend into antelope, sheep, goats, and other groups. Whilst the collection of trophies and, no doubt, the competition between hunters which it engenders, has existed throughout history and probably into prehistory, it is only through the creation of a wealthy international coterie that it has reached its current level. Collectors come from most nations and from all trades and professions including especially, at the highest level, business and politics. As might be imagined, therefore, the value of an individual trophy can reach many tens of thousands of dollars. Enclosed deer have the potential to outgrow wild deer because they can be better fed and because their breeding can be directed scientifically towards large antlers.

Lay people will find trophy hunting distasteful but provided the shooting is carried out humanely perhaps there are no welfare issues. To cause the death of an animal will be reprehensible to many but it does not of itself constitute a welfare issue.

This is not the place to pass opinions on husbandry systems but I do consider that it is the role of this book to make people aware of the practices adopted worldwide.

Whenever animals are exploited by humans, controversies will emerge. Given the accelerating rate of climate change, that train coming down the track cannot be ignored. Issues of welfare are now compounded by the need to reduce emissions of carbon dioxide and methane. Research to date suggests that deer are unlikely to differ significantly from other ruminants in the levels of methane they produce (Perez-Barberia et al. 2020; Riddell 2018).

I have been struck by just how much widespread ignorance there is of every aspect of deer management: both within those old deer parks and the newer enterprises as well as of the markets on which they depend. Whether new or ancient, all deer enclosures are managed in ways which have distinct local or national approaches and legislators spend much time and effort in trying to understand and control these diverse systems.

For example, in meeting European regulators in Brussels, I was surprised to find that they simply refused to believe that the venison from farmed deer could ever achieve a higher price than that from wild deer hunted traditionally and treated reverentially. In this case, the almost religious approach to hunting wild deer completely coloured the perceptions of these bureaucrats whilst for supermarket buyers in England the advantages of having venison from deer killed in an abattoir at a certain weight or age and hygienically processed seemed obvious and would be correctly reflected in the price. In England, the hunting of deer is less deeply engrained in the culture than it is in many European countries and the concept of rearing red deer for meat production on English farms is more acceptable although even here it has encountered objections amounting to mild hostility from traditional hunters.

In England, we have a significant number of deer parks in which the deer are maintained entirely for amenity purposes: people like to visit the parks and admire the deer which remain effectively wild despite the need to supplement their grazing during the winter. Surpluses are culled by rifle and the meat is sold but this is secondary to the value of the animals as beautiful adjuncts to the landscape (Fletcher 2011). Some of my veterinary colleagues working to reduce tuberculosis in cattle wanted to test deer within deer parks and were not willing to believe that there was no system known to man which could reliably and repeatedly round up these wild animals and force them through a handling system without major losses and injuries.

In some cases, these administrators attempted to compel the park owners to carry this out regardless of the inevitable compromise to the welfare of the deer. Many of these deer are retained within parks which are surrounded by very imperfect barriers which may be wooden fences or walls that are centuries old. The deer remain because they are 'hefted', that is they remain voluntarily within the bounds of a range with which they are familiar, so that any disturbance is likely to cause them to escape into the surrounding countryside potentially spreading tuberculosis should the deer be infected. The parks, with a sex ration of 1:1, are entirely different from farms where the ratio of adult males to females might be as little as 1:50. On farms, the antlers are normally removed each year to prevent injuries to the deer or the farmer or his fences, whilst in parks the antlers are, of course, left on.

The corona virus pandemic has seen a massive increase amongst those using deer parks for recreation often including the walking of dogs. These dogs, sometimes taken off the leash, have had a tragic effect on the deer in the parks with high mortality especially amongst newly born or unborn calves or fawns. This has resulted in some parks debating whether it is worth retaining their centuries old herds of deer.

For the Chinese and other southeast Asian nationalities, the value of the deer they keep is almost entirely vested in the perceived medical value of the growing antlers as well as the tails, the male genitalia, the sinews, the foetuses, the blood, etc. The meat comes a long way behind and is, by comparison, scarcely valued at all.

For the American sportsmen and women, the size of the hard antlers is all important. They are looking for a trophy they can hang on the wall or have stuffed by a taxidermist to appear in their museum. Whether the deer are wild and free ranging or within a park is not always fundamentally important.

Different again are the values placed on the techniques of hunting wild deer. In Scotland, the traditional stalker will treasure the difficulty of making a tactical approach into the wind in open country with little cover so as to place a carefully deliberated shot and achieve a humane kill. The antlers for him are important but only in the context of that long remembered stalking experience. The trophy is not the priority but rather how they got it.

By contrast, the Spanish hunter will take pride in the ability to shoot moving deer as they are driven past in the *monteria* and again, whether the quarry is within a large enclosurè or free ranging is often of secondary importance.

In many, perhaps most, countries around the world the growing 'velvet' antlers are amputated from living male deer for sale into the traditional Chinese medicine trade. However, legislation in Britain, introduced largely at my instigation, made the removal of growing antlers illegal on welfare grounds in 1980. Many other European countries have followed suit but throughout Asia, deer of various species are kept solely to provide velvet antlers and other organs valued by the Chinese for medicinal purposes: practises which are often repugnant and almost incomprehensible to keepers of deer in traditional English deer parks.

In Germany and some other European countries, the red deer are so much revered as the noble epitome of the forest that the notion of 'farming' them for venison as is practised in New Zealand or Britain is anathema. Instead, it is the fallow deer which are retained in small paddocks and kept for their meat.

In France, their deeply engrained culinary tradition has placed little value on the meat of the red deer, especially that from the stag, but instead values the meat of the roe.

I have always found these cultural distinctions fascinating and this book has grown out of my interest in trying to shed light on their diversity. And, of course, the traditions are being eroded and diluted both amongst wild deer and within the deer enclosures through the process of globalisation.

In Scotland, where I live, pressure to kill more wild deer is growing, fuelled by a legitimate interest in encouraging the natural regeneration of woodland but often inflamed by a simmering antipathy towards those who stalk the deer traditionally. The notion of Scottish deer stalking is perceived by many as privileged and is, rightly or wrongly, linked with people's perception of wealthy, absentee landowners. This attitude is far from globally universal. In many countries, hunters are drawn from a variety of social backgrounds and hunting is perceived as a healthy necessity bringing participants closer to the natural world and providing a healthy food whilst performing a valuable role in controlling numbers of those species which are damaging woodland and agricultural crops. When the human population of the Scottish Highlands was high, and especially as it grew in the late eighteenth and nineteenth century encouraged by the introduction of the potato, smallpox vaccination, and changes in social structure, the woodlands declined and deer numbers fell. This seems to be a universal paradigm—as rural human populations grow they reduce the populations of animals on which they depend for food. And, conversely, as the human population either declines or becomes urbanised those same prey species rebound often then being branded as pests or vermin.

In Scotland at the beginning of the nineteenth century, red deer numbers were low and their range was limited (Whitehead 1964). That has changed dramatically and many landowners are attempting to reduce their deer stocks. Some have espoused the 'rewilding' agenda, encouraging natural regeneration of woodland and doing what they can to boost biodiversity. Most objective observers of the Scottish Highlands accept that red deer numbers have been too high and overgrazing, exacerbated by heavily subsidised hill sheep, have restricted the regeneration of some tree species. In consequence, Scottish red deer have become politicised and there is a strong danger that they will be treated as vermin. Against that background, red deer numbers have been falling for a number of years. This will eventually be likely to have an impact on enclosed deer. Visitors to Scotland want to see red deer and demand for venison is growing which encourages the formation of deer parks as well as farms, abattoirs, and venison processing plants. Against the demands of those wishing deer to be almost eliminated to encourage the growth of trees are those who recognise the need to maintain the presence of natural grazing animals in order to maximise biodiversity.

This book does not aim to discuss issues around the management of wild, unenclosed deer. That is a huge and much debated subject throughout the world. Instead, I have restricted the scope to enclosed deer. I have asked colleagues working with enclosed deer to describe the systems employed and the backgrounds to the enclosures in their part of the world and they have all responded generously. I am enormously grateful to each contributor for providing interesting pieces. I have not attempted to restrict the coverage of some subjects because each contributor will have their own perspective. On the other hand, I am very conscious of gaps. My initial ambition to provide global coverage has not been achieved. I would have liked to have been able to provide a more up-to-date account of the huge Chinese deer industry. I would also have liked to have given a fuller account of the state of the reindeer herding societies in northeast Asia because although unenclosed those reindeer must be considered as domesticated. Indeed they can, according to some authors, arguably lay claim to have been domesticated in prehistory although evidence for any early domestication is missing (Clutton-Brock 1999). There are glaring omissions within Europe and Scandinavia for which I apologise. Deer farming is at a very low level in Italy, and, in Denmark and Sweden, the farming of deer has declined but in Norway there is a small yet vibrant deer farming industry. My sincere apologies for the many omissions but my equally sincere thanks to all those who have put up with my badgering and produced a variety of fascinating accounts within the challenging background of the corona virus restrictions.

I would consider this book to have been a success if it raises awareness of the diversity of husbandry systems applied to enclosed deer across the world, improves the welfare of these deer, and especially if it is at least partly effective in preventing deer being treated as vermin.

Addendum

The General Biology of Deer

It might be helpful for general readers, who may be more accustomed to grazing species other than deer, if I explain a little of the general biology of the family of Cervidae.

It is a strange paradox that the principal grazing animals which are farmed in the temperate zones of the world: goats, sheep, and cattle, are predominantly species derived from the least temperate parts of the northern zone: the Middle East. Deer species, many of which originate and inhabit the northern temperate regions, are less commonly exploited as farm animals yet one might expect them to be more suitable for they are often the most seasonal of all grazing animals. Even those deer species which have spread into tropical and equatorial regions are deemed at some stage in their evolution to have had primordial relatives which existed in temperate zones. It is assumed that it was as an adaptation to regions where the temperate climate dictated a highly seasonal environment that the deer evolved antlers which they can cast and grow in accord with seasonal changes in daylength (Fletcher 2013). Antlers are unique to deer but they represent only one feature of the seasonal adaptations which most deer have made.

For grazing animals dependent on grass growth, and for those which browse shrubs and trees, it is essential that they drop their young in the spring when the vegetation is at its most nutritious and can provide a summer of growth before the winter. Their breeding has therefore to be synchronised and this requires a short and competitive rutting season with clear implications for the management of deer in enclosures.

The antlers of the deer have evolved, we believe, as weapons with which the male deer can compete for access to the females. Also associated with the rut is a catalogue of behaviour calculated to advertise the presence of the male through noise and scent, movement and colour, all of which has the effect of intimidating rivals and stimulating females to ovulate and to be receptive to mating.

In the only deer species in which the female grows antlers, the reindeer, the antlers have taken on an additional role. They are retained by lactating females to give them, and the calves they protect, an advantage in competition with other reindeer for access to the holes which they have dug through the snow (Lincoln and Tyler 1994).

From all of this, it must be clear that deer need to time their antler growth to fit the seasons very precisely. It is no surprise, therefore, that the antler cycle is regulated by the changes in daylength. Thus, in most species, the antlers are programmed to begin to harden as the days become shorter with the furry velvet which covers and nourishes the growing antler being shed in time for an autumn rut. This rhythm is mediated by the male sex hormone, testosterone, levels of which rise as the days shorten to reach their height during the rut. Through the winter testosterone levels decline and the antlers, in most species, are cast in the spring when the growth of the new antlers begins (Lincoln 1971).

Changes in daylength also dictate the timing of ovulation and oestrus in most species of deer with the exception of the few aseasonal species inhabiting tropical regions such as the chital or axis deer.

This pattern of a well-defined rut is clear in the gregarious or herd forming deer species, which are the only ones readily adaptable to life in parks and farms, and a knowledge of the antler cycle and rutting behaviour is essential for the husbandry of enclosed deer. These gregarious species of deer are chiefly adapted to grazing rather than browsing and are therefore usually able to cope with farms and parks in which grass is often the chief source of feed, and, perhaps even more importantly, their herding behaviour makes them well suited to being confined in social groups. There are a few regions in the world in which the less gregarious deer, normally adapted to woodland, are subjected to enclosure for commercial reasons. For example, there has been the recent development of farms in North America for white tailed and mule deer. These enterprises are orientated towards the breeding of deer for hunting and for trophy development.

Another group of animals that is commercially exploited despite not being gregarious is the Moschidae or musk deer (Moschus spp.). I have not included any of the seven species of musk deer in this book because despite their popular name they are not properly deer at all. They differ from the Cervidae in lacking antlers and lacking pre- or infra-orbital glands whilst they do possess a gall bladder. The cervids all have four teats but the Moschidae have only two. Crucially, and of tragic commercial significance, the male musk 'deer' possess a caudal scent gland which is very highly valued. The musk glands or 'pods' were selling in 1974 for four times their weight in gold (Green 1978). For many centuries, musk was used as a scent but the use of musk in perfume was made illegal in 1979 by the Convention on the International Trade in Endangered Species of Wild Flora and Fauna (CITES). Nevertheless, today trading continues within the traditional Chinese medicine trade and the most numerous species, the Himalayan musk deer, is declining and listed as endangered (Timmins and Duckworth 2015). Today, the trade quantity of natural musk is controlled by CITES but illegal poaching and pressures depleting forest habitat continue to reduce the population.

There were hopes that Chinese musk farms where captive male musk deer were 'milked' of musk without killing them might alleviate the demand but hunting pressures which kill females and males indiscriminately together with a decline in their habitat were thought in 2015 to have seen the population decline by 50% during the last 21 years (Timmins and Duckworth 2015).

Antlers are the chief reason that many people around the world keep deer whether for trophies, amenity or as a vital part of the traditional Chinese medicine trade. Biologically antlers are, of course, extremely interesting. They are the only example of organ regeneration in mammals and they grow very fast—a centimetre a day in even those deer with quite modest antlers. As such they have attracted attention and been adopted as symbols of regeneration in most cultures in those regions where deer are common. The speed of their growth and associations with vigour and vitality have also no doubt contributed to the role of velvet or growing antler in Chinese medicine. In that system, velvet is dried and sold as a general *vade mecum* for all complaints but is perhaps especially valued as an aphrodisiac although, of course, paradoxically the velvet grows only when testosterone levels are at their lowest.

The hard antler by contrast was of immense value in prehistory to peoples before the use of metals as tools in shaping stones or as tools themselves in a variety of forms from combs to fish hooks. I have argued that it might have been in order to facilitate the collection of the highly valued antler that deer might first have been enclosed. Red deer cast their antlers in the spring when vegetation is coming into growth so if not found and picked up very quickly the cast antlers may become overgrown and lost. But the antler casting season is also the time when deer are most short of food. It would be very simple for people to have collected browse that the deer could not reach such as ivy, and gathered the deer by providing feed at particular feeding sites. The possibility of containing these deer into temporary parks at the feeding sites certainly exists (Fletcher 2011; Vera 2000).

In red deer and their relatives, as well as in fallow deer, the females in the northern hemisphere come into oestrus in September or October and, if they don't conceive, will continue to return into oestrus every three weeks until the late winter (Guinness et al. 1971). In these species twins are unusual.

Lactation will normally continue through the summer and into winter but farmers find that by weaning calves at the end of the summer they can stimulate hinds into oestrus a little earlier and keep the hinds in better condition as they enter the winter. For the deer farmer, there is a significant advantage in having calves born early to give a long growing season.

Auchtermuchty, UK John Fletcher

References

Clutton-Brock J (1981) Domesticated animals from early times. Heinemann and British Museum (Natural History), London

Darwin C (1868) The variation of animals and plants under domestication. John Murray, London

Fletcher J (2011) Gardens of earthly delight – the history of deer parks. Windgather Press, Oxford

Fletcher J (2013) Deer. Reaktion Books Animal Series, London

Green MJB (1978) Himalayan Musk Deer (Moschus moschiferus moschiferus) in endangered, vulnerable and rare species under continuing pressure in IUCN threatened deer programme. In: Proceedings of Deer Specialist Group of International Union for Conservation of Nature and Natural Resources

Guinness F, Lincoln GA, Short RV (1971) The reproductive cycle of the female red deer (Cervus elaphus, L.) J Reprod Fert 27:427

Knox M (2011) The Antler religion. Wildlife Society Bulletin

Lincoln GA (1971) The seasonal reproductive changes in the red deer stag (Cervus elaphus). J Zool Lond 163:105–123

Lincoln GA, Tyler NJC (1994) Role of gonadal hormones in the regulation of the seasonal antler cycle in female reindeer, *Rangifer tarandus*. J Reprod Fertil 101: 129–138

McConnel DL, Loveless MD (2018) Nature and the environment in Amish life. Johns Hopkins University Press, Baltimore, MD, p 119

Perez-Barberia FJ, Mayes RW, Giraldez J, Sanchez-Perez D (2020) Ericaceous species reduce methane emissions in sheep and red deer: respiration chamber measurements and predictions at the scale of European heathlands. Sci Total Environ. https://doi.org/10.1016/j.scitotenv.2020.136738

Rackham O (2006) Woodlands. New Naturalist Library. HarperCollins, London

Riddell H (2018) An investigation of greenhouse gas emissions that arise from venison production in Scotland. Unpublished BSc thesis, SRUC and Edinburgh University

Timmins RJ, Duckworth JW (2015) Moschus leucogaster. The IUCN Red List of Threatened Species 2015: e.T13901A61977764. https://doi.org/10.2305/IUCN.UK.2015-2.RLTS.T13901A61977764.en. Accessed 16 July 2021

Vera FWM (2000) Grazing ecology and forest history. CABI Publishing, Wallingford, Oxfordshire

Whitehead GK (1964) The deer of Great Britain and Ireland. Routledge and Kegan Paul, London

Whitehead, GK (1993) The Whitehead Encyclopedia of Deer. Swan Hill Press, Shrewsbury, UK

Acknowledgments

I have called in favours from a large number of friends in compiling this book and I am grateful to them all. I approached people from a wide variety of backgrounds, some academic and some with truly invaluable practical experience of working with deer. They have obliged me and I am truly grateful. Over and above the contributing authors, I owe a huge debt of gratitude to Annette Klaus and to Srinivasan from Springer whose patience throughout the ridiculously long gestation of this book during the Corona virus pandemic has been truly impressive. Thankyou!

Contents

Part I

Farming Red Deer in New Zealand

Farming Red Deer in New Zealand: Industry History, Structure and Administration

Tony Pearse

Abstract

This chapter describes the history of deer farming in New Zealand and the development of its management structure including the raising and use of industry levies, as well as the application of DEERSelect to optimise genetic improvement. The development of export markets for venison and velvet antler will be discussed further. In conclusion, readers will learn about the management of the velvet industry, including marketing initiatives and regulations for welfare and quality.

Keywords

New Zealand · Deer · Deer farming · Venison · Velvet · Deer Industry
New Zealand · Velvet antler industry · Welfare · Velvetting regulation · Stag
management

1 Introduction and Industry Establishment History

In a year dominated by reaction to the worldwide COVID-19 outbreak that saw the most disruption to our international markets and ways of doing business, the New Zealand deer industry quietly celebrated 50 years since the formal start of deer farming as a new livestock farming enterprise.

The first license to farm deer was issued by the NZ Forestry Service to Michael Giles of *Rahana Station* in the Taupo region of North Island in 1970. The industry expanded rapidly from that point driven through off-farm investor taxation incentives and a new style of farmer seeking diversification from sheep and beef

T. Pearse (✉)
Deer Industry New Zealand, Wellington, New Zealand
e-mail: tony.pearse@deernz.org

production. The first farmers were typically motivated by innovation, supported by a new opportunity in city farming investor partnerships made possible by attractive taxation deferral rules, and the lure of the potentially high returns of farmed deer, already established with the young feral venison trade export markets.

The interest in deer farming was supported greatly at that time by government initiatives to encourage diversification in livestock farming, including incentives that promoted an evolution of city business investment in partnership with an emerging generation of young farmers. The first farms of scale were often established high country farms that bounded the public estate where wild introduced deer were well established and simply enticed onto farms into capture blocks or recovered by trapping and released into fenced areas as the first steps into a new profitable livestock farming industry.

An export market was already well developed in Europe for wild shot venison. This was sourced as a valuable product as a consequence of a government-managed effort to cull and substantially reduce the introduced wild deer population where large herd numbers were damaging the complex flora within New Zealand's native forests. Individual commercial businesses began carcass recovery and exporting venison as game meat to traditional seasonal European game markets as well as venison to Europe, and new markets in Asia also evolved for velvet antler, and Asian edible co-products, used in traditional oriental medicine including deer tails, skins, tendons and sinews and stag pizzles (the male genitalia).

In order to develop consistent and quality supply to these relatively lucrative new markets, New Zealanders developed aerial helicopter-based techniques for capturing and transporting live deer to newly established deer farms across the country which began farming deer alongside their other domesticated species on extensive pasture farming systems. Early systems essentially started with learning the biological needs of deer in captivity across the 4 seasons by trial and error and experience with sheep and cattle farming systems. Often the greatest challenges involved working out design and functions of lead-in raceways and holding pens and entry to basic handling systems.

In the 1970s Government had also invested in the establishment of what quickly became a world-leading deer biology and farmed practical deer research institute at Invermay Agricultural Center, originally within the Department of Agriculture. Located adjacent to the city of Dunedin the Deer Research Group also quickly formed links with the University of Otago in the important fields of understanding deer diseases and the value of identifying superior animals and concentrating on applied genetics on the key production traits.

Industry support for the research programme led to the rapid development of a large body of leading research outputs covering deer production, seasonal biology, velvet antler growth and investigations into the properties of velvet antler. This contributed greatly to the growth and sophistication of NZ deer farming and strengthened existing contacts with Sir Kenneth Blaxter and similar work in early deer farming systems, nutrition and physiology at the Rowett Institute in Aberdeen, Scotland.

Recognising the need for research, support and representation the New Zealand Deer Farmers' Association was formed in 1975 to lead the development of industry-friendly regulations, to liaise with Government political leaders and important departments like Agriculture, and the Department of Conservation that managed the public lands, and to foster and support a rapidly growing industry.

The NZDFA is formally an Incorporated Society that at the outset operated based on an annual voluntary subscription fee to fund a national NZDFA President and 8-person Council as the voice of farmers and a political lobby group plus investing in deer industry research and communications.

The NZDFA provided a vehicle for the insatiable demand for information from new entrants to the growing sector. That connection quickly moved into the formation of then 35 active regional branches of the NZDFA which became the outlet for face-to-face deer farmer gatherings and sharing of advice, new practical deer farming ideas, as a social connection and the young deer industry's political lobby group.

The young industry also founded significant deer Breed Societies. The first founded in 1986 was the fiercely independent NZ Elk and Wapiti Society which remains an active and influential body today. (https://www.elkwapitisociety.co.nz). (Cited in "Wapiti behind the Wire" The Role of Wapiti In Deer farming In NZ, David Yerex, GP Publications Ltd Chapter 10 pg 61, 1991).

The Society championed the role of Elk/Wapiti in niche market velvet antler production, world record trophy antler and its commercial venison production potential role in producing selected sires that readily bred with base commercial red deer hinds to provide large fast-growing crossbreds. These met target carcass weights for the European chilled season autumn peak consumption 6–8 weeks earlier and heavier than the red deer breeding herd of the time was able to achieve and in days of venison export made a significant contribution and maintains a successful niche today.

The NZDFA also had the vision with Government support to seek a new status of a further overarching Industry Body to establish the NZ Game Industry Board (1985) as a Statutory Marketing Authority established under the Primary Products Marketing Act 1953 and the Game Industry Board Regulations 1985.

The NZGIB was charged with promoting and assisting in the orderly development of the deer industry and products derived from deer. Its formal powers were simply to be able to deduct a levy per kg carcase weight sold of processed venison (collected at slaughter at the Deer Slaughter Premise DSP) and kg of frozen velvet antler sold by farmers to velvet processing and export specialist companies. Levy rates are reviewed annually by the Board.

The Board collects information about the game industry and uses this information to help its members. Specific information which is collected includes data about industry production and export volumes and prices, support for research into the processing area, development of industry grading standards and product specifications and market research, and strategy development and targeted promotions.

The Board representation comprises four appointees nominated by the New Zealand Deer Farmers' Association, three nominated by the New Zealand

Deer Industry Association, one of whom represents velvet antler processors, and initially one government nominee, although that was discontinued in the late 1990s. In addition, the Game Industry Board had a Research Trust.

The role and influence of the NZ Deer Farmers Association too continued to grow as the representative voice of the deer farmers. The NZDFA was formed to assist in maximising sustainable benefits for all deer farmers and to provide a linkage to the agricultural industry and the public.

An industry structural review in the mid-1990s led to the Government passage of the Commodity Levies (Farmed Deer Products) Order 1995, charging producers primarily for the responsibility for paying levy on deer product both per kg of venison and velvet antler weight and was paid by liable buyers and remitted to the NZDFA and recovered from the liable producer.

2   Application of the Levy Funds

These funds were predominantly used to support on-farm production research where projects were submitted and evaluated and allocated to the research organisations on contract or co-funded with the Ministry of Agriculture's research division or the universities, and in industry communications. Political activity remained supported by annual subscriptions. The levy was always somewhat controversial to some, and in turn, encouraged the establishment of the NZ Levy Payers Association within the Industry as a group that demanded NZDFA transparency and were a force to be recognised in both budgeting debates and scrutinising NZDFA expenditure.

In 2002 the NZDFA and the GIB administratively merged to incorporate with Deer Industry New Zealand with the NZDFA relinquishing its power to collect a compulsory levy granted under the Commodity Levies Act. In turn, the NZDFA Council was reduced to a 4 members Executive Committee, committed to the management of producer interests for the DFA and all deer farmers. Its role remained just as the formation goals of the NZDFA to assist in maximising sustainable benefits for all deer farmers and to provide a linkage to the agricultural industry and the public.

With the passing of the Commodity Levies (Farmed Deer Products) Order in September 2001, all deer farmers producing venison and/or velvet for sale continued to be members of the Association and contribute by levy on the basis of the weight of venison and/or velvet sold, however that was a 12-month transition to the present structure.

With the change from the NZ Game Industry Board to Deer Industry New Zealand effective from 1st October 2002, the Deer Industry New Zealand office now carries out the administrative operations of the NZDFA. This change in organisational structure has provided new strength at the NZDFA Branch level and resulted in the DFA Commodity Levy being set to nil from 1 October 2002.

The Council was reduced to a four-person Executive Committee led by a full-time professional executive, the Producer Manager. The Executive Committee Members are directly elected by members for a two-year term. The Deer Industry New Zealand Producer Manager manages the associations' programmes and provides the link between the branches and the Executive Committee and the Deer industry New Zealand executive team and the DINZ board.

The DINZ Board mission remains the same and is a unique governance body with a 50:50 Board of 4 members drawn from the venison and velvet processing and marketing and 4 producer appointed representatives. This Board is funded by a statutory levy on the products of venison and velvet antler. (Per Kg whole carcase weight & per Kg of frozen velvet antler). For both the levy is paid 50:50 between processor exporters and the farming sector and funds an agreed industry strategy and the activities of DINZ.

3 Governance and Industry Management Directions and Policy DINZ 2020–2025

Deer Industry New Zealand (DINZ) has four strategic priorities as the centre pieces of its mission and industry direction and these are relevant across all the activities involved with the DINZ Board and its executive team. These are:

- Premium positioning of NZ deer products
- Market development and diversification
- Sustainable on-farm value creation
- Cohesive and respected industry

The Executive team within DINZ is headed by the CEO (currently Innes Moffat) reporting to the DINZ Board. DINZ features a diverse range of portfolios that evolve to suit the changing dynamics of the agricultural sector but has an increasingly close association with both the producer sector and the product processing and marketing industries.

Key DINZ staff include a Velvet and Venison marketing manager and a Quality Assurance manager with roles that oversee the National Velveting Standards Body (NVSB) and Industry Quality Assurance schemes. The velvet removal scheme is essential for the maintenance of the industry's ability to practice velvet removal. This is a formal partnership between veterinarians, industry trained and accredited deer farmers and meets the Ministry for Primary Industry's expectations for good animal welfare.

DINZ portfolios also include a science and policy role guiding and commissioning relevant projects with research providers, as well as the myriad of policy issues that are affecting food producers in New Zealand.

A rapidly evolving area for New Zealand deer farmers is environmental management and environmental stewardship. Central government policies on freshwater management, biodiversity protection and greenhouse gasses significantly impact deer farmers. DINZ has engaged an environmental stewardship manager to represent deer farmers' interests at a political and on-farm practical level.

The industry has in recent years also invested significantly in activities to encourage more effective deer farming and improve productivity. A manager of farm performance oversees a programme of practice change activities designed to improve deer nutrition, health, reproductive performance and mitigate environmental impact.

Those remaining in the industry today are absolutely committed to deer and many have large operations with breeding hind numbers in the 2–4000 range with equal enthusiasm for large-scale velvet antler operations.

Venison production remains the core of the business although the commitment to velvet antler production continues to grow on the back of the new marketing initiatives into the healthy functional food market in South Korea and increasingly expanding in China.

4 Producer Engagement

In 2002 the Producer Manager role was created within DINZ to bring a producer perspective directly to the DINZ board and staff and to in effect act as a CEO of the NZDFA.

Today the support for the NZDFA has grown to include communications and event management functions. Central to DINZ and DFA connection to the sector remains the national industry conference (now its 45th year). Electronic and print communications are maintained through www.deernz.org the vast information bank and learning resource, the Deer Hub (https://www.deernz.org/deerhub), the bi-monthly *Deer Industry News* magazine (www.deernz.org/publications) and regular electronic newsletters and information.

5 2000-Present: Times of Significant Change

In the mid- early 2000s, the industry peaked at close to 5000 farmers with approximately 1.5 million deer on farm across the full breadth of NZ farming terrain. Many were small operations often associated with lifestyle blocks or as a diversification on traditional sheep and beef farming operations where deer would be 10–20% of livestock numbers. Velvet antler production drove the interest and passion for most deer farmers.

Markets for venison and velvet were confined to traditional channels for game distribution and were price and volume sensitive.

The key velvet market was the traditional Oriental medical market in Korea which was very sensitive to volumes increasing above 500 tonnes of NZ production and based predominantly on antler from red deer. Elk-Wapiti heads also find a ready demand as a Supreme grade.

With increased volumes, venison returns for farmers were subdued, leading many to question the returns from farming deer on land suited to other options. A dramatic rise in dairy farming had a profound impact on the shape and activity of the dry stock farming in New Zealand, including the deer industry. Much of the fertile plains and downland hill country, where deer farmers were specialising in venison finishing was swallowed up by the rise in dairy farm conversion.

Industry numbers declined significantly. DINZ estimates there would be 1350 deer farms operating today with approximately 800,000 deer. Most deer will be on larger operations, generally located on fertile hill country or in the high country where breeding and velveting operations often run hand in hand.

Some large specialist venison finishing farms remain where farmers buy 4–5-month-old weaners from breeding farms and finish the venison at 9–12 months of age.

Marketers concentrate on expanding chilled markets in the USA, Canada and United Kingdom and Europe, with emerging opportunities in China and other Asian countries. In recent years, the venison returns during the September to November 'chilled season' have ranged from NZ$8.50–$10.50 kg return to the producer for 50–64 kg carcasses.

6 DEER Select and Genetic Improvement

To meet these growth targets and opportunities NZ farmers have invested heavily in stags and selection for high growth rate, improved carcass yield, resilience to parasites.

Deer Select is the national database of genetic merit of performance-recorded animals in linked herds. Virtually all main stud herds producing venison sires (Red deer, Eastern European red deer, and Elk /wapiti) record and use breeding values as part of an objective approach to genetic progress. DINZ has a Deer Genetics manager to manage the programme and work with deer herds to improve the accuracy of recording and the publication and promotion of genetic selection based on objective breeding values.

For venison, the two key breeding objectives are guided by Deer Select indices—one for a breeding/replacement/finishing system, the other for a straight finishing (terminal) system. In future more maternal traits will be added to the breeding/ replacement index,

Venison yield per carcass can be recorded by ultrasound or CAT scanning eye muscle area, which has been shown to correlate to meat yield.

The leading studs are also measuring CARLA, (**CAR**bohydrate **L**arval **A**ntigen) representing an animal's ability to produce an antibody that helps protect it against internal parasites.

7 2021: Venison Production and Markets

Deer Industry New Zealand had three clear priorities for the venison sector heading into the 2019–2020 year. Supported by venison marketing companies, the key activities were:

Fig. 1 Export destinations and volumes of venison by tonne during the 2019/20 season. Source: Deer Industry New Zealand Annual Report 2019–2020

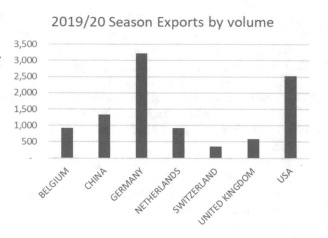

2019/20 Season Exports by volume

- increase sales of the Cervena branded venison in Europe during the summer
- increase sales to foodservice in the United States
- open opportunities for farm-raised New Zealand venison in mainland China.

The venison market entered the 2019–2020 season with a slight overhang of manufacturing venison in the United States and some concerns about demand for high-priced proteins in the European foodservice market. The chilled season in Europe was not as good as some importers had hoped, with high prices leading to product substitution to other items. However, the situation was manageable, and exporters were looking forward to a successful 2020 (Fig. 1).

In February/March of 2020, the global food service sector targeted by New Zealand venison was severely affected by the outbreak of Covid-19. The pandemic significantly disrupted exporters' and DINZ's programmes, with sales cancelled and events needing to be postponed. Added to this were import issues in China with confusion over the status of venison as a game meat.

The priorities for exporters and their customers, and DINZ, changed rapidly. With the suspension of the restaurant trade, DINZ was focused on:

- Restoring access to China
- Helping marketers find new channels for venison
- Creating material to target customers online or via retail outlets

DINZ chefs were put to work creating digital content and conducting webinars with retail buyers and product developers. Social media programmes were activated to take the message about the quality and availability of New Zealand venison to potential customers' homes. In addition, new meal ideas were provided to chefs as they tackled the challenge of shifting from fine dining to road-side pick-ups.

Rapid reallocation of energy, focus and funds allowed replacement activities to pick up over the second half of the year, but the reduction in the international restaurant and foodservice trade severely impacted returns for New Zealand venison producers. Schedule payments to producers dropped over the year with spring payments on contract peaking a little over $7.00 per kg.

Production Up As a result of the ability to process deer, even through Covid Alert levels 3 and 4, total production in New Zealand for the year was up 8.2% on the previous year due to an increase in numbers and an increase in average carcass weight. Total export volume for the year ending September 2020 was down 2 percent year on year. The value of these exports was down 19 percent, however, reflecting the drop in international meat prices and a reduction in chilled exports to summer markets in Europe and the United States.

With a higher exposure to the food service sector, the impact on venison returns has been greater than for sheep and beef producers.

The industry however is confident that diners will return to venison as restaurants reopen and some of the new initiatives to increase sales through retail outlets and home delivery will broaden demand and leave the industry in a stronger position. But current conditions remain unsettling for farmers who sell young stock for venison production to specialist finishers.

8   The New Zealand Velvet Antler Industry

While the demand for deer velvet antler initially came from the South Korean market, by the 2000s the traditional channels and uses had become constrained in terms of volume demand and returns. In 2011 New Zealand velvet antler production was approximately 430 tonnes. Production had been falling for 3 to 5 years prior to this and prices subdued, even though product quality had been improving.

The NZ velvet industry, led by the scientific evaluation of velvet composition that was undertaken by Velvet Antler Research New Zealand (VARNZ) a joint venture between AgResearch at Invermay and the Game Industry board (now Deer Industry New Zealand) encouraged the launch of the use of deer velvet in healthy food products initially in Korea and latterly in China and Taiwan and an emerging Vietnamese demand. This has created an expansion of new products available to new consumers, well away from the traditional medicines market and attracting a new type and class of consumers.

New Zealand's focus on free international trade has also benefitted the velvet market with improved market access. A Free Trade Agreement was negotiated with China in 2008, and with Korea in 2015. This opened the way for much closer collaborations between velvet suppliers and velvet retailers. The shift to being able to direct export to China saw a reduction in pressure on the Korean market, allowing expansion of velvet production while maintaining, and then increasing price (Fig. 2).

The new dominant consumer in Korea has become the discerning 30- to 50-year-old mother buying for herself and her family. Products that provide protection and improve the resilience of her family are key to purchasing decisions. This is

Fig. 2 Michael Lee—Rokland Corporation, second-generation NZ velvet processor showing the contemporary premium health food products that NZ velvet now goes into

empowered by the research that demonstrate the anti-stress, immune response support and recuperative properties of deer velvet.

In February 2021 the industry took a further step in growing Chinese demand for NZ product with the establishment of a three commercial party Velvet Coalition. Three companies marketing deer velvet antler in Asia—CK Import Export, PGG Wrightson and Provelco—have formed a coalition to develop a market for New Zealand velvet as a health food ingredient in China. The three-way collaboration will give the critical mass needed to make an impact. In all other respects, the companies will continue to compete vigorously with each other for sales and farmer supply.

DINZ's experience in these new initiatives recognises that two of the most highly prized ingredients in traditional oriental medicine: deer velvet and red ginseng— have in recent years enjoyed an explosion in demand from South Korea for use in branded natural health products to combat fatigue and boost immunity. This has been a driving force in the increasing sophistication and new use and positioning of NZ deer velvet in South Korea.

The Velvet Coalition has a vision to replicate this phenomenon in China, where locally produced and imported velvet is still mainly used in traditional medicines. The Coalition has appointed a Shanghai-based business development manager to direct the groundwork. Their role will be to identify and work with a small group of brand-name companies that are willing to develop and promote products based on velvet from New Zealand. The three companies will collectively provide most of the funding, with a matching contribution from NZ Trade and Enterprise (NZTE) and administrative support and some funding from DINZ.

The remarkable growth of the velvet antler industry and its strong connections in the key markets remains a highlight of industry evolution and continues to fire the enthusiasm and passion within the NZ deer farming industry and underpins solidly the status and direction of the NZ industry today.

That is better expressed by the recognition that the NZ export receipts from both venison and velvet antler sources are roughly equal in export value and the strength and opportunity of velvet antler in traditional and new functional food use underpins the on-farm production, genetic improvement management and QA systems across the NZ deer farming landscape.

The velvet industry's current priorities and actions have been concentrated on the growth of the New Zealand velvet (ingredient) brand position in South Korea. That is enacted by working closely with a few selected partners in Korea to promote New Zealand velvet as a premium ingredient and continues to pay off. Most Korean companies report good online sales growth over the previous year, offsetting some decline in sales through physical stores due to COVID-19 lockdowns.

With ongoing market promotional activity, there were another 20 velvet health food products launched onto the Korean market in 2021.

The second part of the overall programme is to protect New Zealand velvet's position in the traditional sector. DINZ launched a new Oriental Medicine Doctor (OMD) scholarship initiative to get closer to leading universities that have tradition- ally taught that Russian velvet is the best. The New Zealand Ambassador to Korea presented the scholarships, which resulted in a wide media coverage.

The major next step is to implement a similar Korean health food strategy in China through encouraging companies to invest and develop products containing New Zealand velvet with a company recently achieving successful registration of a healthy functional food product that contains velvet. Two more companies are investing in velvet product development.

The market strategy continues to seek new geographic markets for New Zealand velvet with the industry's closest Korean partner, Korean Ginseng Corporation, launching successfully its new convenience velvet product '*Choen Nok Everytime*' into Taiwan.

9 Velvet Statistics for the 2019/20 Season

Velvet production increased from 803 tonnes in 2018/2019 to around 840 tonnes for the 2019/20 season. However, some importers reported that increasing velvet production (nearly doubling in 8 years) resulted in the decreased prices paid to producers this year.

While the average velvet price came back from the previous year's historic high of $ NZ 132.50/kg to $ NZ 120.00/kg, DINZ estimates the industry has achieved a second year of farmgate values of more than $NZ 100 m (see Fig. 3).

For processed New Zealand velvet the industry's current unique trade advantage with Korea continues to deliver improved benefits. On 1 January 2020, New Zealand processed velvet only had 12 percent duty compared with 20 percent duty paid for frozen velvet or velvet from other countries. The advantage was successfully negotiated by New Zealand officials for the New Zealand–Korea Free Trade Agreement (FTA). The FTA was implemented on 20 December 2015 and 1.3% of duties are removed every 1 January until New Zealand processed velvet eventually becomes duty free.

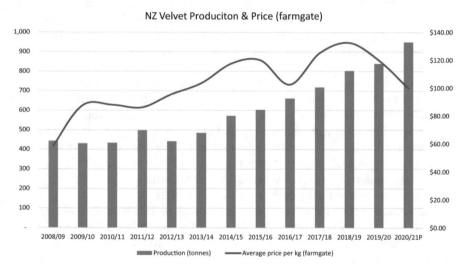

Fig. 3 Increases in both price (NZ$) and volume (tonnes) of velvet antler harvested from 2008 to 2020 demonstrating the trend towards increasing value and volume as trade agreements with China (2008) and South Korea (2015) were negotiated (DINZ 2016)

10   Velvet Production and Management on Farm

Venison or velvet antler? For most deer farmers the removal of deer antler from breeding stags, for highly productive selected velvet specialist herds and for young venison stags (9–15 months of age) is a carefully managed balance of maximising the genetic and nutritional needs for growth of both liveweight and antler potential.

The profitability and specialisation of both systems often leads NZ deer farmers into specific velvet focused production systems by choice, or fast finishing premium venison supply from predominately young animals.

While diversification of venison markets into the growing North American and Chinese and other Asian markets is relatively recent, the UK and European autumn and early winter venison consumption remains dominantly important to NZ. The seasonality contrast with these markets' peak demand coincides with supply of young 9–12 month old males from late August to November which overlaps with the southern hemisphere velvet growth season in spring and early summer.

Today's genetic advance to meet this peak season of venison demand with supply from young males coincides with adolescence, including pedicle initiation and the early growth of yearling or spiker velvet antler. The NZ revised Code of Welfare, Deer (2020) and the Transport of Deer regulations prohibits the transport of deer in growing velvet antler longer than 110 mm in length including pedicle of any age. This essentially demands that all deer farmers at some point will have a management need to have velvet antler removed from breeding stags, young prime venison animals at 9–12 months of age or mixed age velvet stags.

This has led the industry to evolve a well-structured targeted mixed-age specialist velvet herd and industry initiated and independently monitored velvet removal protocol and formal structure. Government and public expectations also required this to be capable of independent audit and formal reporting across the industry to meet all welfare expectations, pain-free removal and all hygiene, identification and cold chain storage demands.

11   Velvet Antler Industry

11.1   Structure and Processes: Stag Welfare and Velvet Removal Programmes

11.1.1 The National Velvetting Standards Body (NVSB) Programme
From the early 1990s, it was clear that the production of velvet antler would be a mainstream contributor to the interest and expansion in deer farming and a highly productive and profitable part of the industry.

The young industry had some considerable advantages in supporting this new venture. Across the country leading veterinary practices had already been involved in understanding the safe tranquilisation of deer being darted and captured in the wild deer recovery phase of establishing this new industry.

The skilled understanding of the role of chemical sedation, recovery and anaesthesia were a critical part of developing the new velvet harvest industry and meeting the animal welfare regulations of that era. With a rapidly expanding stag herd and a concentrated very active harvest season, it became critical and obvious that trained accredited farmer involvement in the process was imperative.

The industry began negotiations with the Ministry of Agriculture, the New Zealand Veterinary Association and the New Zealand Deer Farmers Association and the NZ Game Industry Board (NZGIB).

As the concept evolved, discussion and inclusion of the Royal NZ Society for the Prevention of Cruelty to Animals (SPCA) also assisted with the development of a set of standards, meeting public expectations, practices, training, certification and independent audit that would allow qualified farmers, under indirect veterinary supervision to carry out this controlled surgical procedure. In addition, post-removal animal care and recovery were an integral part of the process.

The New Zealand Game Industry Board (now DINZ) is funded by compulsory levies on both velvet antler and venison, also moved to support the process with the appointment of a National Quality Standards manager and resources to administer and run the programme now termed the NVSB in the early 1990s. The NVSB continues successfully and mostly unchanged today although under consistent review to Government (via Ministry for Primary Industries) and the constituent bodies, NZ Deer Farmers Association and NZ Veterinary Association, plus formal reporting to the DINZ Board via the Quality Assurance manager.

The National Velvetting Standards Body (NVSB) is a formal structure with a governance committee comprising two farmer representatives nominated by deer industry levy payers and approved by the NZDFA's Executive Committee and two veterinarians nominated and approved by the New Zealand Veterinary Association (NZVA). Appointment on to the committee is for a period of 3 years with the administrative work to support the NVSB undertaken by the Deer Industry New Zealand (DINZ) Executive specialist, subject to funding approval from the Board of DINZ. The costs of running the NVSB's registration, supervision and audit programmes are directly recovered from velvetters applying for annual registration and who are registered.

11.1.2 NVSB Form and Function

The NVSB's mission statement is to implement the Code of recommendations and minimum standards for the welfare of deer during the removal of antlers (and applies to both antlers in the velvet stage of growth and hard antler and any regrowth removal).

The NVSB implements a national velvet removal training and certification programme and provides appropriate management of the programme such that farmers wishing to velvet their own animals under veterinary supervision do so within the framework as established in the Code, having been trained and certified to an internationally acceptable standard. The NVSB ensures that the programme operates efficiently and effectively, and reports to the constituent bodies on an annual or on request basis.

Functionally NVSB acts in these critical areas:

- Provides a contract for execution between velvetters and their supervisory veterinarians which is reviewed and renewed annually
- Ensures appropriate training of veterinarians and arranges for veterinary assessment of farmers wishing to become certified
- Effects the distribution of training manuals to veterinarians that are the resources for both a written examination and for on-site observation and approval of a farmer properly velvetting 3–5 stags
- Maintains a management system for recording critical information
- Monitors and records the progress towards certification of individual farmers and communicates results to veterinarians and farmers systematically
- Communicates the development and implementation of the programme and other related issues to deer farmers, veterinarians, and the relevant animal welfare regulators (Ministry for Primary Industries, MPI)
- Mediates and arbitrates in situations requiring independent judgement and resolves technical and administrative issues as they occur
- Establishes policies and procedures to achieve maximum farmer and veterinary compliance with the programme
- Conducts an annual in season audit of farms, facilities and process in a formal programme in collaboration with Deer Industry New Zealand ('DINZ')
- Reviews and reports on the performance of the programme, and modifies it where necessary after consultation with DINZ, MPI, NZVA and the National Animal Welfare Advisory Committee ('NAWAC').
- Provides annual reports to the Chief Veterinary Officer MPI on the performance of the working parties.

11.1.3 Certification
The velvetter can choose to be certified for any or all the following methods for removing, or preventing velvet growth:

- Chemical restraint (with local analgesia)
- Local anaesthetic (with physical restraint)
- NaturO™ ring procedures for spikers (yearling stags) (including the cable tie method)
- NaturO rings for fallow spiker deer.

(NaturO rings are rubber rings which provide analgesia by compression of the antler base)

11.1.4 Initial Registration
To register on the NVSB programme, the velvetter needs to sign a supervisory contract with a supervising veterinarian) and pay the initial joining fee to the NVSB. Material is then supplied to the applicant including the Farmer Velvet Antler Removal Manual ('The Manual') which covers the practical and additional

knowledge related to velvet growth, anaesthesia, correct surgical procedures, post-operative care of the stag and velvet product traceability, hygiene and storage. The applicant must then pass three stages of evaluation:

The first stage requires passing a multiple-choice theory test. The applicant completes those sections of the test relating to the velvet removal techniques he or she wishes to use. The completed test is marked by the NVSB.

Stage Two requires a visit conducted by the Supervising Veterinarian and covers facilities, a practical assessment, and oral questions. Before this visit, the velvetter will undertake training from the Supervising Veterinarian to attain competence to at least the minimum standards specified in the Manual. The practical assessment includes the Supervising Veterinarian observing at least 3 stags being velvetted.

If, despite passing the evaluation, the Supervising Veterinarian nevertheless considers that the welfare of the animals requires further supervision of the appliance, the veterinarian has a duty to undertake more supervisory visits at the applicant's expense, although oral questions and formal assessments need not be repeated.

Stage Three requires an in-depth assessment conducted by an independent assessing veterinarian and again includes assessment of facilities, practical assessment, and oral questions.

The applicant must be competent to at least the minimum standards specified in the Manual. The assessment includes the veterinarian observing at least three stags being velvetted. The applicant's Supervising Veterinarian organises the assessing veterinarian visit, and this may be a combined stage two visit and stage three visit at the Supervising Veterinarian's discretion.

11.1.5 Formal Veterinary Supervision

Due to legal and ethical requirements, it is necessary for a formal written contract to exist between the farmer wishing to carry out velvetting and his or her Supervising Veterinarian. Essentially, since lawful velvet removal by a non-veterinarian involves joint responsibility being taken by velvetter and veterinarian for the welfare of stags during the removal of velvet antler, what that joint responsibility means in practical terms is set out under agreed terms in the formal contract.

The terms cover the supply, storage and use of prescription animal remedies and is supplied by the NVSB for signature by the velvetter and veterinarian for that year only as the full process is an annual obligation.

The contract relates to velvet removal on one property only. If a velvetter moves properties, changes the Supervising Veterinarian, or a Supervising Veterinarian moves practices, a new contract must be signed to ensure the programme remains valid.

In the case of a farmer who is responsible for the direct care of deer on more than one property there must be a separate contract for each property and each property will need to be assessed in consecutive seasons. A separate certificate is issued for each property.

Receipt of the signed contract by the NVSB confirms registration of the farmer for this programme and the relationship between farmer and Supervisory Veterinarian and requires the payment of an annual fee.

11.1.6 Annual Supervisory Visit

Each velvetter certified for Local Anaesthetic and Chemical Restraint must have a Supervisory Veterinarian visit at the start of each season to assess both theoretical and practical competence, using the same criteria as the initial training. The veterinarian's duty to visit more frequently during the velvet season continues where he or she considers this necessary for the welfare of the animals.

11.1.7 Velvetting Records

To fulfil the requirements for the supply, storage and use of registered animal remedies and to ensure the welfare of stags during the removal of velvet, the programme requires the keeping of accurate records relating to velvet removal. Records must include details of the following:

- Stag numbers
- Drug usage
- Deaths and unusual reactions
- Use of NaturO rings
- Details of the identification tags used

A NVSB Velvet Record Book is supplied for this purpose to a new velvetter and annually thereafter, before the season commences. Annually by 31 March, each velvetter must have returned his or her completed Velvet Record Book to the supervising veterinarian, as well as any unused drugs, including partially used containers. It is the supervising veterinarian's responsibility to advise the NVSB of the velvet record book return and drug reconciliation.

The minimum penalty for failure to comply with NVSB programme requirements is revocation of certification for at least one velvetting season.

11.1.8 Audit

Audit of the NVSB programme entails monitoring of people and processes to verify whether those acting under the programme are fully compliant with its requirements. Systematic monitoring of the programme encourages compliance and therefore enables the deer industry to prove that a high standard of animal welfare is being maintained by its farmers.

Examples of non-compliance which that audit is designed to discover and deter are non-certified people velvetting their deer, certified people who have temporarily and voluntarily withdrawn from the programme (these people are said to be 'in abeyance') continuing to velvet their deer, using velvetting drugs for purposes other than for velvetting and mis-reporting or non-reporting of velvetting drug administration.

All velvetters are eligible to be audited no matter what technique they are certified for.

The NVSB selects velvetters for audit, approves well-experienced and highly qualified veterinarians as independent auditors and allocates the selected velvetters to auditors. The auditor arranges the audit with the velvetter. The audit will include facilities, practical assessment, and oral questions. The auditor is required to inspect the reconciled Velvet Record Book and the velvetting drugs on hand.

The NVSB and Supervising Veterinarian are notified of the outcome of the audit by the auditor, including any corrective actions to be taken by the velvetter and Supervisory Veterinarian.

The annual fee paid to the NVSB contributes towards the costs of the annual audit of the programme.

11.1.9 Assessment Criteria

Assessment criteria are used for supervising, assessing veterinarian visits and audits, and recorded systematically now electronically and verified by GPS coordination at the property location.

A major non-compliance is a contravention of any legal requirement and/or something that directly jeopardises the safety/welfare of the stag and/or safety/welfare of the velvetter.

A major non-compliance is a contravention of any requirement relating to:

- Drug storage
- Drug record-keeping
- Drug hygiene
- Contingency responses
- Alleviation of stress
- Competency to carry out reversal technique
- Effective analgesia
- Effective velvet removal
- Tourniquet application
- Operator safety
- Test for analgesia

A minor non-compliance is a circumstance where the requirements of the programme are not met but in a manner that does not contravene any legal requirements and/or directly jeopardise the safety/welfare of the stag and/or safety/welfare of the velvetter.

Under a facilities assessment, major non-compliances must be addressed immediately, and minor non-compliances must be addressed prior to the next Supervising Veterinarian visit.

This has become vastly more important since the 2017 introduction of the MPI introduced and DINZ (via NVSB administered) Regulatory Control Scheme which now monitors shed hygiene and velvet storage facilities during the removal process and extends into the cold chain supply system and frozen product storage and transportation of velvet. Finally, the frozen or dried product is certified by MPI on export who also ensure that the destination countries importation requirements are all met for this NZ origin product.

11.1.10 Abeyance

It is possible to place certification in abeyance once only for a period of up to 3 consecutive seasons. After 3 consecutive seasons in abeyance, certification is revoked, and the entire certification process must be completed again before the velvetter can recommence velvetting under veterinarian supervision.

Termination of certification takes place in accordance with the provisions of the supervisory contract. Upon termination, the velvetter will be required to surrender such things as all drugs and identification tags on hand.

11.1.11 Modern Advances and New Regulations

(a) *2017 Regulatory Control Scheme*

The National Velvetting Standards Body (NVSB) de-velvetting programme has been regulated by the Ministry for Primary Industries since August 2017 and is now a mandatory requirement operating as The Regulated Control Scheme for Deer Velvet Harvest (RCS). (See: www.mpi.govt.nz/document-vault/19379). Its prime purpose is to ensure that deer velvet intended for export complies with market access requirements throughout its harvesting, handling, storage and transport.

On farm this has meant a substantial lift in standards relating to a clean zone with the deer yards used exclusively for the velvet removal process and similarly new specifications for the transport of antler from deer sheds to freezers and subsequently for sale directly off farm or via transport to depots and final NZ grading and sales destinations.

The core elements of a defined clean zone legally require under the RCS for:

- Farmers to supply a floor plan marking all clean zones within the shed and keep this with records for auditing purposes. This is a legal requirement and part of the audit process.
- All surface areas where velvet may come into contact (e.g., floors, racks, tables, benches, scales, and containers are to be non-porous and washed, disinfected and cleaned.
- Clean Zones clear of any visible contaminants, such as mud, dust, rodent droppings, machinery, animal health treatments, tools
- Floor must be able to be washed and disinfected, non-porous, not covered with a build-up of mud, dust, or faecal contamination
- Wall surfaces meet the new requirements including raw timber surfaces must be covered or coated to allow for washing or cleaning
- A supply of water is available for washing the facilities to enable the clean zones to be washed out according to the RCS

- A suitable approved (to the correct code) disinfectant/cleaner is used and appropriately applied, and application of the disinfectant will meet requirements
- All animal remedies are stored outside out of clean zones and the zones are free of any other compounds and animal remedies. Can be stored in cabinets within clean zones provided they are kept closed and washable.
- For mechanical constraint in deer crushes all surface areas are free of any rips, tears, holes, and rough surfaces are eliminated and free of contamination 12.
- Rodent control measures are in place

The RCS requires all consignments of velvet going off-farm to be accompanied by a Velvet Status Declaration (VSD) to identify the velvet moving through the supply chain. The VSD books are administered through the NVSB and DINZ are distributed and sent to all velvetters registered with the NVSB and all known velvet buyers.

When selling velvet for export, velvetters need to sign off a VSD stating that the harvest, handling, storage, and transport of their velvet has met all the requirements of the RCS and that becomes a legal statutory declaration.

Auditing of velvet facilities to ensure they meet the RCS requirements is undertaken by MPI recognised persons that includes all current NVSB auditors who have become recognised persons under the Act. Other recognised persons such as MPI compliance officers can also audit velvetting premises.

For storage of frozen velvet on farm, at collection depots, commercial transport and in grading, collection, and processing centres prior to manufacture in NZ or export, similar expectations are in force

- Freezers must be capable of reaching a minimum temperature of $-15\ °C$. and are clean inside and outside with the areas around all freezers are able to be kept clean and clear of any build-ups of mud, dust, blood, rodent droppings, and control measures are in place for any rodent contamination.
- Freezers used to store velvet are free of any other product and only clean receptacles used for transporting or transferring velvet antler.
- Any velvet stored in freezers must also be tagged with the NVSB official tag and fully recorded in NVSB Velvet Record Book and the Velvet Status Declaration Form correctly filled out and is a statutory declaration.

12   Velvet Antler Identification and Traceability

12.1   Modern Velvet Antler Systems Regulations, Traceability and Best Practice

From the early days of the industry velvet antler had always had some form of identification linking the product to the property of origin. On the establishment of the NVSB in 1993, the use of individually unique numbered plastic zip-lock tags issued by the NVSB through the on-farm veterinarians has been in place. Each tag was individually identified to both the property and the veterinarian practice that supervised the removal process. Tagged traceable velvet was also required at export for verification with the Ministry for Primary Industries of New Zealand origin and its compliance with the NVSB and MPI regulations of the time, plus any additional verification for the Korean and Chinese importing markets.

Since 2017 with the advent of technology (bar codes and QR scans), waterproof tear-resistant NVSB branded paper tags bearing a unique code identify the NZ farmed origin and link with the NZ Industry and the implicit veterinary supervision and compliance standards related to drug with- holding periods, allocation and overview from veterinary practices. The driver for this level of traceability has been the growth and incorporation of verified NZ production and use in the rapidly expanding healthy functional food markets in Korea and now increasing interest in China, which use NZ origin velvet antler as the exclusive ingredient.

The success of this initiative has prompted further sophistication and refinement in identity. 2021 sees the introduction of the exclusive NZ banded VelTraK™ Tags and inventory for all velvet antlers.

VelTrak is a fully electronic, web-based system that enables velvet to be tracked and traced each step of the way from the farm to the market (and vice-versa) built around the Ultra High Frequency (UHF) embedded chip with a unique code and signature in each tag. Each tag also carries an individual bar code, which can be read on farm with a simple inexpensive barcode scanner, or the digital numbers recorded in farmer records.

It proves to customers that the velvet they are buying is produced on quality assured NZ farms and meets our stringent animal welfare and food quality standards and complements the recent activity on farm in upgrading facilities and process developed as part of the introduction of MPI's Regulatory Control Scheme (RCS legislation) that underpins both the welfare and hygiene standards now expected in deer sheds and velvet storage freezers and collection depots.

DINZ has developed VelTrak to help lock in the price premium that NZ velvet now enjoys in South Korea over velvet from competing countries.

It will also help our velvet exporters and DINZ to grow the market for velvet-based health foods across Asia building on the successful new business in South Korea through provision of proof of integrity and traceability needed to protect the reputation of their brands. This level of traceability is also expected to be a major selling point when marketing NZ velvet to similar companies in China, Taiwan and elsewhere.

In practical terms, VelTrak is a web-based application (a website) that can be accessed from a desktop PC, laptop, tablet, or mobile phone and does not require a deer farmer to install any special software on these devices to use VelTrak. The reading and inventory development will be done by the velvet buyers or packhouse scanning velvet tags with a UHF rapid read scanner which can read and record in bulk a large number of individual velvet antler sticks.

For 2021/2022 velvetting season and beyond, all farms, vet clinics and businesses involved in velvet removal, procurement, warehousing, processing, and packing for shipment will need to be registered with VelTrak where each user will have a unique email address as an ID for logging into the database.

Fig. 4 Designated black VelTrak tag made from synthetic paper at a recommended retail price of $NZ 0.50. Source: Deer Industry New Zealand, Industry VelTrak Electronic and bar coded velvet tag Id system Operational August 2021

From the 2021/22 velvet season, each stick of velvet will need to be tagged by the farmer with using a designated tag (Fig. 4).

Tags are part paid for by farmer and part from velvet levies and will be allocated to farmers by their veterinary clinic, which will have recorded them on the VelTrak website. The tags will next be scanned and recorded by the velvet buyer, or the receiving warehouse, using a UHF scanner.

Traceability at each stage is restricted to a one-up one-down level from farm to velvet buyer to processor to exporter and importer.

12.2 On-Farm Production Records

Velvet antler weight at a young age (2 year old) cut at the correct stage of growth for maximum market appeal and processing quality is a proven highly heritable trait and part of the DEER select genetics recoding system for those breeders and commercial herds who wish to contribute their data to DEER Select. On farm the now-mandatory National Animal Identification and Tracing system for lifetime tracking and tracing (NAIT) utilises a low frequency electronic chip with a unique animal ID. Velvet producers can now access software that links that NAIT ID to a visual ID tag and records velvet weights, grades and if required a velvet value plus an animal liveweight seamlessly.

In addition the industry, working in partnership with producers, buyers and marketers have co-operated to produce the New Zealand Industry Agreed Velvet Grading Guidelines as an illustrated recording of all commercial grades for red deer and a specific additional chart for wapiti/elk. This chart is reviewed by an industry panel covering all sectors periodically and will be updated for the 2021–2022 season, grades are determined by both weight and clear beam circumference measurement just below the trez tine, and further defined by the presence or absence of the bez tine. An ideal quality grade has rounded tops on the main beam and is

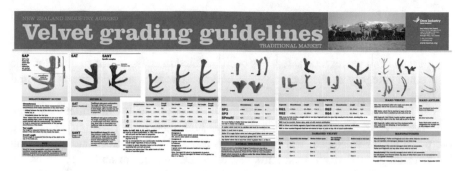

Fig. 5 Industry agreed Velvet grading guidelines. Source: Deer Industry New Zealand, National Velvet Standards Body Resources, Grading Charts Velvet Antler Update 2021

harvested before any indentation or early growth of royal tines begins and judges this growth stage to be the most biologically active, with soft non-calcified tine tips and a soft internal honeycomb structure in the main beam. Processor/exporter input into the grading system has been a key factor for these highly heritable traits (Fig. 5).

13   Nutrition and Velvet Production Management

In the 40 years since New Zealand farmed deer industry began as a new venture in pasture-based livestock production, deer farmers have been driven by the rapid genetic gains in velvet antler production and in greater understanding of product end use in traditional oriental medicine. As knowledge has grown, the value of relationships between our countries, through science, market development and the communications between industry organisations have led to a new international understanding of the customer expectations and the value of the product. This has led to advances in farming production and processing and distribution chains for velvet antler and its products that support today's customer's expectations.

The NZ velvet industry has continued to invest in quality assurance programmes through developing relevant grading standards and continuing improvements in medical and food product quality assurance programmes and velvet supply chain protocols. This drive results from the rapid increase in end use in the healthy food business for velvet antler and velvet extracts and ingredients. This market is now 20% of NZ velvet's end use and is growing rapidly in Korea with a growing interest in China.

To support product integrity, the NZ industry now has a mandatory government-initiated electronic identification animal identification traceability programme for deer (and cattle). This can be linked to the Deer Industry New Zealand velvet antler Identification and origin verification programme that provides traceability from animal and NZ farm of origin through to the market. Authenticity can also be supported by forensic use of isotopic signature if required in any proof of origin and provenance request.

A new programme (Velvet Status Declaration) that requires declaration of deer health status and velvet food safety has been initiated in 2014, as part of an assurance of product integrity as the use of velvet antler expands in a healthy food use and market.

On farm, the ongoing emphasis on production of quality of antler has been embraced by NZ deer farmers. This emphasis highlights aspects around enhancing the potential for biological activity, reduced ash and calcification, and employing better feeding standards to match the rise in genetic quality of antler. Much of the refinement in feeding and nutrition systems comes from a greater understanding of deer nutrition and antler growth needs gained through international collaboration of scientists and national deer farmer and industry groups and the open exchange of knowledge and industry understanding.

The NZ velvet industry believes that the value of velvet antler in healthy food products beyond the Traditional Oriental Medical market is an exciting growth opportunity for all velvet antler industries, as discerning consumers gain access to these products and enjoy their benefits. In turn, the farming profitability of velvet antler production continues to consolidate and offers a strong foundation for the ongoing future of NZ deer farming, which will quickly extend to other industries, particularly in China and Korea.

The NZ industry acknowledges and appreciates the value of a wider association and the potential for cooperation and combined market development in the spirit of both incorporating co-operation, but remaining clearly in commercial competition.

13.1 Feeding for Optimum Velvet Production

The genetics of an animal determine the potential productivity of that animal with adequate targeted nutrition to express that potential. That expression is ultimately challenged by how well the animals are fed across the variability of the seasons and the stag reproductive cycle. Finally, how the environment of the animal is managed to balance its seasonal demands, its age group and social herd structure and the day-to-day management on deer farms have as much influence on the expression of genetics as nutrition might have.

In the velvet industry the application of assisted reproductive technology (AI and ET), combined with investment in the top antler genetics available, and the high heritability of velvet antler weight (heritability estimated at 0.45–0.5) has led the remarkable progress in productivity across all age groups as expressed by velvet weight at the harvesting stage of growth as defined by the industry's velvet antler grading charts.

This is demonstrated by a review of the progress made in red deer and elk/wapiti as seen in the National Velvet Antler competition where the best heads in the country compete for national honours. Figure 6 shows the average weight for key classes including the 3-year-old class since 1982, when the competition began. The key of

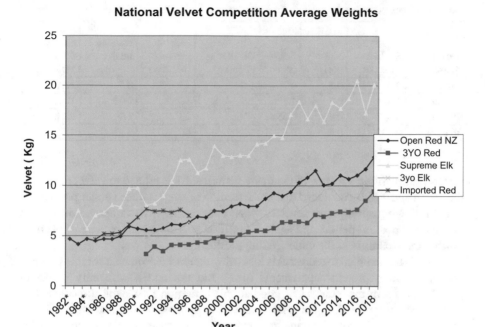

Fig. 6 The New Zealand national velvet competition results from 1982 to 2019 demonstrating the trend of increasing genetic merit across age and breed in the national velvet antler herd

course has been the excellence of feeding and management from these leading producers to deliver this genetic potential.

Management of velvet stags for optimal production depends on judging the growth stage for harvest which is strongly influenced by the expression of the genetics of velvet and antler growth rate as it is a rapid change from ideal to overgrown. Much research has focused on feeding the stag during its adult life, trying to gauge what to feed and when.

It is well known that any restriction in feeding stags for maintenance or growth phases, particularly from autumn until spring has a significant and negative impact on antler growth in New Zealand pastoral-based feeding systems. This has not changed. The simple message is "to feed well… but overfeeding will not deliver anything other than what the stag is genetically capable of".

To avoid the problems of low pasture growth during late autumn until early spring it is recommended to provide targeted supplementation. Recent research has also highlighted that it is not how much weight an adult stag loses in autumn that counts, but how fast he regains that condition in spring. This means that ensuring plenty of feed is available in the early spring, around casting time, is important to maximise velvet production.

Table 1 Concentrate diet formulation for wapiti stags in China (Liang et al. 1993)

Growth phase	Soybean cake (kg)	Maize (kg)	Bran (kg)	Distiller's grain(kg)	ME intake (MJ/day)	Dry matter intake (kg/d)	Diet energy density MJME/kg
Pre antler	1.0	0.5	0.7	1.5	45	3.7	12.1
Antler	1.45	0.7	1.1	1.5	58	4.75	12.2
Rutting	0.8	0.45	0.65	–	23	1.9	12.1
Recovery	0.95	0.45	0.65	1.25	40	3.3	12.1

Dr. Jimmy Suttie's review of the many historical trials related to velvet growth in the 1970s and 1980s concluded that underfeeding of stags reduced antler production in the following spring most severely when poor nutrition was imposed during the late autumn post rut period, and again if nutrition was compromised in early spring. Further trials confirmed the underfeeding production loss as well.

Work that attempted to redress this loss using enhanced protected soluble proteins instead of other protein supplements (peas) did not add significantly to velvet production. From that concept of good supplementation in general the concept of strategic enhanced feeding for velvet production evolved and started with utilising nonpasture diets in combinations and at specific key times of the seasonal antler development cycle.

In practice, the concept of strategic feeding at specific stages of the antler growth cycle to realise both potential beam dimensions and weight has been advanced adapting the pastoral situation along similar patterns to the Chinese pen feeding systems (Suttie et al. 1996).

Specifically formulated diets combining protected protein (18%), protected fat (8%) and organically bound trace minerals amplifying copper and selenium (Bioplex TM 500, Cundy Technical Services) showed a variable but significant mean weight increase of 7% in commercial herds and strong improvement in antler grade.

Encouraged by the Chinese system that changes the balance of protein and energy throughout the year, further developments in strategic feeding are indicated. During the antler growth period, protein levels of up to 25% are targeted. Recent research with Chinese sika deer supports New Zealand research which indicates that dietary protein concentration of between 16 and 18% may be optimum and cost-effective.

Table 1 shows a concentrate diet formulation for a production Wapiti stag *(Malu)* in China (Liang et al. 1993). Diets like this are often the result of local availability and may overstate the requirements for protein feeds. Roughages including ensiled corn, tree leaves, podded plants, forage grasses and roots and melons fed first and complemented by fermented, steamed, or expanded concentrates to improve feed conversion efficiency and balance the cost of grains and roughage sources (Gao et al. 1996).

While additional protein has not improved the velvet production of well-fed stags, it is important to remember what that means. Often stags in New Zealand are fed on silage during the winter and well into the spring. Silage is a product that has already been fermented and this means that although crude protein concentration may look adequate, often the digestion process does not yield enough protein to meet the stags' requirement. This is especially important in early spring when antler growth is initiated around button drop.

Moving to specialist silages that are based around legumes and chicory can help increase the true protein supply to the animal.

In today's winter-feeding regime, there has been a dramatic increase in the role of brassica crops (swedes, turnips and kale/choumoellier and fodder beet) presented as self-fed crops in situ. Such crops are generally strip or block grazed behind 4–5 wire high temporary electric fences as a controlled grazing practice. Supplementation of balanced energy and protein-rich diets via a proprietary manufactured grain-based nut and in recent times offering Palm Kernel Expeller (PKE) is a common supplement; although, there have been no formal velvet growth trials with controls reported to verify this, it has become popular farmer folk lore.

In recent years, the basic winter forage diet for the AgResearch Invermay herd of mixed-age stags on grass pasture baleage (or silage) was changed to a specific crop baleage fed ad lib for the winter and pre casting period in combination with 1.2 kg of whole-grain barley. With both red clover and lucerne sources, wastage was almost nil in comparison with pasture silage. Velvet weights averaged 4.07 kg compared with 3.78 kg the previous year.

14   Management of Velvet Stags and Social Impact

Most deer farms with significant numbers of stags run a herd structure where the mixed-age stags, typically 4 years and older, are run and managed as an elite highly selected group of top performers. Typically these are the first to cast their antler buttons in early-mid August and 55–70 days later are ready for harvest at the optimal stage for value, biological activity, and conformity with the well designed, informed and frequently reviewed industry velvet grading guidelines standards.

In line with that, young spiker and 2-, 3- and 4-year-old cohorts are predominantly managed individually within their own age groups and respond well to that familial social structure and lack of social disturbance. That pattern also fits with the seasonal variation of antler growth with age and allows targeted feeding for each age group. This is particularly important in the critical feeding period post rut where there is a short window in late autumn and early winter to recover lost body condition.

Management of a stable group of like aged stags is also advantaged if it is possible to provide adequate space per animal in the post-rut recovery to minimise competitive aggression and disruption to the social order, particularly if supplements are being fed post-rut during May and June. After button drop, the stag needs a diet that is well balanced to provide enough energy and protein to ensure good velvet growth. Often spring pasture is fine for this period, but it must be remembered that the stag needs enough at this time. The pasture needs to be of good length and unsoiled.

This period begins in early August. Often stags are still on pasture silage through August, and this will mean that the added protein requirement will not be adequately met.

A final factor in velvet growth is the mineral intake of the stag. Little research has been done to verify the concentrations of minerals in the diet. Some Chinese research with Sika deer suggests that the dietary concentration of calcium and phosphorus be 0.89% and 0.52% of the diet respectively for maximum velvet antler growth (Wang et al. 1997).

Ca and P levels in the diet had effects on Ca contents in antler serum at the antler-harvesting stage, but no effects on P contents. However, much of the calcium and phosphorus required comes from remodelling of the stags' bones and so the supply of more calcium has little overall effect on velvet weight. The amounts of other minerals that are essential for cell proliferation, such as copper and zinc, have not been specifically investigated although the role of adequate copper levels at least has been considered very important.

15   Management of Velvet Stag Herds

Management factors also play an important role in optimising production. The key to advancing velvet growth is in the 3- to-4-week post-rut recovery period and in the pre-casting period. Beneficial nutrition effects are enhanced by forming stable cohorts of similarly aged stags from as early as yearling age and to avoid adding in new animals or mixing groups.

During the rut, non-breeding stag groups should be located with as much space as sensible feed conservation allows and as far from active breeding groups as is practical to reduce fence pacing, aggressive behaviour, and extreme rut-related weight loss. Rapid post-rut recovery with targeted feeding is readily achieved. Concentrates should be offered on an individual animal basis rather than by group feeding in troughs or lines. Bulky forages must be of the highest quality possible to counteract the rumen fill limitations to intake.

16   Further Considerations: Genetics and Nutrition

We know that velvet weight can vary significantly from year to year, and can be different from farm to farm, using similar genetics. What might be driving these variations in the expression of genetics?

Recent research from Spain cited the role of growing the stag calf in expressing antler growth (Gomez et al. 2006). They suggest that the lactational performance of the hind, and hence the growth rate of the calf to weaning, will influence later pedicle development and antler growth. Calves well fed during this period-initiated pedicles 5 days earlier and grew more antler than calves with slower growth rates. This was particularly related to the amount of protein in the milk.

Table 2 The conversion of digestible crude protein requirements into crude protein requirements for New Zealand pasture feeding conditions (from Gao et al. 2003)

Stock class			Period	Protein requirements	
				% CP[a]	% CP[b]
Calf			Post-weaning	28.0	16.5
Male					
	Yearling		Winter	18.0	11.5
			Antler growing	22.4	24.5
	2–3-year-old		Winter		9.0
			Antler growing	19.0	19.0
	Mature		Winter		10.5
			Antler growing	17.0	15.5

[a]Crude protein requirements of reported Chinese diets
[b]Crude protein requirements assuming a digestibility of 0.75 for pasture

Chinese experience also points to a high protein requirement during the first year of life. Gao et al. (2003) summarised those experiences (Table 2) and concluded that protein was most important during the first antler growing cycle, recommending a diet containing 24.5% crude protein. During the second year, the requirement reduced to 19%, and then to 15.5% for adults.

These studies highlight the impact of growth and nutrition before the stag begins to develop antler. This process of early nutritional 'priming' of future production potential has been measured almost to the beginning of foetal development. This means that the nutritional conditions that the foetus is subject to during the pregnancy of the hind may impact on the expression of genetic potential later in life.

What else might we have to be aware of when trying to maximise the potential of our genetics?

Spanish researchers have investigated mineral nutrition to some extent, but mainly regarding the final strength of the antler once hard, rather than influencing velvet production. Probably of most significance is their work on the impact of late winter frosting on antler growth and strength (Landete-Castillejos et al. 2010). The occurrence of significant late winter frosting (mean minimum temperatures of -2.4 and $-2.3\,°C$ in July and August) appeared to promote an increase in silica in the feed available with an associated reduction in manganese.

This reduction in manganese was associated with reduced antler weight and strength. They conclude by suggesting that the manganese is required very early in the development of the antler to provide a framework on which the calcium then develops.

This area of nutrition needs research in New Zealand to determine appropriate levels.

Maximising the genetic potential of velvet stags starts with good nutrition of the pregnant hind. This sets the future response to nutrition. Achieving a good weaning weight of the stag calves through feeding the hinds well will then begin to express that potential. As the stag approaches pedicle initiation, the protein content of the diet needs to be relatively high and is best achieved on high-quality pastures. Every spring from just before casting the stags should be gaining weight as quickly as they can on a well-balanced diet, with younger stags getting diets that have a good protein content.

Strategic supplementary feeding is critical to express the genetic potential for velvet production and can be supplied in an increasing variety of hard feeds, fodder crops and high ME and high protein-based conserved forages in the key post-rut recovery phase.

It starts early right from the good feeding of pregnant hinds that is necessary to "prime" the future velvet production of their progeny. The milking performance of hinds will have a big effect on the weaning weights of their fawns and this in turn will greatly affect velvet yield as the stag fawns develop. Top operators aim for growth rates in NZ red weaners of at least 180 g/day in autumn and 140 g/day in winter. This will result in average rising yearling weights of around 78–84 kg going into the spring. In that period offering rising yearling stags spring feed allowances of 4 kg DM/head/day of high-quality pasture (at least 10 cm high) to achieve at least 300 g/day weight gain and a yearling weight of at least 100 kg (NZ reds) is optimum.

In feeding stags of 1–2 years of age in summer and autumn when standard pasture rations are available there is no need for special feeding. However, in drought years stags will benefit from supplementation, typically with good hay/silage/baleage or concentrates.

In winter, good hay/silage/baleage with supplements of crops and/or grain is needed with a transition in early spring to graze stags on good leafy pasture. Spring pasture is very high in protein and a great feed for velvet production, provided it is 10–14 cm in length and unsoiled. Specialist silages based on legumes and chicory are excellent winter feeds for stags.

Good management when feeding out autumn and winter rations to stags is to space the rations well apart on the ground, to reduce fighting and bossing. This is an advantage if the age groups are maintained as stable mobs until at least 4 years of age. It is not necessary to feed stags high value/cost supplements on top of an adequate quality feed supply as it has little impact on velvet yields.

- In early spring, 3–4 weeks before the first button casting, providing stags with a lift in quality feed pays dividends. The penalty for not doing so is delayed casting and reduced velvet yields. While under-nutrition in spring can depress velvet antler yields by up to 20%, there is no need to feed expensive supplements if good pasture is available. Spring pasture is an excellent high protein feed provided it is at least 13 cm long and unsoiled. The major antler growth period (up to 2–3 cm in overall length cm/day) occurs in the spring with under-nutrition during this period potentially depressing velvet antler yields by up to 20%.

Stags need a high protein ration from button drop. New-season pasture is an excellent high protein feed but in NZ pasture of this quality is not always available in early August. Fodder crops such as swedes and fodder beet are a good alternative, supplemented with high-protein feeds like processed feed nuts or Palm Kernel Expeller.

In summer once velvet regrowth has been cut, there is no need for high-quality feed and while it is important to keep velvet stags in stable social groups and use them to tidy up pasture ahead of lactating hinds and keep stable groups, with the genetics available today there is significant and high-value regrowth. Up to 30% more weight will be produced for this specific market and many velvet herds are managed with this in mind. Under NZ's Deer Code of welfare, on-farm hard antler must be removed pre-rut by early March with the standard practice being to do this at the point of stripping while stags are comparatively docile in groups and upon yarding.

References

DINZ (2016) Management for Profit 02, Feb 2016 DINZ deer facts series Passion2Profit strategy, primary growth partnership joint venture (DINZ and the Ministry for Primary industries)

Gao X, Fuhe Y, Li Chunyi, Stevens, DR (2003) Progress on the nutritional requirements of deer farmed for velvet production in China. Deer nutrition symposium. Grassland research and practice series 9. New Zealand grassland association: 69–72

Gao X et al (1996) Recent development in feed utilization and nutrition of antler producing deer in China. '96 International symposium on deer science and deer products: 71–77

Gomez JA, Landete-Castillejos T, Estevez JA, Ceacero F, Gallego L, Garcia AJ (2006) Importance of growth during lactation on body size and antler development in Iberian red deer (Cervus elaphus hispanicus). Livest Sci 105:27–34

Landete-Castillejos T, Currey JD, Estevez JA, Fierro Y, Calatayud A, Cearero F, Garcia AJ, Gallego L (2010) Do drastic weather effects on diet influence changes in chemical composition, mechanical properties and structure in deer antlers? Bone 47:815–825

Liang F, Wang Q, Weng T (1993) Deer farming for velvet production. The 4th ARRC International Symposium: 115–122

Suttie JM, Webster JR, Littlejohn RP, Fennessy PF, Corson ID (1996) Increasing velvet production by improved nutrition. Deer branch of the New Zealand veterinary association. Proc Deer Course Vet 13:149–153

Wang F, Xiuhau G, Shundan J (1997) Studies on the optimal Ca and P levels in diet at the antler growing stage in 3-year old sika deer. Acta Zoonutrimenta Sin 9:35–38

Yerex D (1991) "Wapiti behind the wire" the role of wapiti in deer farming in NZ, Chapter 10. GP Publications, Wellington, p 69

The Management of New Zealand Farmed Deer

Tony Pearse

Abstract

The Chapter provides an overview of the management of New Zealand farmed deer at weaning and mating and includes descriptions of the industry's schemes for genetic improvement. Modern deer farming improvements are elaborated with an explanation of the intent and delivery of the Passion2Profit project in which Deer Industry New Zealand (DINZ) collaborated with the Ministry for Primary Industries to improve both on farm and in venison markets returns of the performance of the deer farming industry.

Keywords

Weaning · Passion2Profit · Welfare · Velvet antler removal · Regulations · Genetic improvement

1 Weaning Management: Pre- and Post-Rut

There are several weaning management preferences related to timing, pre-weaning management, and preparation to consider in the NZ deer farming calendar, although the key objectives, minimising stress and maximising growth are the drivers for each farm operation.

The common factor is good planning in advance and reducing stress at the time. It is worth taking the time to plan to wean and to get it right.

A major change in recent years has been the demise of specialist finishing venison production systems typically located on highly productive land suitable for a few livestock-intensive farming activities. This was a common specialist use for finishing

T. Pearse (✉)
Deer Industry New Zealand, Wellington, New Zealand
e-mail: tony.pearse@deernz.org

young deer for peak season venison demand where autumn and winter conditions and feed reserves were not limiting.

Traditionally these finishers would actively purchase weaners at 4–5 months of age for venison finishing quickly within an age span of 9–14 months of age. Such operations manage weaner growth and finishing operation on highly productive land that may also accommodate both lambs and beef cattle finishing as part of that yearly rotation.

Venison production is seen as a premium operation for the NZ export-focused spring chilled season over the August-end–October period. As land use has changed (dairy farming conversions where both milking platforms and rearing young stock have flourished) along with other animal land-use changes, deer finishing profitability has been more volatile in terms of profitability and has ceased in many of these operations.

Consequently, many more hill-based deer farms have now converted to breeder and finishing systems often in somewhat more challenging terrain and have changed their management and feeding systems accordingly. In general, that has meant some degree of establishing improved pastures which may utilise clovers, and new cultivars of plantain and chicory to encourage high autumn growth rates of 160–240 g/day and may also include overwintering of brassica crops or fodder beet.

These are fed with some form of controlled grazing system either a 5-strand temporary electric fence supported on about1.8 m insulated standards at 5–8 m intervals defining a narrow strip of fresh crop or autumn saved pasture which might be shifted daily or up to 2 or 3 times a week. Alternatively, others prefer to fence a block of about7–10 days grazing duration.

In terms of weaning either pre- or post-rut, the key to success is minimising the stresses involved and their impact on post-weaning growth rate and any risk to stress-induced disease based around the physical separation and unexpected weather changes in autumn and early winter.

The New Zealand industry has not advanced any great measure of intensive housed finishing systems for weaners to finish in time for the September chilled season. That intensification is not supported by the Cervena® venison programme and in NZ conditions is not cost-effective.

Although the advantages of shelter and a partially controlled environment certainly moderate the impact of wet cold winter conditions, the high cost of infrastructures and concentrate feed sources to feed adequately to meet the chilled market peak returns leave this alternative only marginally profitable.

Modern management systems incorporate weaning techniques and the implementation of planned weaner health programmes and the ability to preferentially feed and manage young stock is the dominant activity, although very large (3000–4000 weaners) finishing farms do exist as part of premium supply contracts with venison companies.

An early technique from the first days of deer farming was to familiarise the weaners to their weaning block while running with the hinds some days before the separations and return them to that area on weaning. This has re-emerged as the so-called Aitken technique as a lesson from one of the recent focus farms projects. At its core for success is the management commitment to complete all the potentially stressful processes some days/weeks prior to separation so that on weaning it is the simple process of separation from the dam on the day.

To minimise weaning stress pre and post activities might include:

- Familiarising weaners with herding and yarding with their mothers in advance if practical. Yarding larger groups requires patience and calm handling and it is very important to avoid overcrowding in the available space inside yards, far better to have herd breakdown options in smaller areas adjacent to yards and handle smaller groups quickly and efficiently.
- Avoiding bad weather when weaning and returning newly weaned fawns to a familiar paddock afterward. Good shelter is a priority in the selection of weaner paddocks.
- Feeding newly weaned fawns on clean, good-quality pasture. If pasture covers are low, supplementation with grain or other high-quality feeds is a common practice. Such feeds are best to have had a short period of introduction pre-weaning while fawns are running with the hinds.
- That familiarisation with supplements pre-weaning also has the benefit of rapidly putting on 0.5–1.5 body condition scores of the hinds prior to them being drafted into selected mating mobs.
- Increasingly and somewhat as a response to more frequent and prolonged dry summers and occasional severe droughts on NZ's eastern provinces on both islands, some sort of late summer brassica or cereal crop may also contribute to late lactation and pre-weaning growth rates and maintaining hind condition.
- Best practice preparation for weaning encourages completing all deer health and potential painful or stress-inducing procedures vaccinations, parasite control drenches and formal Nataional Animal Identification and Tracing programme (NAIT) stock ID ear tagging pre-weaning if practical, allowing just separation from hinds and capturing the important weaning weight as part of important KPI's on the day.
- The industry's genetic evaluation programme, Deer Select breeding values recording requires a weight as of 1st March as the proxy for weaning weight in breeding value analysis.
- In addition, this pre-weaning period allows familiarising fawns with any supplements or autumn forages they will be eating after weaning while they are still running with their dams.
- If possible, a small number of unrelated adults possibly dry hinds can be released with the weaned deer into their weaning blocks to act as mob leaders for gateways, and future mustering and yarding.
- If a yersiniosis vaccination programme is in place ideally a pre-weaning first vaccination while on the dams is advised sufficiently clear of the 6 weeks period required for the second vaccination.
- In recognition of the stress of separation and activity associated with handling and unfamiliar routines there is a formal standard related to transportation of weaners on the day of weaning. Where deer are transported at weaning then the young

must proceed directly from farm to farm immediately following weaning and the total duration of yarding and transport shall not exceed 6 h. These deer, when weaned <10 days, are prohibited from being transported on the same unit as their mothers to avoid stress to both groups.

Pre weaning best practice advises an inspection of facilities and pressure points, for example, holding areas, yards and raceways, fences, gates and flooring, heavy traffic areas and load out ramps should be inspected for and risk areas (broken boards, sharp projections in flooring), leg traps related to gates and hinges that might risk foot damage or bruising. Under stress, some diseases like foot abscess ((*Fusiformis*) or yersinosis can rapidly emerge as a post-weaning risk.

To further reduce that physical risk of injury or adding further stress and time spent in facilities, practically it is easier and less stressful to keep mob sizes smaller and reduce time spent in the yards.

A smaller group of proactive farmers are keen to identify the dams of fawns that are poorly grown, often malnourished and may have pale fluffy? coats. A simple technique is to smear a coloured crayon (from a ram marking crayon or similar) and release these poor fawns back into the hind group overnight or for a few hours and identify any hinds with trace colour on the udder and selectively cull those.

1.1   Determining Parentage

On stud properties, the role of DNA matching and confirmation of parentage benefits from a significant investment by AgResearch and the Invermay-based DNA service, GenomNZ.

At the top end of DNA pedigree and parentage confirmation services, GenomNZ offers breeders a SNP (single-nucleotide polymorphism) based parentage test, containing tens of thousands of markers in the process known as genotyping-by-sequencing (GBS).

That technology confers the ability to verify precise genomic relationships, breed composition and level of inbreeding for every animal as well as confirming parentage through molecular-level analysis.

With the advancement of technology, AgResearch Invermay based genetics laboratory, GenomNZ now also advise the use of ear punches that collect an ear tissue sample on application of the visual ID tag.

These small samples are stored in Tissue Sampling Units (TSUs) and are taken on farm by using an applicator (The Allflex® TSUs and applicator) to punch a small piece of tissue from the animal's ear directly into the sampling unit. This sample type has been shown to provide both higher quality DNA and potential on-farm benefits. Where the TSU is purchased with the identification tags the numbers/identity are linked and removes any need for additional manual recording.

While the Genomic testing service is most used at the stud breeding herd level and primarily related to confirming parentage there may also be advantages in leading-edge commercial herds to select replacement breeding stock and create increased confidence in selecting young stags with high-value velvet antler weights and genetics or trophy antler traits more effectively.

At the stud level stag sales catalogues feature reporting breeding values for many commercial traits related to growth rates, carcass yields (e.g., a BV ema (eye muscle area) amongst other carcass traits), early conception dates, 12-month weight and antler weight.

Given the early industry of cross-breeding with elk (wapiti) genotyping using more GenomNZ new technology, (Genotyping by Sequencing (GBS)) the laboratory also offers deer breeders analysis of breed composition, genomic relationships and can determine the level of inbreeding within the herd as some of the early current test results.

The laboratory also offers Genomic breeding values with a goal in the medium future to determine both the lethal and desirable variants that exist within the stud breeding herd.

GenomNZ has a long-term goal to continue to develop and offer Genomic tools developed specifically for accurate selection for robust, high yielding NZ deer.

In addition, the large red deer-based deer research herd maintained by AgResearch at the Invermay site—is a critical resource as a research herd where it is possible to invest the time and resources recording difficult to measure traits that are commercially important. Today that herd recording effort is concentrating on immune function traits and disease and parasite resistance in deer. The programme is called Tomorrows Deer and the aim is to identify the best genetics for robust healthy deer that perform well.

2   The Post-Rut Weaning Option

On extensive deer farming operations or smaller lifestyle blocks, it is an option to leave fawns with their mothers right through mating until late autumn.

It may also be advisable on other farms when the facial eczema risk is high.

By weaning when the fawns are older, they are likely to experience less stress than they would if they were pre-rut weaned and pre-winter growth checks may also be avoided. On the downside, it is commonly thought late weaning is likely to delay conception (and therefore fawning) by 8–10 days. That may simply reflect the lack of a flushing impact that pre-mating weaning can generate with the lactation burden reduced. Late weaning suits a 'minimal' style of deer farming where the labour input is low, and animals tend to look after themselves. Many such operations are on large blocks of difficult country where mustering difficulties make pre-rut yarding impractical at a busy time of the year. Certainly, such weaners appear to experience much less separation stress in May or June, although equally bad weather in early winter and transfer onto hard feed or less quality saved pastures can impose additional stress and potentially a yersiniosis risk if those late weaned animals are compromised with lack of feed or poor weather.

As winter approaches, there are growing risks to the unweaned fawns from yersiniosis and parasites if some sort of animal health programme is not in operation.

Transportation of late weaned fawns in winter to new properties also poses risks. It is also important to avoid any predicted bad weather when planning to wean and make sure the weaners have good feed and shelter. Although anecdotally farmers who practice later weaning believe that the separation stress is minimal and any growth check also minor.

3   The Hind

Many farmers consider Body Condition Scoring (BCS) hinds at weaning (see Fig. 1). If their BCS is 4 or greater indicating either poor lactation or a lost fawn, conception rates are also likely to be around 10% lower. Many of the lower BCS hinds may have produced some of the best fawns because of their high milk production and it is recommended that, if you can, you should identify the mothers of the best fawns associated with lifetime deer identification.

A BCS wall chart (Fig. 1) was posted to all NZ deer farmers in late 2014 and the use of this technique is encouraged as a strategic tool in management, especially after drought or demanding summer lactation, or difficult winters when recognition of the

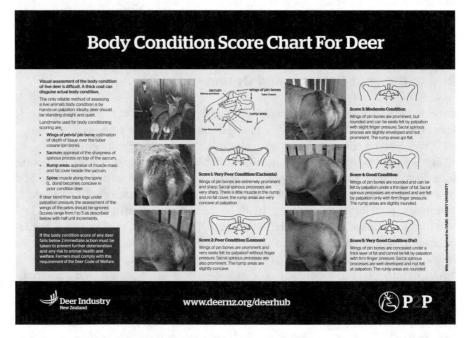

Fig. 1 Body Condition Score Chart (DINZ 2016). Best practice mating management (adapted from Deer Facts: Mating Management 2016) Ref: Management for Profit 02, Feb 2016 DINZ Deer facts series Passion2Profit strategy, Primary growth Partnership joint venture (DINZ and the Ministry for Primary industries)

loss of condition can be rectified with targeted supplementation prior to critical season pinch points like mating or pre fawning.

4   Mating Management

The whole process of weaning management and mating management are closely associated for most intensive deer farms under the fundamental principle that reproductive success, generally at is its most basic live fawns on the ground drives the profitability of deer breeding (see DINZ Best Practice Mating management: https://www.deernz.org/sites/dinz/files/DeerFact_BestPracticeMating_updated_Web.pdf).

At its broadest good mating management outcomes are expressed as higher annual conception rates, concentrated conception dates, and ultimately more viable fawns on the ground and survival to weaning, and seasonal factors including management of summer lactation, drought management and overall nutrition including supplementary feeding has a major bearing on the timing and spread of fawning dates.

The reality is that feeding non-pregnant hinds is wasteful both in terms of direct costs, as well as the loss of the potential invested in the breeding programme. Studies have shown that if the fawning rate on the average deer farm was increased by 10%, net earnings per hind wintered would increase by more than $NZ 30/head, or $NZ 15,000 in a 500-hind herd.

4.1   Pre-Mating Treatment of Hinds

The biggest influence on the rate and date of conception is the Body Condition Score (BCS) of the hind (See Fig. 1). This important Body Condition Score chart originated in the late 1980s through the work of supervisor Professor Peter Wilson and Dr. Laurent Audige during his Ph.D. work at the Veterinary School at Massey University and has become a key industry standard and skill set for NZ deer farmers since then and has been adapted and republished as part of the Passion To Profit resources and distributed to all known deer farmers and on the DINZ Website (Audige et al. 1988).

The detailed studies of Audige and Wilson reinforced that a simple eyeball assessment of condition is not accurate, but a formal scoring after hands-on assessment of the muscle and fat (or lack of) over the hindquarter region over the pelvis is consistent and accurate and defines management actions that are required. The skills involved are easily learned and applied.

In addition, the Body Condition Score (BCS) is incorporated into the formal Code of Welfare for Deer, (administered for Industry with the Ministry For Primary Industries (MPI). Under the Standards of the Code, deer farmers are required to seek veterinary advice if their deer have a BCS of <2 on the 5 point scale. Deer in this condition are considered to be emaciated and unfit for transport and a high welfare risk (Code of Welfare (Deer) 2018).

At its simplest Body Condition Score (BCS) is a quick, visual, on-the-spot method for assessing the health, well-being and nutritional status of adult hinds. It is simply an eyeball and recommended palpation) assessment of the amount of muscle and subcutaneous fat over the rump and torso of individual hinds when they are in the pens. Palpation is preferable during winter coat phase since a thick winter coat can visually hide poorer body conditions.

For hinds in moderately good condition or better, BCS is a reflection of the amount of subcutaneous fat and condition but for hinds of poor condition that carry very little fat, it also reflects loss of overall muscle mass and may indicate an underlying parasitism, disease or social pressure resultant from bullying. These characteristics are mostly easily assessed over the rump and hip bone, but also over the length of the spine.

BCS for red deer hinds is assessed on a five-point scale, with 5 representing 'very good condition' (fat) and 1 representing 'very poor condition' (cachexia or emaciated). However, most farmers assess using half-scores (effectively making it a 10-point scale, e.g. BCS of 3.5). While there is some consistent variation between individuals each tends to be quite consistent in their own assessment and judgement.

The value is that at key times of the year (late lactation, pre-weaning, late winter) assessing a BCS of hinds can influence positive early decision making related to general nutrition and any need to strategically add high energy or protein supplements.

Considered estimates suggest that a change in BCS of one unit is equivalent to a gain or loss of 8–10 kg in a red hind. Occasionally extreme losses can occur. Substantial and rapid loss in BCS in individual hinds can signal major health issues or severe biological challenges. For example, if a hind undergoes a change from 4.0 to 1.5 over a matter of weeks or months, this usually indicates a severe health issue (e.g. clinical Johne's disease). However, BCS losses of this amount also occasionally occur due to the demands of lactation over periods of severe drought, particularly for old hinds.

These and less severe losses in BCS particularly in late lactation in drought/poor feeding conditions often create doubt in farm management and the decision to wean or not farmers who may be supplementing with grain, Palm Kernel Expeller (PKE) or baleage/quality hay, often fear to wean if there is little quality pasture to wean onto. However, as long as the weaners are properly introduced to whatever supplements are on hand taking that lactational burden off hinds and weaning early allows animal health programmes to be initiated with young stock and precision feed supplementing to meet growth and health needs.

The advantage to the hind is profound and they quickly gain condition and weight and successfully mate and conceive earlier and predominately in the first cycle.

Practically on farm experience good hind shows that good feeding pre-mating— shifting the BCS up by 0.5–1.0 points from say, 3 to 4, greatly increases first cycle conception rates perhaps an advance in conception date by 5–7 days and improves overall fawning outcomes.

As the impact of climate change and increasingly dry summers impact negatively on pasture quality and quantity from January to March and is becoming increasingly important. As hinds will strip condition of their body weight to maximise lactation the modern hinds are under increasing lactation demand growing high growth rate offspring sired by stags that are also highly selected for rapid growth rate and body conformation.

This quick, simple and reliable assessment of deer condition is based on extensive research and farmer experience. DINZ encourages deer farmers to use it as part of good herd health and management. The scoring of hinds as a management tool is encouraged because of the impact that seasonal changes in body condition have on deer productivity.

Pre-rut weaning should be completed at least 2 weeks before mating (late February/early March) to allow hinds time to settle after the removal of their fawns. Mixing of hind groups, particularly hinds bought in from another farm, should be done well before mating.

Key Management Points
- Mature hinds: Feed hinds well in later lactation, as the breeding season approaches. If green pasture is not available, quality supplements are required along with a cautious approach to introducing new feed sources, especially grains.
- Wean hinds in late February/early March and assess Body Condition Score (BCS). Preferentially feed hinds with a BCS below 2.5.
- Yearling red hinds should be at least 80 kg and wapiti hybrids at least 130 kg at mating. A target threshold of ~80% of their final mature weight at 15 months of age is recommended to ensure oestrus occurs early.
- Join the stags with hinds by 10 March.
- Use stag:hind ratios of between 1:30 and 1:50 in most situations. Some stags can handle more than this, but non-pregnancy is a risk.
- Yearling hinds can be run with well-grown yearling stags at a ratio of about 1:8–1:10. These groups should be set up well in advance in mid-February.
- Look for signs of non-performance and exhaustion or any injury signalled as the non-completion of mating and replace such stags.
- Remove stags at the end of the breeding season (by 10 May at the latest) to prevent late births.

In addition, practical on-farm management includes:

- In severe droughts, weaning as early as mid-February is justified to ensure the best mating outcomes. Early weaning allows social order to be established in mating groups before the stag goes out.
- For unweaned hinds having a fawn at foot does not directly influence successful conception itself but can have indirect effects as an impact on body condition score as the hinds continue to lactate. Unweaned hinds tend to have a later start to breeding.

- Yearling hinds: Red hinds and wapiti cows on NZ deer farms normally attain puberty (first ovulation) at 16 months of age during their second autumn. Those that do so and have reached a threshold puberty weight are highly likely to produce fawns at about 24 months of age.
- Best practice advises weighing and BCS yearling hinds in January so there is time to take action before putting the stag out. To achieve a high conception rate, they should have a BCS of 2.5–4 and be at least 70–75% of the average weight of mature hinds in the breeding herd. There is a wide variation in hind genetics in the industry, but the genotype of the 'average' hind in a typical commercial NZ deer herd is now estimated to be 30–40% eastern European genetics, with an expected mature body weight of 125–130 kg.
- In such a herd, the minimum pre-mating liveweight should be 80 kg and the average pre-mating liveweight should be close to 95 kg.
- Yearlings that have some wapiti in them (say 50%) need to be at least 130 kg, although using this crossbred is not as common a practice today with the introduction of the larger productive easter red deer as sires. Most wapiti sires across red hinds are a targeted terminal sire operation where early slaughter for the chilled season is in reach for both young crossbred stags and hinds.

4.2 Pre-Mating Treatment of Stags

- Mature stags: Feed sire stags well in summer. If they have a good BCS during the rut they will focus on mating rather than feeding.
- Body Condition Score young stags early in the season (January) and check for lameness and injuries.
- It is a requirement under the 2018 Deer Code of welfare to remove hard antler including re-growth from all stags (with the exception of trophy stags) before the rut and by no later than 1 March.
- Yearling stags: Feed stags used for breeding as yearlings. Join stags with hinds by 10 March at the latest, then monitor their performance closely while keeping disturbance to a minimum.

4.3 During Mating

Disturbance of established mating groups is always a risk but may be necessary if a breeding sire is noticeably lame or otherwise injured, although often these can often be quietly worked out of a mob and replaced. In multi-sire mating situations, a dominant stag can hold a large number of hinds and good management requires simply keeping a watchful eye on mating behaviour. Most modern mating practices would involve introducing backup after the first 3–4 weeks with ideally removing and replacing the sire rather than introducing a backup in competition. It is not uncommon for stags to lose as much as 30% of summer body weight during the

3–4 weeks of active rut, and requires some vigilance and post-rut care and quality feeding.

While it is acknowledged that some sire stags have successfully mated 100–120 hinds during the rut, such high ratios carry a considerable risk of pregnancy failure, especially in single-sire mating situations. Even though the rut is very intense in terms of overt stag behaviour, sire stags have limits to their libido and mating capability. Also, individual stags vary enormously in their mating ability. The Industry guidelines for mating success suggest:

- Safe adult stag:hind ratios range from 1:30 to 1:50 depending on stag age as 2 and 3-year-old stags generally have lower mating capability than older stags.
- Safe yearling stag:hind ratios range from 1:8 to 1:12, highlighting their lower libido and their inexperience.

If oestrous synchronisation procedures are used to tighten calving spread, it is very important to reduce stag:hind ratios for mating (e.g. 1:10–15) as stags have considerable difficulty serving multiple oestrous hinds in 1 day.

5 Post-Mating Care of Breeding Stags

With such dramatic weight loss over the rut at recovery of as much condition as possible pre-winter is important. These are expensive animals and well worth investing time and feeding well. If space permits ideally stags would be run in small groups and feed high quality high energy feed to recover some of the body condition and weight loss in the short window that is available pre-winter. Some care is required as inevitably re-combining dominant stags will also introduce social hierarchy and fighting in contest to re-establish the social order.

5.1 Sire Selection

5.1.1 DEER Select

The importance of genetic improvement on a national scale has been prominent in industry leadership with DINZ, NZDFA and the wider industry for many years. To use that opportunity of selecting verified superior breeding sires with growth rate and carcass yield traits quantified is one of the main ways farmers to improve the profitability of their deer farming operations.

The DEER Fact Related to DEER Select cites the features of Genetic improvement as being:

- Permanent: Unlike improved feeding or health, improved genetics influences the performance of an animal for its lifetime.
- Cumulative: Improvements made in this generation are added to those made in previous generations and then passed on to the next.

- Sustainable: Improvements can continue to be made so long as there is genetic variation as a basis for selection.

The most visual and dramatic progress has been expressed in the rapid advancement of velvet weights and mass and in trophy stags breeding where today new world record heads are been created and exceeded on a year to years basis. That success has been assured with the high heritability of velvet traits with recent recalculations suggesting heritability estimates are around 0.48–0.50 for velvet weight and beam and will be strongly influential in improving productivity.

Equally and significantly the venison industry has benefited from targeted improvements in carcass weights, slaughter pattern timing in terms of faster growth to meet the traditional northern hemisphere chilled venison markets with 9–12 month age yearlings. Base breeding stock have steadily increased in weight, lactational ability, carcass muscle yield, body and muscle conformation and overall performance.

To assist and promote genetic gain the NZ industry invested heavily over the past 20 years in DEER Select as a formal genetics improvement programme housed within Deer Industry New Zealand and supported by the research and analytical expertise within AgResearch and its Deer Research group. This is an industry-funded initiative and is managed by contracted expertise engaged by DINZ (The DEER Select Manager).

Deer Select's goal is to provide farmers with an accurate, unbiased way of identifying superior breeding animals (normally sires) for use in their herd. It functions as an extensive database based on pedigree records and performance data collected from performance-recorded animals in participating stud herds as expressed through calculated Breeding values (BVs or estimated BVs) for key productive and commercial traits.

Deer Select uses the well-established SIL (Sheep Improvement Limited) genetic engine to calculate how much of the performance variation between individual animals is genetic. It then ranks these animals in order of their genetic merit for particular traits.

Animals can be ranked across multiple herds when the stud herds are genetically linked through common sires, mainly through bought-in stags or semen. This allows the effects of farm/climate and management to be excluded from the estimation of genetic merit. The industry encourages and if needs be invests in ensuring cross herd linkages are maintained.

In terms of operation differences in feeding can be accounted for by the use of mob codes within a farm, providing there are verified genetic links between the mobs—such as by having the progeny of one sire in both mobs accepting that some of the variation around measured traits are environmental in nature or unknown and cannot be corrected for, for example, a disease challenge.

After accounting for known environmental effects, the performance of an individual, as well as the performance of its known relatives, are used to predict the individual's genetic merit for a trait. Where a trait is not expressed in both sexes, for example velvet weight in hinds, a breeding value can be estimated based on the performance of relatives of the opposite sex. It is a given that for animals to express their genetic potential they will require improved feeding across all seasons and stages in growth and pro-active animal health management.

The industry leaders have promoted the value in recoding and promoting reliable parameters related to genetic gain for many years, but it has been the last 12 years of concerted effort that have made the difference. Average carcass weights for young animals have risen from a static 56 kg (males and females combined) to 60 kg over the past 5 years, a key aim of the Passion 2 Profit programme.

Within DEER Select over time 30 foundation and key stud herds have contributed vast amounts of data recording for analysis. Base data and breeding values calculations are updated 4 times per year and published on the Deer select website and distributed direct to the breeders. DEER Select provides reporting with both a venison production primary focus and has a primary emphasis on sires described and defined by the breeders as being 'European and composite sires'. A similar analysis is also available for 'English sires' where both analyses reporting and traits are directly comparable.

The information presented is for growth, meat yield and reproduction (Fig. 2). Both the estimated breeding values (eBV) and the accuracy of each estimated breeding value (acc%) are reported to the breeders.

Key traits measured and analysed are primarily based around venison production as the primary focus for European and composite sires and for English sires.

A similar DEER Select service operates for Elk (wapiti) and crossbred sires breeders across 7 major herds. Currently the published EBVs between red and wapiti sires are not directly comparable but extensive work is underway to allow all sires and breeds to be comparatively ranked on a single Breeding Value calculation across all traits and is expected to be in place by late 2021 and will certainly illustrate the extensive genetic gains that have been made by the stud industry across many important and commercial traits.

Key Points
- Deer Select is the genetic evaluation programme that enables farmers to accurately choose replacement sires, based on genetic merit.
- The main genetic traits Deer Select evaluates are those relating to growth, meat yield, velvet weight and some maternal qualities.
- Commercial farmers are encouraged to buy sires from stud breeders who provide Deer Select estimated breeding values (eBVs) for sale stags that match their breeding objectives.
- Most leading studs expand on their breeding objectives and the EBV calculations for traits are published alongside the stags pedigree, breeding background and performance in their sales catalogues.
- Modern catalogues also can have calculated and have published Deer Selects lists and in individual sales catalogues a series of economic indices.
- Economic indices a provided by Deer Select to enable buyers to work out which stags will give them the best return on investment. A breeding index consists of a

Code	Trait Description (Developed in 2015)
W12EBV	Weight at 12 months (kg)
CWEBV	Carcass Weight (kg)
WWTEBV	Weaning Weight (kg) (Indicator also of maternal milking ability)
AWTEBV	Autumn Weight (kg)
MWTEBV	Mature Weight (kg)
CDEBV	Conception Date (days)
EMAceBV	Eye Muscle Area
CARLAeBV	Saliva immune response to internal parasite challenge (units/ml) Research BV
R-EARLY KILL	Growth eBVs x economic weights + Meat eBVs x economic weights + Conception eBVs x economic weight
TERMINAL Index	Growth eBVs x economic weights + Meat eBVs x economic weights
VW2eBV	2 Year old velvet weight
MVWeBV	Mature velvet weight

Fig. 2 Current Traits and Codes used by DEER Select

set of breeding values multiplied by an economic weighting reflecting the value of a one-unit increase in the trait in a typical deer production system.

- By multiplying each eBV by its financial contribution, the individual with the best combination of eBVs will have the highest value. Indices are expressed as $ venue advantage per hind calving As prices change over time and production systems evolve, the economic weightings of each trait in each index are updated.
- The NZ deer industry uses two indices—one for a breeding/replacement/finishing system, the other for a straight finishing (terminal) system with the expectation that in future more maternal traits will be added to the breeding/replacement index, better representing the overall merit of sires that breed replacements.
- The two current Deer Select indices components are (a) *Replacement & early kill* = Growth eBVs × economic weights + Meat eBVs x economic weights + Conception eBVs × economic weight and (b) *Terminal* = Growth eBVs × economic weights + Meat eBVs × economic weights.

6   The Deer Progeny Test

When Deer Industry New Zealand introduced the 5 year productivity improvement programme in 2007 the strategy was aimed at making productivity gains for venison growth and reproductive traits for farmed deer.

On farm the reality was there had been little productivity progress in venison growth rates and carcass yields as much as breeders had tried to make progress. On an industry-wide basis uptake and real improvements was minimal apart from the success with highly heritable with antler traits.

The industry at that time also began its Focus Farm programme which was seated in the NZDFA branches, as a venture to improve uptake and application for new knowledge and management techniques. The Focus farm project established a byline *'more, heavier earlier'* as venison production main theme. It is recognised as an industry challenge where biological production was the big challenge. Simply put, it is the disparity in normal growth of young stock taking 12–15 months to a finished natural weight at 14–15th months old, when market demand was looking for quality carcasses 4–6 months earlier. It was already established that the economics of overwintering young stags for a second winter was uneconomic, and also at risk of having carcass weights at the top end of preferred venison cut sizes.

The *'More, heavier earlier'* Focus Farms programme was based around facilitated field days and defined objectives to improve both weight and timing of premium supply with farmers learning from farmers and implementing improvements on the Focus Farm over a 3–4 year time frame.

This project had a theme of more deer on the ground referring to improved survival and reproductive success and to have them grow faster, earlier to move more animals into the chilled season and heavier simply a reflection of improving profitability. The focus farms underpinned the uptake of new genetics verified by breeding values and performance. Uptake of modern on farm-technology including weighing, pregnancy scanning, and new management and strategic feeding techniques were amongst the practical leanings and very much captured farmer innovation and new management techniques related to performance improvement. On an industry-wide basis, a secondary aim was to motivate and increase the uptake and use of objective breeding values or eBVs.

It was the slow uptake that prompted the industry led by AgResearch and Deer Industry New Zealand and a selection of top farmers, geneticists, veterinarians and deer farm consultants to establish the national Deer Progeny Test programme which has had such success (Ward et al. 2014). It had five principal aims:

1. To encourage and improve sire linkage across recorded breeding herds in DEER Select which had only been achieving ~70% connectedness (This was a major driver of the project)
2. To provide a platform to evaluate breeding values across the breeds of red and wapiti both maternal and terminal breeding operations

3. To evaluate new traits, especially those related to venison production (carcass yield and growth rates) and generate genetic correlations to allow the best understanding end-use of selection goals
4. To provide a starting point for the evaluation of maternal traits under commercial breeding environments
5. To establish a well-described phenotyped DNA resource population for future genomic tool development

The DPT was a three-year breeding programme run across three different farms with two birth years of progeny. Three years of artificial insemination (AI) were completed and overall involved 2417 inseminations, using 35 different sires and 1581 hinds, with an average rate of progeny weaned to hinds inseminated of 68% (range 77–64%).

The project received a substantial boost when Alliance Group Limited, a major NZ venison exporter, who had previously been involved in a central progeny test for the sheep industry, (where AGL was involved in evaluating meat through analysing carcass yields and quality meat, quality attributes) also committed to support the Deer Progeny Test with that background experience. AGL also had on hand a trained taste panel. Deer Breeders were invited to submit an EOI to participate and with most donated semen from key sires for the breeding programme on the three properties used.

All maternal (red) male and terminal (wapiti crossbred) progeny were slaughtered at approximately 11 months of age to measure carcass traits while maternal females are retained into the breeding herd to measure maternal traits. Over the 3 years of the DPT, 1640 animals survived to weaning and subsequent recording and analysis and parentages were confirmed via blood or tissue samples.

Seasonal live weights and temperament scores were collected on progeny and dams, while ultrasound eye muscle area, parasite antibody levels and foetal age were recorded on progeny. A subset of core traits was also measured and submitted by partner herds who contributed AI sires. Slaughter occurred on a single date for each farm and a variety of meat yield and quality traits were assessed. DPT data was uploaded to DEERSelect and publicly communicated via Sire Summary reports. The Industry today continues to publish 4 times a year updated sire summary reports with adjustments to the key trait EBVs as records are systematically analysed.

A bonus from the programme was confirming that success rates of the artificial insemination programme averaged across the participant herds ranging from 62% weaned to 77% over the 3 years.

The three base herds were all larger-scale commercial farms covering both hill and high country locations under excellent management DPT has increased the percentage of linked herds recording growth on DEERSelect from 70 to 100%.

This means DEERSelect breeding values and economic indexes produced from red deer can be validly compared across herds.

The average genetic merit for weaning weight (WWT) in all Deer Select herds in 2015 was +8 kg compared to the 1995 average.

The genetic merit of the top individuals was 2–3 times this, a huge difference when it comes to selling weaners. Average genetic merit for 12 month weights (LW12) across all Deer Select herds in the same period increased by +11.6 kg, or 6 kg/head carcass weight. The elite top sires in 2021, selected entirely for their role as venison sires have 12 month weight EBV values in greater than +30 kg and significant improvements in carcass trait values enhancing total carcass weight and venison yield, particularly in the loin and hinds legs cuts.

7   Passion to Profit   P2P

In 2012/2013, a frank overview of the deer industry's productivity performance recognised that while there were standout performers in terms of productivity and profit, as an industry in spite of the wealth of knowledge and application of technology, management evolution and performance increases and growth were static.

A strategic review was required and a new era in industry growth and profitability beckoned. It was recognised that potential for productivity improvement was there and achievable. In addition, a next generation of young deer farmers were also emerging as potential industry leaders and a new fresh face of the business of deer farming and new markets beckoned.

A wide-ranging body of skills and experience was captured in review team drawn from across many disciplines in and outside the industry and developed an application for a seven-year project for co-funding with the Government's Primary Growth Partnership (PGP) administered through the Ministry for Primary Industries but with a clear whole of the venison industry target and inclusion.

The application's business case concluded in its executive summary that the New Zealand deer industry required a transformational change to improve profitability to levels competitive with alternative land uses. Land use competition was especially fuelled by NZ's dairy boom on best land classes and creating competitive challenges for sheep, beef and deer farming. For the venison breeding and finishing industry retaining critical mass of active farmers was an emerging concern that had already been impacted by land-use change with the loss of specialised venison finishers on the better classes on land.

The systems were not currently in place to ensure that market demands were being responded to by producers or that our distinct offering in terms of the back story and quality of New Zealand based. Grass-fed naturally raised product was being adequately communicated to consumers.

It was clear that without change, New Zealand would miss the opportunity to derive full market value from exported farmed venison products, which are almost unique to New Zealand. Without delivery of greater value to markets and a corresponding increase in deer farming profitability, returns to New Zealand would remain sub-optimal and land use change would also continue to marginalise this pioneering symbol of primary industry diversification.

Velvet antler production and marketing was not included in this 7-year P2P initiative.

The Government co-funding was sourced through a national government commitment and vehicle in the form of the Primary Growth Partnership (PGP) which also has similar programmes in the other livestock industries targeted to stimulate and encourage transformational change in production profitability and increased market returns and market share.

The outcome was that Deer Industry New Zealand (DINZ), and the Ministry for Primary Industries (MPI) became co-investors in the Passion2Profit (P2P) programme in partnership with the NZ Deer Farmers Association (NZDFA) at both national and regional levels. The industry initiative was supported also by the Deer Branch of the NZ Veterinary Association with the agreed goal to create transformational change in the production and marketing of New Zealand farm-raised venison.

Clearly, the bottom lines remained as ensuring that the profitability of the New Zealand deer industry needed to improve so it remained more competitive with alternative land uses. This was critical as the active deer farming numbers had decreased dramatically since a peak in the early 2000s which had industry numbers including lifestyle blocks at close to 5000 individual farms. And in 2013 were estimated at 2000–2300 at best. Most of the lost numbers were either smaller operations near cities in the lifestyle block category or on larger farms now in family succession where the succeeding generations opted for dairy support or sheep and beef farming as they were not confident about deer farming futures or had little experience wit deer as farmed animals.

In 2010–2012, the low relative profitability of venison farming was impacted largely because it remained over-reliant on the commodity game meat market in continental Europe. This market has a narrow window of demand in the northern autumn/early winter when chilled venison cuts have an average value 20–30% greater than the rest of the year and is 6 months out of phase with the natural growth and productivity of young deer.

On-farm, many producers have been unable to produce young venison in the optimum weight range for this narrow market window. As a result, much New Zealand venison was being frozen and stored for many months before being traded as an undifferentiated commodity into this market.

In response, the more pragmatic deer farmers converted to more profitable farming options and the national deer herd fell from its peak of 1.7 million deer to under 1 million.

However, those that stayed and, in many cases, expanded their commitment agreed that without wholesale change, the deer industry would continue to shrink and with it, the ability to fund research, market development and other services needed to ensure its long-term viability. This was feared as a major loss and risk not just for those who are committed to deer farming, but for Brand New Zealand and for those other farmers for whom deer are a potential avenue for diversification.

As major working objectives, the vision and planned programme agreed that the changes needed are unlikely to be achieved through a business-as-usual approach with small changes made each year. A game-changing investment was needed to allow many initiatives to be implemented simultaneously—the only way they are likely to be successful.

Only with co-funding from the government's PGP fund was such a strategy possible. Two core areas of investment were identified:

7.1   Marketing Premium Venison

The P2P aims to increase the amount of NZ venison sold in chilled form, year-round, at higher prices. This was aimed at diversifying the consumer base for venison, thereby spreading market risk, extending the consumption period and generating greater sales of higher value chilled venison.

The five largest venison marketing companies—Alliance Group, Silver Fern Farms, Duncan NZ Venison, Mountain River Venison and Firstlight Foods agreed to work together to design a collaborative marketing programme and together became cornerstone investors. The marketing programme scope was to go far beyond the scope of existing promotion programmes for NZ venison. It also involved the development of common brand values, product specifications and positioning by these companies for their use in new markets and new market niches.

The critical actions included developing a single set of product and production quality standards to underpin the positioning of NZ venison in both developed and developing markets with company marketers working together to open new markets where NZ venison would be positioned as a luxury, non-seasonal meat but clearly identified with each company's branding reputation and market connections.

This programme was initially directed to explore opportunities in new markets such as mainland China. Venison was relatively unknown in Chinese cuisine, but the creation of demand for premium venison products in a country as large as China had the potential to increase year-round demand and reduce reliance on the European traditional and somewhat commodity-based game meat market.

The second stage was to test New Zealand's ability to position the finest venison as a new, exotic, non-seasonal grilling concept in one or two smaller markets in Europe with the products positioned to be distinct from 'game' meats.

7.2   Market-Led Production

The P2P key mission and activities evolved to develop systems to help farmers respond to current and future market demands. It was clear that there were very successful leading farmers using their own innovation and had a high awareness of the potential for improved performance. These early adopters followed experiences and advice coming from research, extension and productivity system change also being used in other livestock industries.

As has been a proven practical application, leading farmers have always success-fully integrated the best scientific knowledge on deer feeding, animal health, and genetics into their farm systems. This enabled them to profit from delivering deer earlier but in the weight range and at the time that the traditional market demands. The industry has had a strong focus on employing these technologies or systems successfully.

Some farmer-led precedence existed based on work at the Invermay Research Center, Dunedin in determining a role for a terminal sire production system for venison. The early New Zealand deer industry example is classically seen in the enthusiastic adoption of cross-breeding red deer dams with the captured Fiordland wapiti (~75% Elk genes from the original wild release in 1905 (Cervus *elaphus nelsonii*)) and from the mid-1980s the imported Canadian elk sires from Saskatchewan and Alberta (*Cervus elaphus manitobensis*).

These fast-growing crossbreds under good autumn and winter management could meet the 55 kg-60 kg slaughter weight in the 6–8-week peak season in September and October and peak prices returns.

However, the industry was somewhat divided on the role of elk wapiti crossbreeding and sought and imported new genetics from the UK and Denmark, Sweden and Germany initially. Industry leaders in the stud business then ventured into eastern Europe (Yugoslavia, Hungary Cervus *elaphus hippelaphus*) and imported sires and dams via quarantine and testing in the UK then onto New Zealand to establish new larger red deer strains on farm for both increased venison growth and carcass weight. In addition, the focus was also weight and dimension to the quickly emerging trophy antler production sector.

That investment from the leading studs carried the industry improvement programme through the 1980s and 1990s and onto the next step with the develop-ment of the Passion to Profit (P2P) initiative.

The Business case for incentivising on-farm productivity increases in 2012 was straightforward.

The programme advanced three distinct themes that underpinned the potential for substantially better performance, Animal Health, Feeding/Nutrition and Genetics.

To assist in the 'packaging' and delivery of solutions and technology for each of these categories a formal Reference group drawn from industry expertise and specialists plus a broad motivated farmer base, and also linked with the research and technology units. A further feature of the reference group was that it was well connected with industry media and allowed a wider communications strategy and information flow from each theme group.

Other elements have included:

- Benchmarking: An expert working group of large-scale farmers has been formed as part of Advance Parties to develop productivity benchmarks for commercial farmers allowing some meaningful comparisons to be made between farms.
- Expertise development: The Programme set about identifying farm consultants, veterinarians, and other specialists with expertise in deer and where necessary filling regional gaps. Special training and accreditation were to be explored.

- In part that was to replace the large gap as a forum for scientists' veterinarians and specialist expertise left as the industry declined in numbers. Significantly too and the quiet loss of the annual focused technical conference run by the Deer Branch of the NZ Veterinary Association. The technical reporting across the whole range of deer health and productivity, and industry initiatives were reported annually in voluminous, and information-rich Proceedings of a Deer Course for Veterinarians, championed for all its productive life by Professor Peter Wilson from Massey University from 1984 to 2015 with the last sole deer focus technical conference. (Now termed Cervetec).
- The Deer Branch NZVA still exists and plays a very important role in the industry with a significant number of key veterinarians as facilitators involved in the P2P as Advance party facilitators or as specialist expertise in the leading three theme groups.
- The programme and industry also recognised that much of the science about deer reaches only a minority of highly innovative farmers, farm advisers, veterinarians, and other rural professionals. Better communication systems were set up to enable this knowledge to be better packaged and disseminated to farmers and those who advise them.
- Tailored farmer communications: Individual farmers are now provided with information that is relevant to the performance of their operation. For example, how their farm performs relative to peers in average slaughter date or average carcass weight. DEER Pro. [https://www.deerpro.org.nz]
- The final activities were a complete rewrite and update of the Deer Industry New Zealand's website and productions of the very comprehensive 'Deer Information Hub' the deer industry's web-based contact with deer farmers (https://www.deernz.org/deerhub).
- For deer farmers, the practical productivity jewel in the crown has been 7-year project that annually produced a series of short 2–4 pages well researched and professionally written 'Deer Facts'. These sit in an electronic form on the Website and are printed and distributed to all known farmers with the industry 'Bi-monthly Deer Industry News', A deer farming resource kit. These now number over 50 and have been peer-reviewed by a panel of industry farmers researchers and writers for accuracy, balance and relevance and are openly available to all on the website.
- The Deer Facts key sections cover Best practice, Biosecurity, Breeding, Environment, Handling, Health, Management for Profit, Nutrition and contain 4–8 publications each [https://www.deernz.org/deer-facts]

The multi-year programme was started with the intention of improving deer farmers' ability to improve production efficiency and meet existing and new markets' requirements and funded at roughly half of the full P2P investment. The parallel goal was to create collaborative opportunities for marketing companies to work together to position New Zealand venison in new market segments that will increase returns to New Zealand.

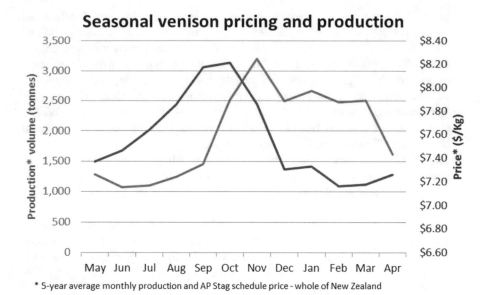

Fig. 3 Seasonal venison pricing and production calculated using 5-year averages over the whole of New Zealand

The productivity challenges for venison production and production mismatch for the majority of all northern hemisphere peak venison (both fresh chilled and frozen market) are well expressed in the schematic below (https://deernz.org/dinz-activity/ p2p-deer-industry-initiative/what-passion2profit).

The P2P programme has transformed the farming innovation and breeding, feeding and on-farm and animal management skills that were identified as part of the earlier Focus Farm's programme and its 'More, Heavier Earlier' underlying themes in those key farm management programmes (Fig. 3).

The P2P programme also developed a range of new knowledge and management application to achieve improved deer farming efficiency and performance. It also introduced a national Quality Assurance scheme underpinning product quality related to pH, tenderness, shelf life and age that allowed marketing companies to provide better market-led feedback to producers (Fig. 4).

The current P2P programme finishes its funding round in September 2022 and there is strong interest from farmers to continue the P2P and both broaden and deepen the programme within the Deer Industry. Design and consultation are of a potential Passion to Profit 2 (P2P2 are underway with all stakeholders to the current P2P programme.

There is a range of on-farm market-led programmes that have been developed, and all are guided by an advisory group of practice change and thought leaders from both within and outside of the deer industry.

Fig. 4 An overview of P2P at a glance

The primary engagement groups are known as *Advance Parties* who are very active in a wide-ranging area of self-determined topics depending on the needs of each group.

These include strategies for farmer engagement using well supported small groups of motivated (8–10) farmers) providing peer-to-peer learning and support with the aid of a skilled facilitator and importing subject matter expertise when required, who work together to identify opportunities to improve profits on their deer farms.

The Advance Party members demonstrate the improvements they make to their wider community to encourage wider adoption of farming practices that can improve the profitability of deer farming.

A key aspect in the success of the project is that in history deer farmers have been collaborative by nature and want to learn from each other. They have always been thirsty for information from the outset and have shared with honesty and inquiry, what mistakes or challenges they have faced and how can these be prevented or is success is proven how can these be incorporated into other farm settings and objectives. History records that most deer farmers tend to learn best where they can see changes, relate them to their own farms and are able to try them. Advance Parties are a new way to support change using these attributes and preferences.

Advance Parties create change by learning and trialling different techniques as part of a supportive group to see their effect on profit and sharing the results with the wider deer farming community.

The purpose of an Advance Party is to be a catalyst for change by demonstrating new or different deer farming methods or technologies to inspire change for increased profit.

Members agree to some individual productivity improvement projects on farm and its progress and outcomes are openly shared as members are committing to shared personal and farm business development, including their data, methods, plans, results, problems and reward and successes.

The benefits from Advance Parties are not limited to the farmers in the group. They are a means of testing and refining profit gain opportunities and demonstrating those methods (and their limitations) to the wider deer farming community.

The name Advance Party is used to differentiate the group from a 'Discussion Group'. An Advance Party is not a traditional discussion group. An Advance Party is not primarily about 'discussion'.

In more recent years this successful mode of engagement has been extended to new groups known as *Deer Industry Environment Groups* focused as the name suggests on environment topics and working together to complete and keep active a Farm Environment Plan (FEP) for each deer farm. The production and maintenance of a formal Farm Environmental Plan is evolving as a formal requirement for farmers at the local body Government level and an emerging national policy from the Ministry for the Environment based around enhancing water quality, reducing nitrate leaching and conceravation of soils.

Between both Advance Parties and Environment Groups, some 500 deer farmers are involved in achieving progress within their farm systems. There are 28 Advance Parties in operation and 15 Environment Groups.

The P2P programme also has a focus on developing depth in deer systems knowledge within the rural professional and supply networks that support deer farmers and runs regular *Rural Professional Workshops* throughout the country to build knowledge and confidence for rural professionals.

There are also four specific theme groups *Animal Health, Improved Breeding, Feeding, and Environment*, who have between them produced a wealth of resources, apps, charts, workbooks, and spreadsheets to support deer farmers.

The P2P programme has developed significant resources to help deer farmers and the Rural professionals that support them, and during the past year there has been a focus on how they could be integrated for even greater benefit.

For example, a 2020 project around improved management of intensive winter grazing enabled a combination of resources and expertise from the P2P feeding, environment and animal health and welfare to be successfully integrated into one programme.

It has been recognised that a risk to wider industry and typically the less engaged deer farmers, who may only have deer as a smaller part of a mixed livestock operation, is that these practically may not have ready access to the published material and reporting that the Advance Parties and theme groups regularly promote.

As is typical in a smaller compact industry too, Advance Party members are often also the same committed people who are the life force of energetic regional Deer Farmers Associations local branches. To some extent that energy and activity within the P2P and Advance Parties has superseded the DFA's regional role in field days and on-farm activity for members.

To address this and extend the activity and practical management gains outside the Advance Parties each group is encouraged to develop themed regular *Regional Workshops* which are held throughout the country in association with the NZDFA to showcase the progress that Advance Parties in particular have been making and to share that learning with deer farmers who are not members of groups.

The P2P programme has also supported the development of *Industry Assurance programmes* to support the increasing development of quality systems that underpin the deer industry and market needs.

In addition to the farm-related P2P programmes, the wide range of market-led programmes on farm support an equally vibrant P2P marketing programme which has a range of collaborative programmes involving all of the venison marketing companies working together to expand the depth and breadth of markets for premium venison products. The 5 companies regularly report their progress and initiatives at industry fora and industry communications and these are often highlights of the annual conference and at the regional NZDFA events.

The involvement of deer farmers and those that support them is pivotal to the success of the programme as is the partnership, support, and funding from MPI, DINZ and DFA.

Priorities for the programme now involve:

- Ongoing support to embed changes in farmer skill and confidence, so they continue to improve performance and increase efficiency. This means continuing to support uptake of improved management practices through the P2P practice change activities. An increased focus on integrating across projects will bring greater synergy for farmers across the range of market-led initiatives.
- Enlivening further the Quality Assurance programme to ensure it provides value to producers and their customers. This means adding in market-led requirements as part of initiatives to provide an integrated farm plan structure and assisting with the adoption and auditing of standards.
- Encouragement and support for farmers to document and adopt improved environmental stewardship to ensure they meet their own, societies and market expectations.
- The design and development of an expanded successor programme that can continue the progress that the P2P programme has achieved to date.

In terms of productivity across the industry directly related to the P2P activity average carcass weights have risen by 4–5 kg over the last few years from 55 to 56 kg and been maintained at 59.5 to 60 kg. This approximate 7.5% increase in dressed weight and the ability to hit the chilled market timing at 9–11 months of age has been greatly enhanced through a combination of targeted genetics related to elevated seasonal growth rates, better carcass yields (through a focus on selection for the Breeding Value for enhanced eye muscle area, a proxy for total carcass saleable muscle yield) and a wide appreciation and application of DEER Select breeding values including earlier conception dates and growth rates and weight at 12 months of age. These gains in productivity are expected to be ongoing and permanent.

Fig. 5 P2P logo

The industry now awaits the emergence of the P2P2 and a new era in industry unity, growth, productivity and profitability (Fig. 5).

References

Audige L, Wilson PR, Morris RS (1988) A body condition score system and use for farmed red deer hinds. N Z J Agric Res 41:545–553

Code of welfare (Deer) (2018) Issued by the Minister of Agriculture under sec 75 &76 of the animal welfare act 1999. National Animal welfare Advisory Committee, Ministry for Primary Industries (MPI), Regulation and Assurance Branch, Animal Welfare, Wellington, p 39

DINZ (2016) Management for Profit 02, Feb 2016 DINZ Deer facts series Passion2Profit strategy, Primary growth Partnership joint venture (DINZ and the Ministry for Primary industries)

Ward JF, Archer JA, Asher GW, Everett-Hincks JM, Mathias-Davis HC (2014) Design and implementation of the Deer Progeny Test (DPT). In: Proceedings of the New Zealand society of animal production, vol 74. Napier, pp 220–225

Diseases of Farmed Red Deer in New Zealand

Colin Mackintosh

Abstract

This chapter provides a brief background to the deer farming industry in New Zealand and the emergence of diseases of economic importance within farmed deer. The diagnosis and control of the diseases is described.

Keywords

New Zealand · Deer farming · Deer diseases · Bacterial disease · Viral disease · Parasites · Fungal disease · Tuberculosis · Paratuberculosis · Mineral deficiency · Yersinia

1 Introduction

In the late 1800s and the early 1900s, seven species of deer were introduced into New Zealand (NZ) for hunting and released into the wild in various places in the North, South and Stewart Islands. The species were red (*Cervus elaphus*), wapiti (North American elk, *C. elaphus canadensis*), fallow (*Dama dama*), sika (*C. nippon*), rusa (*Rusa timorensis*), sambar (*R. unicolor*), white-tailed deer (*Odocoileus virginianus*) and moose (*Alces alces*). By the mid-1900s, their numbers had grown substantially and they were causing extensive damage to native forests. This led to Government funded deer culling and later to large-scale commercial hunting and recovery of venison for export in the 1960s. This resulted in dramatic falls in the number of wild deer. As a consequence deer started to be live-captured and released into fenced enclosures, leading to the start of deer farming in the early 1970s. The main species farmed were red, wapiti and fallow deer. Initially a lot was

C. Mackintosh (✉)
AgResearch Ltd, Mosgiel, New Zealand

learnt about the capturing, fencing, yarding and domestication of deer. Over the following years, there was a steep learning curve about the feeding, management, restraint, health and disease issues of farmed deer. There were few precedents in the world relating to farming deer and much of the research into deer diseases and the development of treatments and prevention strategies has evolved in this country. Population densities in the wild in NZ were relatively low and food was generally plentiful. Consequently they appeared to be very healthy and had good reproductive rates. However, their capture and adaptation to a farming environment resulted in wastage due to misadventure, trauma and stress related diseases. Exposure to sheep led to Malignant Catarrhal Fever (MCF), while increased stocking density and grazing on pasture soon led to parasite problems, especially lungworm. In the late 1970s, bovine tuberculosis (Tb) became a major problem. Trace element deficiencies, especially copper, became apparent in the early 1980s and then Johne's disease (JD) started to become a problem in the late 1990s. Concurrently, a wide range of other herd or individual animal diseases of a more sporadic nature, similar to those seen in other domesticated livestock, have been encountered.

Considerable research and development has been carried out over the last 50 years by researchers at AgResearch Centres at Invermay, Ruakura and Lincoln, the NZ Government Laboratory at Wallaceville, Massey University and Otago University, diagnostic laboratories, drug companies and Ministry field veterinarians, as well as veterinary practitioners. These advances led to the Deer Industry and veterinarians introducing animal health programmes to improve animal welfare and reduce wastage. However, clinical and subclinical losses due to disease are still encountered on many New Zealand deer farms.

In 1984, a group of veterinarians started the Deer Branch of the New Zealand Veterinary Association. Since then the Branch has held an annual conference and published proceedings, which are available for purchase from the NZVA website (https://www.nzva.org.nz). The New Zealand Deer Farmers Association, which was started in the 1970s, has a website (https://www.deernz.org) that has a wide range of production and animal health information.

Today (year 2021), there are approximately 900,000 farmed deer in New Zealand and they are predominantly red, red x wapiti or pure wapiti deer and relatively few fallow deer.

This review provides a brief summary of the most important diseases diagnosed by veterinary practitioners and/or published in laboratory surveillance reports in the following categories:

Bacterial: Bovine tuberculosis, paratuberculosis (Johne's disease), yersiniosis, leptospirosis, clostridial infections and fusobacteriosis.

- Viral: Malignant catarrhal fever and deer parapox.
- Parasitic: Lungworm, gastrointestinal nematodes, *Elaphostrongylus cervi*, flukes, ticks and cryptosporidiosis.
- Fungal: Facial eczema and ryegrass staggers.
- Trace element deficiencies/toxicities: Cu deficiency/toxicity, Se deficiency/toxicity and iodine deficiency.

2   Bovine Tuberculosis

2.1   Introduction

Bovine tuberculosis (Tb), caused by infection with *Mycobacterium bovis*, has been found in farmed and wild deer worldwide. In NZ the first cases of Tb were diagnosed in farmed deer in Canterbury in the South Island in 1978. In the 1970s and the early 1980s, Tb spread widely throughout the country because deer farming was rapidly expanding, deer were bought and sold, there was no compulsory Tb testing, no movement control restrictions and large numbers of wild deer continued to be live-captured from areas that were subsequently found to be Tb infected due to infected wildlife vectors. In 1985, a national voluntary Tb control programme for deer was implemented in addition to the compulsory national Tb control measures in cattle. This became compulsory in 1989. The Government, through the farm industry-led Animal Health Board (AHB), started implementing widespread control of wildlife vectors, primarily possums (*Trichosurus vulpecula*), introduced from Australia, and ferrets (*Mustela furo*). A National Pest Management Strategy (NPMS) with extra government funding was implemented to increase pressure on wildlife vectors. The number of infected cattle and deer herds started to fall in the late 1990s. Technical and managerial improvements, including the development of accurate blood tests, tighter controls on movement of stock from infected herds and compulsory ear-tagging of cattle and deer to identify their herd of origin, enabled more reliable tracing of infection sources. By June 2007, there were 148 infected cattle and deer herds in New Zealand, with a herd infection prevalence of rate of 0.39%. A new body, OSPRI, was set up to replace the AHB, with increased funding and an updated NPMS. Since then TB has been eradicated from 1.6 million hectares since the plan started in 2011 and the annual infected herd period prevalence rating declined to 0.09%, with 43 infected cattle herds and 1 infected deer herd by June 2016. The aim is the complete eradication of Tb from all cattle and deer herds by 2026.

2.2   Transmission

It is accepted that the primary route of infection between dairy cattle is by the aerosol route. Deer-to-deer spread of Tb may occur by several routes, depending on the location of lesions shedding Tb organisms. This may be via aerosols from infected lungs. However, discharging external or internal abscesses have often been found in farmed deer with advanced Tb and spread via the oropharyngeal route (saliva) may also occur from animals sharing the same food source, through grooming or from exposure to contaminated secretions, fomites or dust particles. Evidence for this route comes from the frequent observation that the most common sites of Tb lesions early in infection are the tonsil and retropharyngeal lymph nodes.

2.3   Clinical Signs

Tb infected deer often showed few clinical signs of Tb until the terminal stages of disease. In the 1980s and the 90s, 15–27-month-old deer killed for venison often had only a single caseo-calcified lesion in the retropharyngeal lymph node and appeared otherwise healthy. Animals with more advanced Tb showed weight loss and some with bronchial disease developed a cough. Some severely affected deer developed peripheral lymph node enlargement, some of which had open sinuses tracking onto the skin surface and excreted large numbers of acid-fast-organisms (AFOs). Clinical disease was often associated with the stresses related to poor management. The interval between the development of overt clinical signs and death was often relatively short.

2.4   Diagnosis

Early research showed that a single mid-cervical skin test (MCST) with bovine tuberculin had a sensitivity of approximately 85%, if applied carefully and interpreted conservatively. Compulsory whole herd testing (WHT) with the MCST became the basis for whole herd Tb control. WHTs were carried out at 6 monthly intervals until a clear test was obtained. Then WHTs were carried out annually. Early in the scheme, any deer with a positive test was slaughtered and all the lymph nodes checked for lesions. As the number of positive tests and the number of infected herds declined ancillary (or secondary) testing with the Comparative Cervical Test (CCT), using avian and bovine tuberculin, or a Blood Test for TB (BTB) was used to retest MCST positive deer to improve the specificity of testing, especially in herds where deer had increased avian sensitivity. Any animal positive to these ancillary tests was killed and necropsied. The BTB was a composite test that measured the reactivity of circulating lymphocytes and antibody levels in blood samples. It had consistently high levels of sensitivity (>90%) and specificity (>98%). It was eventually replaced by a modified ELISA, which measures IgG_1 antibodies, which has a sensitivity of 87% and specificity of 97%.

In the 1970s and the 80s, Tb surveillance was carried out at Game Packing Houses during routine inspection of carcasses of deer that were shot in the wild and recovered for feral game venison, and this helped to track regions where Tb was present in wild populations. Since deer farming started Tb surveillance in Deer Slaughter Premises (DSP) has played an essential role in Tb control. All carcasses are inspected and samples of any suspicious lesions are sent for diagnosis of Tb using histopathology, culture and/or PCR to confirm the herd status. Strain typing of *M. bovis* is performed using DNA-based techniques to track the probable source of infection.

2.5   Control

The control and eradication of Tb in farmed deer is strongly influenced by the presence or absence of wildlife vectors of tuberculosis. In the absence of infected wildlife, the most effective strategy for eradication is to use a comprehensive test-and-slaughter policy, surveillance of animals sent for slaughter and strict control of the movement of livestock from infected properties. In addition, herd management strategies aimed at minimising the risk of further spread and the identification of infected deer that are test negative (i.e. false negative) will accelerate eradication from a herd. Herd depopulation may be invoked if there is a severe outbreak and the prevalence of infection is above 40%. However, the currently available diagnostic tests for tuberculosis allow eradication from infected herds cost-effectively in a short period if managed properly, using the available test protocols, combined with selective slaughter.

In New Zealand, the main infected wildlife vectors are the brush-tailed possum (*T. vulpecula*), the ferret (*Mustela furo*) and wild deer. Although wild pigs can be infected they are generally regarded as "deadend" hosts and not liable to spread Tb to any great extent. The presence of infected wildlife vectors in extensive areas of native forest and grassland has made it impossible to totally eradicate Tb from NZ at present. However, the widespread trapping, shooting and poisoning of wildlife vectors in bush margins around farmed land have enabled Tb eradication on farm.

3   Johne's Disease or Paratuberculosis

3.1   Introduction

Johne's disease (JD) caused by *Mycobacterium paratuberculosis* (*M. ptb*) was first reported in NZ farmed deer in the mid-1980s and is now relatively common. Since then it has also emerged as a problem in farmed deer in the United Kingdom, Germany, New Zealand, Canada, Ireland, USA, Argentina and France. In the 1980s and the 1990s, it did not appear to be causing many problems. However, since 2000, JD has started to cause significant outbreaks of disease and mortality in young red deer, 6–18 months old, as well as sporadic wasting and deaths in older deer.

Strains of *M. ptb* that infect cattle are primarily C type or Type II, while sheep are infected with S type or Type I. Both strains have been isolated from farmed deer but over 95% of strains of MAP isolated from farmed deer in New Zealand have been of the C type and there is some evidence that the cattle strains are more pathogenic in deer than sheep strains.

Young deer appear to be more susceptible to clinical disease than older animals. In an experimental challenge study, 30 three-month-old weaners, 20 yearlings and 20 adult red deer were all dosed with 4×10^9 colony-forming units of a C strain of *MAP*. Ten of the weaner animals developed clinical disease in the year following challenge while none of the yearlings or adults were clinically affected (Mackintosh et al. 2010).

Passive surveillance, principally by the examination of suspect tuberculous lesions at slaughter plants, subsequently resulted in *M. ptb* being identified in over 600 farmed deer on 300 properties. Serological surveillance for paratuberculosis on 627 of New Zealand's deer farms has been carried out since 2000, using blood samples submitted as part of the national tuberculosis eradication scheme. The findings from this project suggested a national prevalence of herd infection of around 63%. Since 2007, abattoir surveillance showed enlarged mesenteric lymph nodes suggestive of Johne's disease (JD) on >95% of deer farms. While outbreaks of clinical JD occurred commonly prior to 2010, regular herd testing with ELISA and culling of reactors since then has reduced clinical outbreaks of JD in New Zealand deer herds to a very low level.

M. avium infection has occasionally caused a disease syndrome very similar to JD.

3.2   Transmission

JD is spread primarily by the faecal–oral route from animals excreting *M. ptb* in faeces onto soil or pasture that other deer then ingest. Faecal contamination of water or the udder may act as a source of infection for young animals. Infected dams may excrete bacteria in milk and infect the suckling offspring. Intrauterine transmission from dam to foetus, as demonstrated to occur in cattle, also occurs in red deer. Van Kooten et al. (2006) reported that in a sample of 9 late-stage pregnant hinds clinically affected by JD, 9 out of 10 (90%) of their foetuses were BACTEC culture-positive for *MAP*. In another study, *MAP* was isolated from 78% of foetuses of a group of 18 subclinically infected red deer hinds (Thompson et al. 2007). Neonatal animals are considered to be most at risk from infection, and adults appear to be more resistant.

3.3   Clinical Disease

Typically animals with JD may develop chronic scouring, weight loss and death. Two clinical syndromes have been recognised in NZ farmed red deer: sporadic cases in mixed-age deer, with an incidence of 1–3% per annum and outbreaks in 8–15-month-old deer, which may involve up to 20% of a group. Sporadic cases of clinical JD, with chronic granulomatous enteritis, occur in all ages and classes of farmed deer. Affected animals typically lose weight over a period of a few months, and the majority develop continuous or intermittent diarrhoea. There is usually low morbidity (<1% of infected animals) but high mortality (~100% die or are euthanased) and no response to treatment, which is very similar to JD in cattle and sheep. By contrast, outbreaks of more acute JD in young red deer, involving 5–20% of the 8–24-month-old animals, have occurred on 5–10 farms a year in the early 2000s (Mackintosh et al. 2004). The affected animals initially "fail to thrive", stop growing and then rapidly start to lose weight and condition. They invariably develop diarrhoea and became soiled with green faecal material around the tail, hindquarters and hocks. In spring they only partially moult their winter coats and take on a patchy or "moth-eaten" appearance. The clinical disease has a course of a few weeks, and it appears that the younger the animal, the quicker the progression to emaciation and death. The differential diagnosis includes yersiniosis (in weaners in winter) and abomasal parasitism. Blood samples taken for clinical pathology tests typically show low total protein and low serum albumin concentrations, and there may be elevated concentrations of acute inflammatory proteins such as fibrinogen and haptoglobin. Cervids typically mount a vigorous immune response to *MAP* infection characterised by high titres of antibody and are capable of shedding high titres of *M. ptb* organisms into the environment as the disease progresses from the paucibacillary to multibacillary state (O'Brien et al. 2013).

3.4   Diagnosis and Control

Currently the most widely used serological test for diagnosis and control in farmed deer in NZ is the Paralisa Test™, while the qPCR is used to quantitatively detect *M. ptb* in faecal samples (O'Brien et al. 2013). Faecal culture can be used on faecal and lymph node samples but takes some weeks. The Paralisa Test or faecal qPCR is used for rapid diagnosis in clinically affected deer. The qPCR is also used on post-mortem samples of enlarged jejunal and ileo-caecal lymph nodes from dead or euthanased animals.

In the face of a recognised infection, the aim of control should be to reduce the level of infection in order to eliminate clinical disease and minimise subclinical effects on productivity. Serological tests suffer some lack of sensitivity because antibody is not produced early in the disease and therefore it is unrealistic to try to eradicate paratuberculosis from a deer herd. On affected farms, the real cost of paratuberculosis should be estimated by adding some or all of the following produc-tion losses: outbreaks of mortality in 8–27-month-old deer, sporadic losses of mixed-age hinds and stags, reduced growth rates in calves and yearlings, increased non-pregnancy rates in hinds, suboptimal antler growth in stags and interference with TB testing. Control measures to reduce these costs include the identification and culling of clinically affected deer, as well as culling infected hinds, because their calves will almost invariably be infected either in utero or shortly after birth.

The use of Paralisa testing of serum samples is relatively economical and effective in control programmes because the culling of positives eliminates the most seriously affected animals that are most likely to transmit infection. However it is very unlikely to eliminate disease in a herd. The faecal qPCR can be used in control programmes to detect some animals that have a false negative Paralisa result but are shedding *M. ptb* in faeces. It can also detect "super-shedders" that are the most dangerous. Herds can be screened for infection by collecting faeces from all the animals and using the qPCR test, either on all individuals or on pools of 5–10 animals. If any of the pools are positive, then individual samples can be tested.

Vaccination with a commercial killed paratuberculosis vaccine has been shown to reduce clinical disease by 60% by approximately 12 months of age, but did not reduce the proportion shedding. However, vaccination has very limited use in NZ because it sensitises deer to the intradermal skin test for bovine Tb. Therefore, it can only be used in deer that are going for slaughter at under 27 months of age and do not require skin testing for Tb, but are examined post-mortem at slaughter for Tb surveillance.

One advantage of the use of culture is that the strain of *M. ptb* can be determined.

3.5   Prevention

Ideally, deer farmers should prevent the introduction of MAP on to their farm by acquiring only uninfected livestock originating from test-negative herds, maintaining a closed herd and avoiding cross-grazing with other livestock species. Use of artificial insemination to bring in new bloodlines and control of wildlife on the property may also be worthwhile. However, these measures are not as easy as they may appear. Establishing the infection status of a herd or source of uninfected deer is very difficult, and avoiding cross-grazing, surface water run-off from neighbouring paddocks and controlling wildlife such as rabbits, rodents, roe deer and mustelids may be very difficult. As well as only sourcing deer from "low-risk" farms, farmers should only purchase deer that are negative to the most sensitive blood tests available and/or to a faecal qPCR test.

4   Yersiniosis

4.1   Introduction

Yersiniosis, caused by *Yersinia pseudotuberculosis* (*Y. pstb*) was first diagnosed in outbreaks of diarrhoea among deer herds in 1978. The disease quickly became one of the most serious and common infectious diseases of young farmed deer in New Zealand. It usually occurs in 4–8-month-old deer during their first autumn/ winter. Predisposing factors include stressors such as weaning, transport, bad weather and underfeeding. Yersiniosis has also occurred in farmed deer in Australia, North America and Europe and *Y. pstb* can also infect humans and causes lymphadenitis with symptoms very similar to acute appendicitis.

4.2   Transmission

The epidemiology of yersiniosis in New Zealand has been reviewed by Mackintosh (1992). Three serotypes of *Y. pstb* (I, II and III) have been identified in New Zealand.

The serotypes are commonly carried in the intestines by a wide range of wild and domestic animals including birds, rodents, rabbits, hares, pigs, cattle, sheep and household pets. Carrier animals tend to excrete more *Y. pstb* in the faeces at times of stress, especially in winter, and the organisms survive in the environment for long periods in cold, wet conditions. The majority of farmed deer are exposed to contaminated food and water during their first autumn/winter and experience sub-clinical infections. Outbreaks of disease occur when animals are stressed and experience a heavy challenge. Genetic factors also appear to influence susceptibility.

4.3   Clinical Disease

Affected young deer usually show signs of depression, anorexia and green watery diarrhoea staining of the hocks and tail hairs. The diarrhoea generally turns dark or bloody. Elevated temperature (pyrexia) is frequently noted early in the course of the disease. Affected deer rapidly become dehydrated and weak. Acute cases may be found dead without any obvious clinical signs.

4.4   Pathology

The usual findings at necropsy include haemorrhagic enteropathy with bloody intestinal contents, especially in the lower small intestine and sometimes in the caecum and colon. There is often oedema of the serosal surface of the small intestines and thickening of the intestinal wall with a pseudomembranous necrotic lining. The mesenteric lymph nodes are often swollen and oedematous.

4.5   Diagnosis

The differential diagnosis of a number of cases of rapid death in weaners during the autumn/winter period includes yersiniosis, malignant catarrhal fever, leptospirosis and parasites, especially lungworm and abomasal *Ostertagia*-type nematodes. Healthy young deer often experience subclinical infection with *Y. pstb* and can carry and excrete the organism in their faeces for some time. Therefore, it is not uncommon to isolate *Y. pstb* from faecal samples taken from normal deer, using *Yersinia*-specific bacterial culture media. In cases of clinical yersiniosis, *Y. pstb* organisms cross the wall of the intestine and spread to the mesenteric lymph nodes via the lymphatic drainage. Therefore, in cases of sudden death in suspected cases of yersiniosis it is worthwhile taking samples of mesenteric lymph nodes for culture for *Y. pstb* culture, along with samples of small intestine and mesenteric lymph node samples for histopathology to confirm the diagnosis.

4.6   Treatment

Acute cases of yersiniosis usually respond well to early intensive treatment with antibiotics, such as tetracyclines or potentiated sulphonamides, and oral or parenteral fluids. Isolation and housing and other nursing care are essential to achieving the best outcome of treatment. During an outbreak, yarding should be undertaken for detection of new cases, which can be treated early. If there are multiple cases of yersiniosis in an outbreak, it is sensible to give mass treatment to all in-contact deer with oral or parenteral antibiotics.

4.7   Control

The best strategy for prevention of yersiniosis in farmed deer in New Zealand involves the use of good management to optimise feeding, provision of shelter in bad weather and minimising stressors. Careful handling of young deer over their first autumn will condition them to interaction with humans. Vaccination of 3–4-month-old weaners with two doses of Yersiniavax, a killed adjuvanted *Y. pstb* vaccine, has been shown to give significant protection against severe yersiniosis and is a valuable adjunct to good management for disease prevention. The vaccine is widely used in the industry and studies have shown that it gives significant protection in experimental challenge trials and reduces morbidity and mortality by more than 65% in field trials.

5   Leptospirosis

5.1   Introduction

Leptospirosis is caused by serovars of the bacterium *Leptospira interrogans*. Throughout the world, a large number of serovars are each carried asymptomatically by a restricted range of maintenance hosts, which are often rodents, but also include other wildlife species and domestic animals. Infection can "spill-over" into a wide range of mammalian species, including deer. Infection is often inapparent, but can also result in serious disease, including redwater (haemoglobinuria), hepatitis, jaundice and kidney disease (nephritis). In New Zealand, serovars *L. pomona*, *L. hardjo-bovis* and *L. copenhageni* have been isolated from clinically affected farmed red deer.

5.2   Transmission

Maintenance hosts for particular leptospiral serovars develop chronic interstitial nephritis and shed leptospires in their urine for long periods of time. Accidental or spill-over hosts shed the organism in their urine for shorter periods. Cattle are the maintenance host for *L. hardjo-bovis,* pigs are the usual maintenance hosts for *L. pomona* and rodents host *L. copenhageni.* Red deer appear to be spill-over hosts for all three serovars, but can shed *L. hardjo* in urine for as long as 8 months. The organism can persist for long periods in the environment, especially under cool, wet conditions. Animals are primarily infected by exposure to leptospires in urine, contaminated water or food and the organisms enter across mucous membranes, especially conjunctiva of the eye, and membranes of the mouth or nose, and through skin cuts or abrasions. Once they enter the body, leptospires spread via the blood to the liver, kidneys and placenta.

5.3   Clinical Signs

Young animals are generally more susceptible to clinical disease than adults. Infections may be inapparent in many cases, depending on the serovar. Clinical signs may range from mild pyrexia, especially with hardjo-bovis infections, to general malaise, pyrexia and anorexia, through to severe haemolytic disease, severe nephritis, haemoglobinuria and jaundice with *L. pomona* and *L. copenhageni* in young red deer. There is clinical evidence of an outbreak of redwater and jaundice associated with serovar *copenhageni.* Leptospirosis has been shown to cause abortions in farmed deer in New Zealand, and reduce weaning percentage in red deer.

5.4   Pathology

Post-mortem signs are highly variable and depend upon the severity of the infection. In farm animals the acute form is characterised by anaemia, jaundice, haemoglobinuria and widespread haemorrhages. Acute nephritis may also occur and most animals that recover develop chronic interstitial nephritis. Meat inspectors at Deer Slaughter Plants commonly find white spotted kidneys, which are likely to be due to leptospirosis.

5.5   Diagnosis

Diagnosis on clinical grounds is difficult because other diseases may cause similar symptoms. A combination of dark-field microscopy of urine, culture or PCR of urine or kidneys, and fluorescent antibody techniques, especially on kidney tissue, are important for confirmation. In freshly dead deer, including aborted late-term foetuses, dark-field examination of aqueous humour may reveal leptospires. Serology on a herd basis, particularly of paired samples taken at an interval of 14 days or more, can also help establish evidence of infection. The microscopic agglutination test (MAT) is the most common serological test and is especially useful on a herd basis. However, serology on individuals is not completely reliable because some acutely infected animals can have negligible titres.

5.6   Treatment

Streptomycin at 25 mg/kg daily for 4 days appears to be effective at treating clinically affected animals in the early stage of disease and eliminating the carrier state. Ampicillin and amoxicillin may also be effective. Tetracyclines may be used during the acute phase but may not prevent kidney colonisation. Electrolyte therapy should be considered, especially if the kidneys are damaged.

5.7   Prevention

Vaccination of young deer against the prevailing serovars has been shown to protect them against leptospirosis. Either a two serovar (Hardjo-bovis and Pomona) or three-serovar (Hardjo-bovis, Pomona and Copenhageni) vaccine is used. It is recommended to give a sensitising dose to young deer at 12 weeks of age, followed by a booster dose 4 weeks later. An annual booster to all breeding stock in late pregnancy strengthens the immunity and promotes good colostral protection in newborn deer.

Deciding whether to vaccinate depends on the risks of infection. Risk factors may include whether dairy cows/dairy grazers or beef cattle are run on the same farm as deer, whether pig effluent may enter the farm, the history of past infections, farming downstream from farms known to have the disease present or operating an open or closed herd.

Vaccination is considered to be a cost-effective option. Research shows a reproductive efficiency increase of only 1.3% would pay for vaccination costs.

Purchased animals may be treated with streptomycin before entry into a herd and then vaccinated. Wildlife sources of infection should be controlled or eliminated. Clean drinking water should be supplied and natural water sources fenced off. Deer should not be grazed with other livestock, especially cattle and pigs, unless known to be free of infection.

6   Fusobacteriosis

6.1   Introduction

Fusobacteriosis (necrobacillosis), caused by *Fusobacterium necrophorum*, can affect wild and farmed deer causing purulent necrotic lesions, especially of the mouth and feet, which can lead to abscesses of the liver and lungs. The pleomorphic Gram-negative organism is a strict anaerobe with a worldwide distribution. It commonly causes foot abscesses in farmed red, wapiti and fallow deer, and mouth lesions in fallow deer in New Zealand, Australia, Europe and North America, and has also been associated with die-offs of wapiti on winter feeding grounds in the USA, and white-tailed deer in Canada.

6.2   Transmission

Fusobacterium necrophorum is a part of normal intestinal flora and is an opportunistic pathogen. The organism is present in faeces and survives well in soil and mud. Entry is usually through damaged skin or mucous membranes and infection often occurs in animals that are debilitated or suffering stress such as overcrowding. Abrasions or injuries of the oral mucosa, due to thistles, foreign objects or teething, can lead to infections of the mouth and throat, while cuts and abrasions to the feet or lower legs can lead to infected limbs, especially in cold, muddy conditions or in animals kept in feedlots. Toxins produced by the bacterium kill cells in the surrounding tissue, disrupting the blood supply and providing the anaerobic conditions suitable for multiplication.

6.3   Clinical Signs

Infections of the oral mucosa, which usually affect young fallow deer under 3 months of age, result in swelling of the head and/or tongue, submandibular oedema, salivation, fever and an inability to suckle or feed. Infections of feet and lower limbs lead to interdigital dermatitis, infected joints, ligaments and tendons, and lameness. Systemic spread to the liver and lungs results in pus filled abscesses and leads to fever, loss of appetite, respiratory distress and death. Diagnosis is usually made on the appearance of the lesions.

6.4   Treatment

Early cases of foot abscess may be treated with injectable antibiotics together with cleansing and dressing of the affected foot. Hydrogen peroxide is useful for washing wounds and abscesses, and the application of oily antibiotic mastitis preparation has been successful in treating deep-seated abscesses in red deer feet. In severe or chronic cases, digital amputation may be required. Treatment of early cases of necrotic stomatitis (severe mouth lesions) involves extensive debridement of lesions and vigorous antibiotic therapy. High doses of penicillin, tetracyclines and sulphonamides have been used successfully. If an outbreak is occurring in a herd, preventive treatment with long-acting penicillin or tetracyclines of unaffected deer has proven to be useful. Concurrent vaccination against tetanus and clostridial diseases may be indicated. Once generalised lesions have developed, treatment is usually futile.

6.5 Control and Prevention

Prevention of fusobacteriosis depends on the removal of predisposing factors such as sharp or protruding surfaces, which may lead to abrasion of the lower limbs. Access to thistles or other plants with prickles, which may injure the oral mucosa, should be prevented. Grain overload should be avoided, and in feedlots, clean dry bedding should be provided. A *F. necrophorum* vaccine is no longer available in New Zealand.

7 Clostridial Diseases

7.1 Introduction

Clostridial organisms cause a wide range of diseases in domestic livestock and wildlife throughout the world. *Clostridium* spp. are commonly found in soil, as well as in the gastrointestinal tract of healthy animals. The organisms are spore-forming anaerobes (living in the absence of oxygen), which are highly resistant to a wide range of environmental conditions. Disease is caused by the production of potent specific toxins excreted by the organisms (exotoxins).

Since the beginning of deer farming in New Zealand in the 1970s, sporadic cases of clostridial disease have been reported in deer. Cases of pulpy kidney (*Clostridium perfringens* type D) have been recorded in young red and fallow deer, and blackleg or malignant oedema associated with trauma or injection sites in adult red stags and wapiti. These animals are prone to injury from fighting during the breeding season and often need intramuscular injections for chemical restraint or local anaesthetics during handling and velvet antler removal. Tetanus (*C. tetani*) has also been reported occasionally by veterinary practitioners on deer farms in New Zealand.

7.2 Pulpy Kidney (Enterotoxaemia)

Enterotoxaemia due to *C. perfringens* type D has been recorded in young red and fallow deer and usually causes sudden death. Animals may be found dead, or may die in convulsions within 24 h of first showing signs of anorexia, diarrhoea and depression. Few post-mortem signs of disease are observed in peracute cases. The carcass is usually in good condition, and an excess of clear straw-coloured fluid is often noted in the pericardial sac. Petechial haemorrhages may be found in the heart. In acute cases, the contents of the small intestine have a watery consistency, and areas of congestion of the abomasal and intestinal mucosa may be observed. Usually the liver is congested and the kidneys are soft and gelatinous. Confirmatory diagnosis requires identification of typical bacteria or toxin in intestinal contents. Given the very rapid onset and short course of the disease, treatment is usually impractical.

7.3   Malignant Oedema and Blackleg

Malignant oedema, also known as gas gangrene, is an acute, rapidly fatal wound infection caused by several different members of the genus, including *C. septicum*, *C. chauvoei*, *C. perfringens*, *C. sordellii* and *C. novyi*. Blackleg, caused by *C. chauvoei*, develops when the spores which lodge in normal animals in skeletal muscle without symptoms, proliferate under anaerobic conditions such as in deep bruising or wounds, either of which may be associated with handling or fighting.

Malignant oedema and blackleg are usually associated with deep puncture wounds or heavy bruising. In most cases, few signs are present other than peracute death. At necropsy, malignant oedema is characterised by gangrene of the skin and oedema of the subcutaneous and connective tissue in the surrounding areas. Affected wapiti had subcutaneous and intramuscular haemorrhage, with emphysema in the extremities, oedematous internal organs, tarry blood and bloody fluid exudate from body orifices. Swelling and crepitus (bubbly gas in tissues, felt when pressure is applied) of affected muscles are typical manifestations of blackleg. Diagnosis requires the isolation of the organism from lesions or identification of the organism using a fluorescent antibody test.

A number of cases of severe septic oedema of the head and face of stags after velvet antler removal have occurred. Some were associated with local anaesthetic containing adrenalin, which reduces blood flow, possibly creating the anaerobic tissue environment that stimulates clostridial spores to multiply. *Clostridium septicum* has been isolated from at least one case. Treatment is usually impractical because of the very rapid onset and short course of the disease.

7.4   Prevention of Clostridial Diseases

Vaccination is the only practical means of protecting animals at risk, in addition to good husbandry and the use of sterile technique when injecting deer. Although not licensed for deer, multivalent clostridial vaccines are commonly used. Generally, pregnant females are vaccinated annually in late gestation, and fawns are vaccinated at weaning. Adult stags are often boosted annually to optimise protection against clostridial complications of wounds or bruising associated with fighting during the breeding season.

8   Viral Diseases

8.1   Malignant Catarrhal Fever

8.1.1   Introduction

In the first 20 years of deer farming in New Zealand, in the 1970s and the 1980s, Malignant Catarrhal Fever (MCF) caused many cases of sudden death and was the most important viral disease of farmed deer. Initially the cause was unknown but was later shown to be caused by a virus carried asymptomatically by sheep, ovine herpes virus-2 (OHV-2). This is related to other viruses that cause similar diseases in other species. MCF, which is similar to Bovine Malignant Catarrh (BMC), has gradually declined in incidence and significance over the last 30 years, although it still occurs occasionally.

8.1.2 Transmission

OHV-2 is thought to be endemic in most sheep flocks worldwide and sheep-associated MCF (SA-MCF) occurs in deer where there is direct or indirect contact with sheep. Ewes experience a recrudescence of infection in late pregnancy and shed virus in lacrimal (eye), nasal, oral and vaginal secretions, infecting the next generation of lambs soon after birth. This may also explain the seasonal increase in cases in deer in late winter and early spring. The disease has also occurred in other circumstances, and direct contact is not necessarily required. Wind-borne infections have been reported and deer have become affected after being carried in a truck that had earlier been used for sheep transport.

Stress also plays a major role in the development of MCF. It is well known from studies in the UK, Australia and New Zealand that the incidence of the disease rises in winter/spring, at a time when conditions are harsh, deer may be in poor condition and disease incidence rises in crowded conditions. MCF occurs more frequently on intensively managed properties than on extensive operations. Clinical disease due to MCF is rare in deer <1 year of age. Stags are three times more susceptible to MCF than hinds.

The disease only affects a small number of susceptible animals in a group, although occasionally a higher morbidity can occur. Some deer species, including Père David's (Elaphurus davidianus), rusa, sika, axis and white-tailed deer are highly susceptible, while some, especially fallow deer, appear highly resistant to MCF.

8.1.3 Clinical Signs

MCF can occur in forms ranging from peracute to chronic disease. The outcome is invariably fatal. In the peracute form, infected animals may die with no prior clinical signs. The most common manifestation of MCF in deer is a very rapid onset of bloody diarrhoea, dark stained urine, marked depression and death within 48 h. Chronic cases last from a few days up to 5 weeks and may develop the so-called head and eye form of the disease, which is the type normally seen in BMC of cattle, and in the early stages is characterised by congestion of the mucous membranes, excessive salivation, lacrimation and nasal discharge. Bilateral corneal opacity develops, there may be haemorrhage into the anterior chamber of the eye and as cases become more chronic the eyelids may swell and eventually become closed with a mat of catarrhal exudate. The muzzle, and often the vulva, develops a dry crusty appearance and erosions of the mucous membranes of the mouth are evident. Lymph nodes are swollen and the superficial ones can readily be palpated. Nervous signs are also sometimes seen, particularly an initial dullness followed by incoordination and hyperaesthesia.

The clinical pathology most often involves an initial rise in white blood cells followed by a marked leucopenia and an increase in acute inflammatory proteins.

At necropsy, the lesions vary somewhat according to the clinical course of the disease. In the peracute forms gross changes may be limited. In less rapidly fatal forms lesions may be seen throughout the entire gastrointestinal tract. These include necrotic erosions, congestion, oedema and frank blood in the intestinal lumen. Lymph nodes are usually swollen, and the mesenteric nodes may be grossly enlarged, with much oedematous fluid around them. Cross section of these and other nodes shows necrosis and inflammation. The most characteristic histological lesion is a lymphocytic vasculitis affecting the arteries, arterioles, veins and venules. In peracute cases they are characterised by invasion of lymphoid cells into all levels of the vessels, and in the most severe cases they may even occlude the lumen. Vascular haemorrhages, epithelial degeneration and lymphoid hyperplasia with invasion of lymphoid cells into non-lymphoid tissues also occur, especially in chronic cases. These infiltrations may be visible macroscopically in the kidney. In chronic cases, there may be numerous nodules around major blood vessels, known as peri-arteritis nodosa. These may be found in many organs, but are most common around the blood vessels to the uterus and intestines, and in the kidney.

8.1.4 Diagnosis
Clinical signs and post-mortem lesions are important means of confirming a case of MCF.

8.1.5 Treatment and Control
There is no effective treatment. There is no vaccine available. Prevention requires that deer do not have direct or indirect contact with sheep.

8.2 Deer Parapox

8.2.1 Introduction
A new species-specific parapoxvirus infection of red deer was identified in New Zealand in 1985. The virus is a member of the parapox genus of the poxvirus family that has specific characteristics that distinguish it from other poxviruses and parapoxviruses. Restriction endonuclease analysis of viral DNA and PCR has shown it to be different from parapoxviruses of sheep or cattle.

Pox viruses in the parapox group have a worldwide distribution, but the specific parapoxvirus of deer has not been reported outside New Zealand. The three situations where severe infections have occurred are infection in stags in velvet, hinds and calves close to calving and recently stressed or transported animals. Deaths have occurred but these are rare. As all deer in New Zealand originated from other countries it is possible that the virus is also present elsewhere but has not been either seen or reported.

Transmission of deer parapoxvirus probably occurs by direct contact with active lesions on infected animals, or with exudates or scab material in the environment. In severe outbreaks in New Zealand the presence of thistles in paddocks is thought to be a major contributing factor. The parapox viruses are extremely resistant to most environmental factors such as sunlight and cold and can exist outside the body for long periods of time.

This is a zoonotic virus (affecting people) and gloves should be worn when handling infected animals and velvet antler. The risk of transmission of infection to humans via the consumption of infected velvet antler is not known, but some methods of processing velvet antler may not destroy the virus. Therefore, affected velvet antler should be destroyed.

8.2.2 Clinical Signs

In New Zealand the most common clinical signs in stags are scabs on the velvet antler, ears, muzzle, face and lip margins. In more severe cases the lesions develop over widespread parts of the body and in one outbreak 21 of 55 recently captured animals were severely affected and died. The disease in young animals results in scabby lesions about the lips, and sometimes, blisters on the tongue and in the mouth. There may also be pox lesions on the hoof coronary band and between the hooves. Sometimes these lesions may bleed. This form of the disease is similar to Foot and Mouth Disease, although there is rarely a fever, as seen in that disease. Lesions on the feet may spread up the leg causing lesions with crusty scabs that also may bleed if damaged. In affected females scabby lesions about the vulva and perineum may be observed. Secondary bacterial infection of the damaged skin may result in severe infection and death. This form of the disease has been seen in deer as young as a few days of age, up to 6–8 months, and because of secondary bacterial infection, can be fatal. Occasionally hinds may show a mild form of the disease with scabs about the mouth and udder. There is evidence that deer can carry this disease without showing clinical signs. Removal of the scabs leaves a red raw surface.

8.2.3 Diagnosis

Diagnosis is based upon the clinical picture and can be confirmed by histological examination and DNA identification of the virus.

8.2.4 Control

Generally severe outbreaks of disease occur when the infection is introduced to a susceptible population. Usually there are few cases of clinical disease in subsequent years, except when new animals are introduced, suggesting that the majority of deer experience subclinical infections in the endemic state. There are suggestions that control of prickly plants may help reduce the incidence of the clinical disease.

Experimental vaccination by scratching with parapoxvirus or orf virus on the inner thigh of young deer gave some protection against subsequent challenge with deer parapox, but it has not been used on commercial farms. In young animals affected prophylactic antibiotic treatment is justified to prevent secondary bacterial infection.

If velvet antlers are affected, management of an outbreak depends on the stage of antler growth. One report describes removal of antlers from those near the harvesting stage, and separation of those stags from unaffected herd-mates. Those that had affected velvet antler in the very early stage were separated and left. Their lesions resolved by the time the velvet was ready to harvest.

9   Parasitic Diseases

9.1   Lungworm

New Zealand farmed red and wapiti deer are commonly infected by the deer-adapted parasite *Dictyocaulus eckerti*. It is a close relative of the cattle lungworm, *D. viviparus*, which can also infect deer, but it is less well adapted to them.

Lungworms usually first infect young deer in the late summer/early autumn when they are 2–3 months old. They are very susceptible because they have not yet developed any immunity. The warm wet environment in autumn favours survival of the parasite larvae on pasture and explosive outbreaks can occur if this parasite is not adequately controlled. Red deer develop resistance with exposure and age, and adult deer are somewhat resistant, although clinical disease can occur in situations of poor management where deer are severely stressed or nutritionally compromised.

D. eckerti has a direct life cycle. Adult worms live in the bronchial tree of the lung, where they mate and lay eggs. The eggs either hatch in the bronchi, or in the gastrointestinal tract after they have been coughed up the trachea and swallowed. Only first stage larvae are passed in the faeces. Under conditions of warmth and moisture the larvae develop on the ground to infective third-stage larvae in 3–7 days. They migrate up the pasture and are eaten. They burrow through the intestinal wall and travel to the lung via the blood supply and lymphatic drainage. They burrow through from the capillary bed into the airways and develop to adults. The pre-patent period (time from ingestion to excretion of larvae in the faeces) is usually 23–24 days. The adults can continue to produce eggs for over 3 months. Cold weather arrests larval development on pasture, and they are susceptible to drying out in hot dry conditions. However, this is a hardy parasite and they can over-winter in cold climates and can withstand a temperature of 4.5 °C for a year.

9.1.1   Clinical Signs
Clinical signs in deer include decreased appetite, loss of live-weight and condition, retarded growth rate, roughened coat and unexpected deaths in a young deer. A soft bronchial cough may be heard, especially after exercise. Deer are usually severely affected by the time these signs are observed. Deer that die of heavy infestations usually have huge numbers of adult worms blocking the trachea and major airways. There is often reddening of the lung and areas of congestion.

9.1.2   Diagnosis

Faecal larval counts (FLC), using the Baermann technique, give a good indication of the adult lungworm burden, but cannot indicate the number of immature developing larvae that have been ingested in the previous 3 weeks. However, a high FLC is a warning that the level of challenge on the pasture is high and treatment is strongly indicated.

9.1.3   Treatment, Control and Prevention

Young deer are most at risk from late summer (Feb) to early winter (June). The degree of risk depends on a number of factors including climate (temperature and moisture), stocking density, weaning practices, pasture type and length, grazing frequency, stress and genotype. Wapiti deer appear to be more susceptible than red deer and take longer to develop a degree of resistance. Farmers should aim to prevent lungworms from limiting production by a combination of appropriate pasture management, anticipating periods of risk and using preventative treatment.

Lungworm is easily controlled by treatment with appropriate anthelmintics at appropriate intervals, either to reduce the risk to the animal and preferably, to break the life cycle. The latter can be achieved by treatment frequencies that prevent the immature lungworm from reaching maturity. This varies, depending on the type of anthelmintic used and formulation.

Benzimidazoles (white drenches) are effective but have no persistent activity and should therefore be used at three-week intervals over periods of risk. Avermectins and moxidectin are highly effective and can be used at longer intervals according to manufacturer's recommendations, because they have varying degrees of persistent activity that prevent the establishment of infection for that period. Levamisole is not effective against lungworm in deer.

Pastures heavily contaminated with larvae as a result of being grazed by untreated animals with lungworm burdens should be grazed by older deer or other livestock, which are less susceptible, or spelled for a long period. Making silage or hay will also reduce larval challenge on regrowth pasture. Early anthelmintic treatment and repeats doses at appropriate intervals should prevent establishment of heavily contaminated pastures. It is important to maintain optimum nutrition and health in relation to other diseases, to maximise the young deer's immune competence, and therefore its ability to resist serious infestation. Recent research has indicated that pasture plants with high condensed tannin concentration, such as chicory, lotus and sulla, may have an important role to play in reducing the need for chemicals for parasite control in deer.

9.2   Gastrointestinal Parasites

9.2.1   Introduction

Most gastrointestinal (GI) nematode parasites of deer are closely related to those of other ruminants and occupy the same regions of the gastrointestinal tract as their equivalent species. The most important genera of parasite affecting deer are

Ostertagia-type in the family Trichostrongylidae. These parasites have a direct life cycle involving the shedding of eggs in the faeces, the hatching and development through larval stages to an infective larva and the ingestion of larvae while grazing. The larvae either continue to develop in the gut or may have a period of arrested development associated with winter. The larval stages of these parasites require high humidity and warmth in the microclimates in forage to complete their development. Red and fallow deer develop a degree of resistance to *Ostertagia*-type parasites, but wapiti deer are more susceptible and are prone to developing moderate to severe burdens of immature parasites in the lining of the abomasum, leading to reduced acid production, raised pH, poor protein digestion and chronic weight loss. Before the cause was clarified the clinical entity of severe chronic weight loss was referred to as the "fading elk syndrome" and is somewhat analogous to Type II Ostertagiasis of cattle. It has also been recognised in wapiti in North America.

There are a number of other species of gastrointestinal worms of sheep and cattle which infect deer, but they usually do not cause problems. However, *Haemonchus contortus* (barber's pole worm), a common parasite of sheep and cattle, causes some problems in the warmer areas of the North Island of New Zealand. With increasing temperatures due to climate change, it may become a greater risk in deer in the South Island.

9.2.2 Clinical Signs

Acute heavy burdens in young deer, especially wapiti, show clinical signs similar to those of parasitic gastroenteritis in other stock. They include weight loss, or failure to thrive, staring coat, soft faeces or frank diarrhoea, and soiled tail and perineum. The syndrome is akin to Type II Ostertagiasis in cattle, where immature larvae affect the abomasum, is characterised by anorexia, weight loss, weakness, dull hair coat, diarrhoea and low blood protein (hypo-proteinaemia), sometimes resulting in oedema under the jaw, referred to as bottle jaw. Severely affected deer may have worm counts as high as 90,000, with fourth stage larvae, which do most of the damage, being approximately 80% of the total.

9.2.3 Diagnosis

Assessing GI parasite burdens in the live animal is difficult because the correlation between faecal egg counts (FEC) and total worm number is poor. This is further confounded because the most damage to the abomasal lining is caused by the immature stages before they produce eggs. Loss of condition and other clinical signs, combined with positive faecal samples, may be taken as positive in the absence of evidence of other specific conditions. Plasma pepsinogen, which is used in cattle to indicate damage to the abomasal lining, does not appear to be useful in deer. The heavy winter coat of wapiti and red deer can readily mask the fact that they are in poor body condition, and suffering severe weight loss.

9.2.4 Treatment

Until relatively recently, the avermectin/moxidectin anthelmintics were quite effective against adult *Ostertagia*-type nematodes in deer and have been the mainstay of the deer industry for the last 30 years. Modern benzimidazoles such as oxfendazole and albendazole were somewhat effective in red deer but appeared to be less effective in wapiti deer. However, the continual use of these anthelmintics, especially pour-on anthelmintics, which have persistent activity but also a long "tail" at low concentrations, has led to the development of anthelmintic resistance in *Ostertagia*-type worms in deer, similar to that seen in sheep and cattle. Consequently, veterinarians are now advising against using pour-on anthelmintics, but advocating the use of combinations of oral and/or injectable anthelmintics to give better combined efficacy and to slow the development of resistance. Oral benzimidazoles should be given at twice the cattle dose rate, together with either injectable or oral moxidectin/avermectin and some veterinarians advise adding oral levamisole to the combination. Recently an oral triple combination drench has been registered for deer in New Zealand called "Cervidae Oral", giving an oral dose of moxidectin 0.4 mg/kg, oxfendazole 18 mg/kg and levamisole 12 mg/kg. This has shown good efficacy against GI parasites and currently overcomes anthelmintic resistance issues.

9.2.5 Control

Control measures involve good husbandry and the principles described above for lungworm apply to gastrointestinal parasites. Indeed, in most situations, parasite control aimed at lungworm will control gastrointestinal parasites concurrently. Where these parasites have been shown to be a problem regular faecal sampling and anthelmintic treatment should be combined with pasture management to ensure that the animals have adequate nutrition. The use of pasture containing plants high in dietary tannins has recently been investigated as a means of reducing intestinal worm burdens and shows promise. Mixed grazing of sheep and/or cattle is used by an increasing number of deer farmers to control pasture and reduce deer-specific parasites.

Because of their increased susceptibility to parasites, it is advisable to graze wapiti on longer pastures and to avoid grazing them with red deer. Unlike adult red deer that often need little or no anthelmintic treatment, adult wapiti usually require regular anthelmintic treatment, often up to four times annually depending on conditions, to maintain their productivity and avoid problems.

9.3   Elaphostrongylosis

9.3.1   Life Cycle

Elaphostrongylus cervi, also known as the "tissue worm", occurs in deer in many parts of the world and it was introduced into New Zealand with deer importations around 1900. It has an indirect life cycle that involves gastropod intermediate hosts, including a variety of native and introduced slugs and snails. The adult worms reside principally in the connective tissue sheaths surrounding skeletal muscles of the shoulders, chest and abdomen of their cervid host, as well as the arachnoid and subdural spaces of the central nervous system. Gravid female worms lay their eggs into the lymphatic or venous systems. Eggs are carried to the lungs where they are trapped in the capillary bed. The first stage larvae (L1s) hatch, invade the air spaces, ascend the trachea and are swallowed and passed out in the faeces. The L1s penetrate the foot of a gastropod and develop through to third-stage larvae (L3s) at which point they are infective if the gastropod is ingested by the deer. The L3s leave the gastropod and penetrate the gastrointestinal wall and migrate up the nerves to the level of the spinal cord. They undergo two moults and then descend down nerves to the muscular tissues or remain in the CNS.

A survey of faeces from deer farms in the early 1980s suggested infection with this parasite was present on about a third of a relatively small number of deer farms. There has been no recent study of this parasite, so its current prevalence on deer farms is unknown, although it is rarely seen at meat inspection in slaughter plants.

9.3.2 Clinical Signs

In New Zealand, it is rare to find any clinical evidence of disease associated with *E. cervi* infections, which tend to be small in number. Heavy experimental infections have caused respiratory signs and occasionally CNS signs in red deer. In Europe, neurological signs are occasionally seen, and the principal clinical sign is an interstitial pneumonia. Heavy infections may cause lung lesions when eggs and larvae cause vascular congestion, collapsed alveoli, fibrosis and petechial haemorrhages. There are usually few gross signs of lesions associated with worms lying in the subarachnoid spaces and around the spinal cord. In the muscle tissues the worms lie like coiled strands of hair and early infections are very hard to see. Later, the immune response causes a build-up of eosinophils, which cause greenish patches on the surface of the meat and is of concern to venison processors and exporters.

9.3.3 Diagnosis

Diagnosis is based on clinical signs and the finding of spiny-tailed L1 larvae in the faeces using a Baermann technique and microscopic examination. The larvae can be distinguished from lungworm larvae by morphology of the tail. However, there are promising developments in the field of serological and DNA-based detection methods.

9.3.4 Treatment

Avermectins and benzimidazoles appear to temporarily stop egg-laying, but their efficacy for killing adult worms is doubtful. Repeated doses may help to prevent the establishment of infection by killing the migrating larval stages. Widespread use of avermectin/moxidectin anthelmintics in the last 30 years appears to have minimised infections in New Zealand.

9.4   Liver Flukes

9.4.1   Life Cycle

Fascialo hepatica, the common liver fluke of sheep and cattle, is the only fluke causing problems in New Zealand and it not commonly found on deer farms except on the west coast of the South Island. *F. hepatica* has an indirect life cycle, involving Limnaea aquatic snails. Wet areas such as swamps and particularly waterways with slowly moving water and frequently irrigated paddocks are essential for snail survival and the movement of some of the larval stages, such as cercariae and miracidia. The eggs can over-winter, but are susceptible to desiccation. The movement of infected sheep, cattle and deer can introduce infection to new areas where there are suitable intermediate hosts.

9.4.2   Clinical Signs

Red and fallow deer are somewhat tolerant of *F. hepatica* and usually exhibit few signs, even when pastured on the same ground as sheep that are dying of fluke induced disease. In most situations the first indication to the farmer that their deer have liver fluke is a report from the DSP, noting their presence.

There is little other information on the effects of fluke infection upon farm productivity, but reports from Scotland suggest it can cause subclinical disease and reduced productivity. Flukes are occasionally seen at meat inspection at slaughter plants.

9.4.3   Diagnosis

In the living animal the most reliable diagnostic tool for diagnosis of mature flukes is examination of faeces for the presence of the characteristic operculate eggs, but it cannot detect early infections.

9.4.4   Treatment and Control

Triclabendazole, at 10 mg/kg has been shown to be effective against all stages of *F. hepatica* in sheep, cattle and goats and appears to be very effective in red deer. It is recommended to repeat the treatment in 8 weeks.

Control depends on effective treatment of infected animals and prevention of re-infection by fencing off wetland areas harbouring the infected snail intermediate hosts.

9.5   Ticks

9.5.1   Distribution

Haemaphysalis longicornis is the only tick of domesticated animals in New Zealand. It affects all mammals including humans, horses, pets, wildlife, as well as deer, cattle and sheep. Until recently this parasite has been restricted to warmer parts New Zealand, mainly the east coast of the North Island, and north of Waikato, with a few in Nelson. However, over the past few years, there have been heavy tick infestations in the Waikato, and south as far as the Manawatu and Wairarapa. Recently ticks have been observed on deer and cattle in Otago and Southland, causing concern that they may become an established cause of production loss in those districts. However, because this tick is temperature sensitive, it is not known whether populations will thrive and multiply to a point where they will affect production in cooler districts. Global warming may have a very significant effect in expanding the geographic range of ticks on NZ farms.

9.5.2   Life Cycle

H. longicornis is a three-host tick, meaning that three stages of its life cycle must engorge on a series of hosts to complete the life cycle. Each of those stages can be on a different animal species. Adult ticks lay about 1000 eggs on the soil in mid-summer. They hatch to larvae late summer-early autumn. Larvae migrate up pasture and attach to a passing host upon which they engorge lymph and blood. At this stage they are about 1–2 mm. After 3–5 days of engorgement, they drop from the host and return to pasture where they moult to become a nymph, which is relatively resistant to cold, and over-winter, dormant in the base of the sward. In the early spring, the nymphs, which are about 2–3 mm, become active and re-attach to a host to engorge over 3–5 days, expanding to about 5 mm, and then drop off. Nymphs moult to become adults, which are about 4 mm long. They attach to a host early-mid summer, and engorge to about 10 mm diameter over 3–5 days and then drop off to lay their eggs. A high proportion of each stage do not survive in the environment if it is too dry, hot or cold, or do not find a host. Thus, annual differences in climate and various farm management practices have a significant influence on adult tick numbers on animals in the summer. There is evidence that in the warmer parts on NZ the life cycle may be more rapid than12 months. Thus, it is not uncommon to find more than one stage of the tick on an animal at one time.

9.5.3   Clinical Signs

Signs will depend on the number of ticks that have survived to adulthood during the preceding stages of the life cycle. In cooler climates where ticks are present, insufficient numbers may survive to become a deer health concern. Usually, ticks will be observed on the animal only when in the adult stage, which are about the size of a small grape when engorged. However, larval and nymph stages also engorge on the host, but because they are considerably smaller, they are not so commonly seen. Because deer farmers often do not examine their deer closely, particularly around fawning when adult ticks are present, infestations may go unnoticed until severe losses occur. *H. longicornis* infestations can cause deaths in newborn fawns if present in large numbers. Up to 160 adult ticks have been observed on ears of a newborn fawn. Each tick consumes 0.75–1 mL of blood, meaning that the calf can rapidly become anaemic, weak and die of blood loss. It is likely that growth and production are inhibited, although the only trial measuring growth showed no effect. The tick also causes direct severe damage to hides, which is not evident until after the tanning process, thus causing considerable wastage and lost opportunity to the leather industry. Tick infestation of velvet antler can cause severe scarring and may predispose to parapoxvirus infection. This results in downgrading of velvet and financial losses.

9.5.4 Treatment and Control

Tick control can be very challenging. The first factor is to establish the severity of infection on a property, remembering firstly that small burdens may not affect the animal, and secondly, that it is impossible to eradicate the tick once it is established. If only small numbers are present and the environment is marginal for ticks, no action may be indicated. Control methods include animal sprays, pour-on medications, impregnated ear tags, removal of ticks from pasture, pasture renovation or spelling, and pasture treatment. A topically applied 1% flumethrin solution has been shown to both be directly effective against the tick and to have a residual action of about 3 weeks on deer facing natural challenge. Animal treatments must be timed to coincide with when ticks are on the deer and can be done for any stage of the tick. However, management difficulties arise with calving hinds when the adult stage is present, because that is when they are calving, and disturbance can cause losses. Further, stags cannot easily be handled during the rut.

Removal of ticks from pasture may be done by high density grazing animals (sheep, cattle or deer) on infested pastures when conditions are right for tick attachment, such as in warm humid weather, allowing attachment. These animals are then treated with a spray or pour-on to kill the ticks. Ploughing pastures and re-grassing, or passing through a crop will reduce tick survival. Leaving pastures fallow for 12 months will mean that there is limited access to hosts, and this will break the life cycle, although if there are a lot of wildlife species that can sustain the ticks, and this may not be totally effective. It is possible to gauge which paddocks are most heavily affected by dragging a towel across parts of each paddock when ticks are active, and counting them. Pasture spraying with insecticide or acaricide has been done commonly, but only kills ticks on the grass at the time. Residual effect appears limited and there may be adverse effects of the chemicals on insects and bees.

9.6 Lice

Both biting and sucking lice, *Damalinia spp.* and *Solenopotes burmeister*, respectively, infest deer. They are not commonly observed in numbers sufficient to cause clinical problems in otherwise healthy deer, but may be present in large numbers when management is poor and animals are under significant stress. They are difficult to detect on normal deer. The main sign of lice infestation is rubbing of the neck, usually with the hind feet, causing loss of hair. This is seen commonly in adult stags, but it is difficult to find lice on those animals, suggesting that the deer does not need to have a high burden before they feel irritation.

9.7   Cryptosporidiosis

9.7.1   Introduction

Cryptosporidiosis is caused by the protozoan parasite, *Cryptosporidium parvum,* and is believed to be a significant cause of diarrhoea and neonatal deaths of farmed deer in New Zealand, Scotland and Canada (Angus 1994; Hicks 1994). Hand-reared deer appear to be particularly at risk, but severe outbreaks can also occur at pasture. Mortality rates of up to 25% have been recorded. Cryptosporidiosis also occurs in a wide range of animals, including lambs and calves, and is a zoonosis.

9.7.2   Transmission

C. parvum appears to have little or no host specificity and is thought to be carried in the intestinal tract by most animals, including rodents and domestic livestock, such as sheep, cattle and deer. Infected animals can excrete millions of oocytes per gram of faeces. Neonates are probably infected by ingesting oocytes from suckling around the contaminated perineum and teats or by nibbling on contaminated pasture plants at a young age. The incubation period is 2–5 days in most species. The parasite multiplies rapidly in the small intestine, where the various stages lie within the lining cells of the gut, just beneath the cell membrane.

The majority of animals experience only a mild subclinical infection, and they appear to be protected by the presence of colostral antibodies, which are absorbed into the fawn's blood stream in the first 24 h of life. Immunoglobulin A in the milk may also give local protection in the intestines during the first few weeks of suckling. However, some animals are severely affected. This may be because they ingested inadequate amount of colostrum, the colostrum may have lacked sufficient antibodies or the animal may have been exposed to a massive challenge that overwhelmed the natural protective mechanisms. Hand-reared animals are often orphans that may not have had sufficient colostrum in the first 24 h of life. They are often kept as groups in pens that can become soiled with faeces very quickly and they tend to suck each other and inanimate objects in the pen.

Outbreaks of cryptosporidiosis in young deer at pasture are unusual and the following factors may have predisposed them to the disease: poorly drained, wet, muddy paddocks; high stocking density in fawning paddocks; poorly settled hinds or interference around the time of fawning resulting in suboptimal colostral intake; a build-up of faecal contamination in a fawning paddock resulting in the late calves getting heavy exposure to a contaminated environment; contamination (direct or indirect via run-off) of paddocks by infected carriers such as rodents, pigs, sheep or cattle; intercurrent exposure to other organisms such as pathogenic strains of *E. coli*.

9.7.3   Clinical Signs

Affected fawns are usually 1–3 weeks of age and may be first seen wandering behind their peers, often with their tail up, and appearing pot-bellied. There may be an explosive liquid white/yellow diarrhoea, although this may not be obvious, and they lose condition rapidly over the next 24 h. Dehydration, recumbency and death follow quickly without treatment. Diagnosis can be made on finding vast numbers of tiny (4–6 μm) oocysts in the faeces.

At necropsy there is often a full stomach of milk and sometimes there is no evidence of scour. Lesions are typically found in the mid to lower small intestine (jejunum, ileum) and caecum/colon. There may be congestion and mucosal haemorrhages. The contents may vary from pasty grey to yellow or greenish custard. The mesenteric lymph nodes may be enlarged. Microscopic examination typically shows atrophy of the villi of the jejunum and ileum, and under high power large numbers of the organism can be seen in the epithelial cells lining the small and large intestine, just under the cell wall so they appear to be attached to the outside of the cells.

9.7.4 Treatment and Control

Treatment of clinical cases is very difficult and frequently unsuccessful. Fluid therapy is essential to control dehydration. Some apparent success has been achieved by dosing with the drug toltrazuril at 40 mg/kg, given orally as a 1.25% solution. Oral colostrum and intravenous infusion of deer serum may also assist recovery. Broad spectrum antibiotics may help to control secondary infection.

Prevention is aimed at minimising predisposing factors. This is especially true of bottle fed calves where the hygiene of equipment, the environment and handlers is of utmost importance. Strong bleach solutions appear most effective. In a farming situation every effort must be made to fawn the hinds in dry, clean paddocks at low stocking density and with minimal interference prior to and around the time of fawning. If possible calves should be given colostrum as soon as possible and small amounts of colostrum can be added to milk feeds for its local effect in the bowel.

It must be remembered that this is a zoonosis and it is sensible to wear gloves and protective clothing when handling calves with scours and to disinfect clothing and equipment.

10 Fungal Diseases

10.1 Facial Eczema

10.1.1 Introduction

Facial eczema is a descriptive name for the photosensitisation that occurs in sheep, cattle, goats and deer due to liver damage caused by the ingestion of a toxin (sporidesmin) in spores produced by the fungus *Pithomyces chartarum*. This fungus grows on dead litter in the pasture under conditions of warmth and humidity, which occur in the late summer and autumn, especially in the warmer districts in the North Island. Ingestion of this material causes acute damage to and blockage of the bile ducts resulting in liver damage, jaundice and photosensitive dermatitis due to the spill-over of photosensitive pigments into blood and tissues. Fallow deer are the most susceptible species, but red and wapiti deer can also be affected if conditions are dangerous enough.

10.1.2 Clinical Signs

Photosensitisation causes reddening, swelling and crustiness of skin exposed to the sun, especially non-haired skin of the ears, nose, lips, perineal area and vulva. Ulcers on the muzzle and tongue have been reported in fallow deer as a result of excessive licking due to irritation of the muzzle. The affected animals invariably seek shade and the skin irritation can lead to excessive rubbing of the muzzle, lower jaw, eyelids and ears. Some animals may appear blind. Mucous membranes, particularly of the conjunctiva of the eye appear yellow as a result of jaundice. Acutely affected animals are distressed and have an elevated respiration rate. In animals that recover from the acute photosensitivity, there is often a prolonged period of recovery dependent on regeneration of the unaffected areas of the liver. Repeated or severe chronic damage can cause permanent liver damage and affected animals will become debilitated, lose weight and may scour. Diagnosis is from clinical signs and elevation of the enzyme gamma glutamyl transferase (GGT) in the blood.

On necropsy, the liver is swollen in acute cases and there may be general yellowing of the carcass. The liver usually has characteristic microscopic lesions. In chronic cases the liver is shrunken and hard. The regeneration of some areas of the liver can lead to an abnormal shape.

10.1.3 Treatment and Control

Seriously affected animals should be euthanased. Mildly affected animals should be removed from dangerous pastures, kept in a dark shed or given adequate access to shade for 4–6 weeks and given supportive treatment with fluids, vitamins and good feeding to allow liver regeneration.

Preventive measures include: (a) Predicting dangerous periods and areas by doing spore counts regularly during the summer; (b) Make the pasture safe by spraying them at critical times with fungicides; (c) Administration of zinc boluses (unproven in deer); (d) Selecting for resistant stock; (e) Pasture and stock management to minimise dead litter in pastures and adjusting grazing so the deer are not forced to graze pastures too low.

10.2   Ryegrass Staggers

10.2.1 Introduction

Ryegrass staggers (RGS) is a nervous disease of deer, sheep, cattle, alpaca and horses. It is caused by the presence of toxins in perennial ryegrass (*Lolium perenne*) produced at certain times by endophytic fungi (*Acremonium loliae*) growing inside the plant. The fungal spores are present in the seed and after the seed germinates and the grass grows, the fungus grows up inside the plant and into the seed head, where it produces spores to infect the next generation of plants. The toxins, called lolitrems, are absorbed from the plant material when it is ingested and cause brain damage resulting in tremors and staggering, hence the name. Wapiti and wapiti/red hybrid deer appear to be more susceptible to RGS than red deer.

10.2.2 Clinical Signs

There is a wide range of susceptibility between species and also between individuals within species. Red and fallow deer appear only moderately susceptible to RGS, while wapiti are highly susceptible. Often the signs are not very noticeable when the animals are at rest or quietly grazing at pasture. There may be a slight head tremor or trembling of the muscles of the neck, shoulder or back. However, if the animals are stressed or put under pressure, during yarding, for example, the tremors and shaking become more exaggerated and severely affected animals can fall over and thrash or have convulsions. They may stand with a very stiff legged stance and shake or nod. Such severely affected individuals can die of shock or haemorrhagic stress enteropathy. Some individuals have also died of misadventure such as drowning or getting caught in electric fences. Diagnosis in dead or euthanased deer requires histopathological examination of the brain, which reveals characteristic "torpedo-like" lesions seen in sections of the cerebellum.

10.2.3 Treatment and Control

There is no specific pharmacological treatment for RGS. The most effective treatment is to quietly move the animals off the dangerous pastures and either graze them on alternative safe pasture (non-endophyte ryegrass or non-ryegrass pasture such as red clover, fescue, brassicas) or hold them in a yard or pen and feed them on lucerne hay and concentrates. It can take 1–3 weeks for the signs to abate. Ensure that affected animals are kept in a safe environment where they are unlikely to fall down banks, drown in open ponds or get trapped in electric fences. Continued or repeated exposure to toxic pasture can lead to some individuals developing permanent chronic brain damage and tremors.

Prevention is dependent on developing safe pastures by using ryegrass cultivars that have been treated with fungicide or have been selected for nil endophyte or for safe endophytes, or alternatively using pastures that are free of ryegrass. Supplementary feeding with hay, silage or grain will reduce intake of the toxin. It would also be sensible to select animals for resistance to RGS.

11   Trace Element Deficiencies

11.1   Copper Deficiency

Many areas of New Zealand are deficient or low in available copper. Osteochondrosis in young deer and enzootic ataxia in yearlings and older deer are the main manifestations of copper deficiency. However, the incidence of clinical disease is relatively low, even in herds where serum samples indicate the majority of animals have very low copper status. Few trials have shown a growth response to copper supplementation in their first year of life in groups under these conditions. However, supplementation may be economic if it prevents clinical disease occurring, even in a small number of animals. Wapiti are more susceptible to deficiency than red deer and have a higher copper requirement. Supplementation of hinds in mid to late pregnancy may be the most cost-effective means of preventing osteochondrosis in young fawns. The long-term effects of copper deficiency on hind productivity is not well understood.

There are a number of methods of copper supplementation including dosing with copper oxide wire particles (copper bullets), copper injections and pasture topdressing. The ease of dosing and the cost-effectiveness of the various methods of supplementation may depend on the degree of deficiency and the management circumstances under a range of conditions. Excessive supplementation may cause copper poisoning.

11.2   Selenium Deficiency

Most of New Zealand soils and pastures are deficient or marginally deficient in selenium. Clinical disease is rarely diagnosed in deer and there is little information on what liver or circulating blood levels indicate "adequate", "marginal" and "deficient" states in deer. As insurance against deficiency it is likely that the majority of farmers supplement their deer with trace amounts of selenium by topdressing pastures with selenium prills, adding selenium to drenches, giving selenium boluses or selenium injections or using pour-on selenium. These are all relatively cheap forms of supplementation compared to the value of the animal.

References

Angus KW (1994) Neonatal enteritis. In: Alexander TL, Buxton D (eds) Management and diseases of deer, 2nd edn. A Veterinary Deer Society Publication, pp 136–140

Hicks JD (1994) A cryptosporidiosis outbreak in fawns. Deer Branch, NZVA, Proc. Course No. 11, pp 246–250

Mackintosh CG (1992) A review of yersiniosis in farmed red deer in New Zealand. In: Brown RD (ed) Biology of deer. Springer Verlag, New York, pp 126–129

Mackintosh CG, de Lisle GW, Collins DM, Griffin JFT (2004) Mycobacterial diseases of deer. N Z Vet J 52:163–174

Mackintosh CG, Clark RG, Thompson B, Tolentino B, Griffin JFT, de Lisle GW (2010) Age susceptibility of red deer (Cervus elaphus) to paratuberculosis. Vet Microbiol 143:255–261

O'Brien R, Hughes A, Liggett S, Griffin F (2013) Composite diagnostic testing to achieve optimal ante-mortem diagnosis of Johne's disease in farmed New Zealand deer: correlations between bacteriological culture, histopathology, serological reactivity and faecal shedding as determined by quantitative PCR. BMC Vet Res 9:72

Thompson BR, Clark RG, Mackintosh CG (2007) Intra-uterine transmission of Mycobacteriumavium subsp. paratuberculosis in subclinically affected red deer (Cervus elaphus). N Z Vet J 55:308–313

van Kooten HCJ, Mackintosh CG, Koets AP (2006) Intra-uterine transmission of paratuberculosis (Johne's disease) in farmed red deer. N Z Vet J 54:16–20

Part II

Farmed and Park Deer Management in Europe Including Their Diseases

Husbandry Systems for Farmed Red Deer in Britain

Alan Sneddon

Abstract

The background to management systems for British farmed deer is discussed including housing, grazing on forage crops, marketing and abattoirs, monthly management programmes, and handling systems.

Keywords

Deer farming · Scotland · England · Deer housing · Forage crops · Management programmes · Handling deer · Deer crushes

Although the pioneering deer farming research in Scotland was done in the uplands at Glensaugh, with red deer grazing mainly unimproved heather (*Calluna vulgaris*), deer farming is now more commonly an enterprise found on improved farmland where ryegrass or ryegrass/clover/herbal leys can be grown successfully, and deer can be maintained at higher stocking densities. An ideal breeding/finishing unit would have a balance of improved pastures and areas of hill/woodland or rough grazing which would provide shade and privacy for calving hinds and shelter/feed sites for outwintering. Some low-ground deer farmers in areas of heavy soil type and high rainfall, in-winter hinds in sheds with a corral attached to allow access outside. This can significantly reduce pasture damage and soil losses during winter and allow for high overall stocking densities to be maintained.

The majority of calves are weaned either pre-rut (Sept) or post-rut (Nov–Dec) and then in-wintered with an allowance of around 2–2.5 sq.m. of floorspace per calf. Some of the most recently developed deer units have been established on already successful mixed livestock farms where typically sheep numbers have been reduced

A. Sneddon (✉)
Hidden Glen Safaris, Crieff, Scotland

© The Author(s), under exclusive license to Springer Nature Switzerland AG 2022 95
J. Fletcher (ed.), *The Management of Enclosed and Domesticated Deer*,
https://doi.org/10.1007/978-3-031-05386-3_4

and farmed deer are seen as a low labour alternative enterprise with a strong and growing domestic market for venison providing confidence to those seeking to diversify. There is a growing interest in year-round forage systems and the use of forage crops (forage rape/stubble turnips/kale/swedes/fodder beet) to reduce fixed costs (buildings and machinery) and improve feed quality and quantity in late autumn and winter when pasture quality and availability is generally declining as growth slows down. Forage rape and legumes such as lucerne are now being grown in the south of Britain on deer farms to provide high-quality summer feed where drought is now becoming a regular issue.

There are still some long-established deer farmers in Scotland who run breeding herds across semi-improved or hill pastures and sell calves to finishers post-weaning but the most common type of system found in the UK is breeding-finishing where calves are retained on farm, generally in-wintered and then grazed on improved pasture and finished at 15–18 months of age. With domestic consumption of venison in the UK increasing rapidly and strong customer pressure on retailers to reduce imports then the UK deer farming industry faces the challenge in the coming years of producing all year-round fresh venison. With this in mind some deer farmers are looking at alternative systems with a focus on providing high levels of nutrition to hinds during lactation and weaned calves in the autumn and following spring to maximise growth rate potential which should enable them to have a proportion of finishing stock at 45 kg or more carcass weight at 8–10 months. With an increasing number of hinds being presented for slaughter then such a system should allow for hinds to reach viable carcass weights at 12–15 months and be killed when stags are unable to be transported due to being in velvet antler. The major UK processor is now offering a significant price incentive for producers to target early finishing thereby extending the traditional Oct–Feb kill season. The major processor of UK farmed venison is Yorkshire based Dovecote Park who built a bespoke, state of the art, slaughter and venison processing facility in 2017 with a throughput in excess of 7500 deer in the 2020–2021 kill season and a projected kill of around 12,000 in the 2025–2026 season. Prior to building their own facility, deer were killed at Round Green Deer Farm who operated an on farm deer abattoir for many years and supplied around 950 carcasses in 2009 growing to around 2500 by 2016. Dovecote currently has a group of around 70 producers who supply venison and all farms have an annual audit from Dovecote Park and also have to adhere to a set of strict Quality Assurance standards which require a further annual audit from an independent administrator.

Most UK farmed red deer are either of Scottish or English park type genetics with a typical adult breeding hind weight of between 100 kg and 125 kg live weight and adult stags between 175 kg and 275 kg. Eastern European (Romanian/Hungarian) genetics have been available to UK deer farms in the form of breeding stags or semen for AI (artificial insemination) for some time. More recently some breeders have imported elk semen and are in the process of breeding, through AI, elk/red terminal sires for the commercial market some of which will become available for the 2021 rut. Eastern European and elk cross adult stags can reach mature body weights of 300 kg to 350 kg. Eastern European and elk cross terminal sires may have a place in an early finishing system as their faster growth rates and ability to provide carcasses of heavier weights earlier have been well proven in New Zealand. This potential higher growth rate cannot be realised without the provision of adequate nutrition and the nutritional requirement for these larger types increases proportionately with body weight.

Prior to 2000, there were very few deer farms in the UK with more than 300 breeding hinds, most having between 100 and 200 hinds and a significant number with <100 hinds. More recently, the trend has been for larger units and of the 25 or so deer farm developments in Scotland between 2009 and present more than half have breeding herds of between 200 and 400 hinds and a few have projected numbers of 400–1000 hinds.

1 . Management Systems

Typically the stag to hind ratio on UK deer farms is around 1–40 and the size of the unit and paddock layout and area determines whether hinds are rutted in single sire groups or in larger groups with multiple sires. There is a growing trend towards using spikers (16-month-old stags) at a ratio of around 1–10 with yearling hinds. This generally results (on average) in higher conception rates in yearlings and is a common practice in New Zealand. Spikers should be introduced to their mating groups late Aug/Sept to allow for socialisation prior to the yearling hinds coming into oestrus. Post-rut, spikers can be retained for breeding or slaughtered for venison. Mature stags are introduced to hind groups in late September and removed by early to mid-November to avoid late born calves. Calving is from early May through to late June or early July. Pregnancy scanning using transrectal ultrasonography from 30 days post stag removal is now routinely used on UK deer farms to identify non-pregnant hinds for culling and to determine foetal age so that hinds can be sorted into calving groups accordingly. Early and late conceptions can be identified and recorded relatively quickly by a skilled ultrasound practitioner (usually a veterinary surgeon) and the farmer can then sort hinds into calving groups according to expected group calving date. This has a major benefit in that hinds with calves at foot from groups with known calving dates can be moved without the fear of leaving hiding calves behind and also late calving hinds can be calved on clean ground so that their calves are not being born onto paddocks with high parasite burdens (e.g. Cryptosporidium).

2   Summary of Monthly Deer Farm Management (Venison Advisory Service)

Mid-August

- Breeding stags de-antlered as soon as possible as velvet is shed. Worm drench and mineral bolus if required.

- Yearling stags de-antlered in batches as soon as velvet is shed. Weigh and sort for slaughter. Target 110 kg live weight for stags, 88 kg for hinds.
- Plan ahead for quality pastures for rut.

Mid-September

- Wean calves if pre-rut weaning is preferred (this can result in earlier conception). Worm, sex, ear tag and weigh calves.
- Identify hinds which have not reared calves and record.
- Sort hinds into rutting groups and introduce stags at a ratio of 1–40. Mob sizes to suit paddocks (multi sire mating works well with deer). Worm drench if required.
- Monitor grazing and introduce silage or concentrate as pasture declines (if required) to maintain body condition through the rut.

Early November

- Remove stags from rutting groups to prevent late calves and put in sheltered paddock. (on smaller units this may not be practical and stags can be wintered with the hinds but there is a risk of very late calves). Feed ad lib silage or hay and 1.5–2 kg concentrate per head per day.
- Post-rut, wean calves if this is the preferred policy as above.
- Set up wintering mobs in selected paddocks and feed ad lib good quality (10.5–11 ME) silage or soft leafy hay to mature hinds or strip graze on crop such as turnips/swedes/rape allowing 2 kg dry matter/head/day along with baleage or good-quality barley straw.
- Winter rising 2-year replacement hinds in a separate mob on ad lib silage or crop as above and 250–500 gm concentrate (on smaller units this may not be practical but care must be taken to prevent bullying of young hinds).

December–February

- Monitor stock and adjust feeding according to weather.
- Scan hinds for pregnancy and identify/record early, late, empty.
- Cull empty hinds.

March–April

- Increase concentrates for in-wintered calves to around 400–500 gm/day (up to 1 kg depending on forage analysis) and separate any that are being bullied.
- Monitor pasture cover and turn out calves on set up pastures in late April.
- Reduce hind feeding as pasture growth starts.
- Check fences and repair in preparation for calving.
- Plan grazing programme and organise ahead for any cultivation of forage crops.

April–August

- Rotationally graze finishing stock where possible.
- Monitor worm burdens and drench accordingly.
- Set stock hinds by early May for calving. Ensure calving paddocks have areas of cover for calves to hide and find shade. Hinds may be rotationally grazed from mid-July onwards (or 2 weeks after the last calving date to avoid leaving calves behind). Ensure adequate pasture quality for lactation to maximise potential calf growth.
- Close up some paddocks for silage or hay production. Top grass seed heads post calving to maintain high percentage of green leaf and improve summer grazing quality.

Note:

Yearling hinds calve later than mature hinds and should not be allowed to become overfat before calving to avoid calving problems.

3   De-Antlering

Transport Guidelines in the UK state that deer in velvet (and female deer in the last month of pregnancy) should not be transported, except in an emergency for veterinary treatment. Deer in velvet are also prohibited from being transported to an abattoir (DEFRA 1989). Further legislation also prohibits velvet antler removal until the antler is more than 50% stripped of velvet at which point the hardened (calcified) antler can be sawn off without any sensation of pain to the stag. This means that on most UK deer farms stags are de-antlered in the autumn when the decreasing daylength triggers a rise in testosterone levels which in turn initiates the calcification process of the antler. Older stags generally come in to hard antler earlier than younger stags so with large herds mature breeding stags might be de-antlered from early August and younger stags and spikers through mid-August and September. Stags can become very aggressive and difficult to handle just a short time after they start to strip velvet so it is recommended that antlers are removed within a few days of stripping where possible.

In some cases this means that older stags should be separated out early from a mixed age group while in the later stages of velvet growth and then brought in to the handling system either individually or in very small groups to avoid the risk of older stags in hard antler causing serious injury to younger stags in the yards. Spikers should be de-antlered at least a few days prior to transport to an abattoir to recover from the stress and potential bruising caused during handling and restraint.

4   Handling Systems and Crushes

Handling systems for deer in the UK vary in design but where breeding stags are to be handled for de-antlering it is essential that this can be done safely with minimal risk of injury to the handlers and the deer. (There should always be at least two handlers present when working with adult deer.) It is desirable when mature stags are moving through a handling system that there are a series of gates that can be closed behind the animals so that the handler is never in a small pen with an adult stag in hard antler. Gates which antlered stags are expected to pass through should be a minimum of 1.2 metres wide so that there is less chance of them hesitating and turning back. Generally stags are restrained in a mechanical crush for antler removal. The three main types of deer crush are outlined below.

4.1   Drop Floor Crush

The drop floor crush was common on many deer farms in the early days and is still favoured by some (especially for fallow deer) but began to be superseded by other types from around 2000 on. The animal to be restrained enters through a gate at the rear into a narrow Y shaped covered passage and a hinged floor is triggered to drop by the operator which suspends the animal along its length so the feet cannot gain purchase on the floor and the body is held in the wedge-shaped sides. Sliding or hinged panels on the upper sides of the crush can be then opened to allow access to the restrained animal. Once a procedure has been carried out then another lever can be pulled on a side of the crush which is hinged along the top releasing the wedged animal onto the ground where it can walk out of a gate at the front. The crush is then manually reset and the next animal can be brought in. The crush can either be positioned over a dug out trench so that the deer enter at the same ground level as the rest of the yards or the crush may have a lead in ramp which the animal goes up via a race giving enough clearance for the floor to be dropped.

4.2   Hydraulic Crush

The hydraulic squeeze crush is commonly found on most deer farms in New Zealand. Essentially they are two padded sides within a frame which move in and out on hydraulic rams to restrain the animal along its entire body length. They also move up and down to enable quick adjustment and allow most deer sizes from a mature elk bull to a fallow buck to be held comfortably. An extended pad along the inside top on both sides holds the animal's neck and prevents it from jumping out. Above the side pads on either side, sliding curtains create a visual screen as the animals enter to stop them trying to jump over the sides and these are slid back for the handler to gain access once the animal is securely restrained. The crush is usually sited within a rectangular pen in a handling system with the animals entering and exiting through sliding doors at each end. Stags with very wide antlers can access easily as long as these doors are wide enough. Weigh scales can be fitted to hydraulic crushes quite easily. The original hydraulic crush was designed and manufactured in New Zealand but they can now be made to order by a UK based company and there are several manufacturers in Eastern Europe.

4.3   Swing Squeeze Crush

The swing squeeze crush was again developed in New Zealand and works on a similar principle to the hydraulic crush but is manually operated and does not require power or a pump. This crush has two height adjustable padded sides along its entire length which hinge from opposite corners of a rectangular box fixed within a pen. For the animals to gain access the entrance side is opened and the deer can walk in and the operator closes the side like a gate. As the gate closes a spring loaded catch locks into a ratchet section in the end of the box which enables the sides to be gradually squeezed up to restrain the deer in a similar fashion to the hydraulic crush. There is a lever to assist this process if required. Once the deer is held then sliding curtains allow operator access from both sides and the deer can be processed and then released by means of the spring loaded bolt on the other side which will when pulled back allows the exit side of the crush to be opened. The swing crush needs to be fixed down onto a level concrete pad and is generally set up within a pen of around 3 m × 5 m in size.

Once a stag is restrained within a crush, one or both antlers can be tied by means of a rope or strap to part of the frame or some other fixture and then the antlers sawn off with a bone saw or cutting wire. It is essential that the stag's head is firmly secured during antler removal to minimise the risk of serious injury to the handler from a stag thrashing its antlers around. Great care must be taken to avoid damage to the coronet or pedicle and the general rule of thumb is to start the line of cut at least 10 mm above the coronet and at such an angle as to maintain this across the entire cut. It is important to keep checking that the cut is not tracking downwards towards the coronet and if it is then it is usually best to remove the saw blade and start a new cut. If brow tines are very low and do not allow for a horizontal cut to be made without damage to the coronet, then the brow tines should be cut off first by means of a vertical cut before cutting through the main beam horizontally. With spikers any basal snags (extra points originating from the base of the antler sometimes pointing rearward or in an abnormal direction) should also be removed. Occasionally an antler will snap just before it is completely sawn through leaving a sharp upward pointing splinter. These should always be removed before the stag is released. Stags once de-antlered should be kept (where possible) in a separate paddock to stags which still have antlers on.

4.4   Weaning

Weaning is the physical separation of calves from the hinds so that the calves are no longer dependent on their mothers for food (milk) and security. This enables the calves and hinds to be managed separately within the farm system. In the UK, most deer calves are weaned either pre-rut in late August to mid-September (3–4 months old) or post-rut in mid-November to December. The choice of weaning date depends on the farm system and feed availability. Leaving calves on their dams through the rut where there is likely to be insufficient pasture cover or forage crop available to maintain or allow an increase in hind body condition over the duration of the rut can result in later conception dates and lower overall conception rates. There will also be a negative impact on calf growth rates as pasture quality declines, therefore autumn feed availability should be a major consideration when deciding to wean pre- or post-rut. Most deer farmers in the UK house their calves for winter or house them for a period (perhaps 2 weeks) before turning them back out onto saved autumn pasture to be brought in again when pasture cover declines and the weather deteriorates. Some farmers are now growing forage crops such as stubble turnips or forage rape which can provide high-quality feed for hinds and calves in the late autumn through the rut and in some cases calves are weaned back onto such crops and then out-wintered. The key to successful weaning and low mortality is to plan ahead, to minimise stress on both calf and dam and ensure that the transition to a new environment (winter housing) and diet does not negatively impact on calf growth and wellbeing.

4.5   Housing at Weaning

If calves are to be housed at the point of weaning, then it is essential to ensure that the transition from a pasture/milk based diet to a forage/concentrate based diet happens gradually and this can best be done by introducing feeding outside pre-weaning so that the calves can become accustomed to eating the new diet with their mothers in the paddocks. Feed levels can be built up gradually and the mobs of deer can be fed closer to the handling yards and through gateways as a sort of training process to make gathering easier. It is best to avoid bad weather, if possible, when weaning so there needs to be some degree of flexibility around the actual weaning date. Raceways and yards should be checked and any repairs carried out well in advance. It is a commonplace for calves to be ear tagged and wormed at weaning and prior to housing so order materials well in advance so that they are on hand for the day of weaning. Some farmers prefer to do this at a separate handling either pre- or post-weaning to reduce stress on the day of weaning. Winter housing should be checked and any repairs and maintenance carried out well in advance. Bedding and forage and perhaps some concentrate should be put into the pens (if possible) prior to the calves going in to minimise disturbance for the first day or so after the calves are introduced. Ensure that drinkers or water troughs are working and there is always adequate fresh clean water available. Calves should be separated from their mothers in the handling pens and then can be sexed, tagged, and wormed in groups of perhaps 4–8 in a small pen with minimal restraint. Depending on the scale of the deer unit calves could then be separated into same sex groups then moved into the winter housing. As a guide the recommended floor space allocation is 2–2.5 sq.m./head. Any very small calves should be housed in a separate pen. Good-quality forage should be available for the newly weaned calves and any concentrate should be built up gradually over a period of 4–7 days to minimise the risk of acidosis. Calves will be quite flighty for the first week or so post-weaning but will gradually quieten down and during the winter can become very tame.

If weaning pre-rut, then the hinds should be moved into paddocks as far away from the calves as possible. Hinds and calves can be quite vocal for a few days post-weaning and hinds can fret and pace up and down fences. Once the hinds settle down after a week or so then stags can be introduced for the rut.

Some farms wean their calves pre-rut into housing and then turn them back out onto high-quality grazing after 2 weeks or so once they have settled down. The calves would then be brought back in for the winter once the weather deteriorates or pasture quality declines. Having been housed previously for a short period calves under this regime adjust quickly to being re-housed.

4.6   Post-Rut Weaning and Outwintering Calves

If there is sufficient quality pasture and or, forage crops, to ensure a good level of nutrition through the rut, then post-rut weaning can be a good option as it is generally less stressful on both hinds and calves and has a lower labour requirement than pre-rut weaning. If post-rut weaning is the preferred option, then it is a good practice to run the hinds and calves through the handling system in late August or early September and calves can be tagged and wormed at this point. Any dry hinds or cull for age hinds can be removed from the herd to be grazed separately. Lactating hinds and calves can then be allocated quality pasture or forage crop and stags introduced by late Sept for the rut.

The transition onto forage crops from pasture should be done gradually over a few days and there should be a grass run off paddock adjacent to the crop where possible. Baled silage or hay can be offered to provide a source of fibre to balance the forage crop. Stags should be removed from the hind mobs by early to mid-November to prevent late born calves and weaning can take place at the same time as stag removal or deferred until later depending on feed availability and weather. On most UK farms post-rut weaning would be done by mid-December at the latest. In areas where liver fluke (*Fasciola hepatica*) is known to be an issue then calves should be treated at weaning. In some areas of Scotland warble fly (*Hypoderma diana*) are becoming an issue and can cause severe irritation in mid to late winter and can also result in carcass damage and in extreme cases rejection at the abattoir. In such areas hinds and calves should be treated with an ivermectin based drench or injection either pre or post-rut.

Calves can be successfully out-wintered but great care must be taken to ensure that the process of weaning is as stress free as possible and that calves have been transitioned onto either forage crop or concentrate/silage/hay based diets for ideally 2 weeks prior to weaning while still with their dams. Paddock selection is also critical and shelter and access for feeding are the main things to consider. In an ideal weaning scenario the paddock the calves are to be weaned into should have a good pasture or forage crop cover and the calves would be grazed in the weaning paddock with their mothers for 2–4 days prior to weaning then put back into this familiar paddock post-weaning. Care must be taken when moving newly weaned calves along raceways and through gateways which requires patience on the part of the

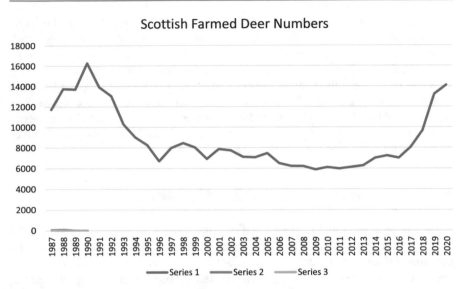

Fig. 1 Deer of all ages which can be gathered, identified, recorded, and handled (Data provided by Scottish Government Rural and Environmental Science and Analytic Services)

handlers. It is a good practice to leave a few quiet hinds in with the calves which will help to keep them settled and make the business of moving them easier. Calves usually settle down within a few days and can be rotationally grazed or strip grazed using electric fences if desired. It is essential that deer are trained to respect electric fences and this can be done by setting up a short section across the corner of a paddock for a few days where deer are unlikely to run through it at speed. As deer are naturally inquisitive they will sniff and rub against the fence and soon realise that it is something to be respected. Once they have gone through this process electric fences can be used very successfully to subdivide paddocks and allow for better pasture or crop utilisation. Deer grazing behind electric fences should be checked daily and the fence itself tested to ensure that it is functioning properly.

4.7   UK Farmed Deer Numbers

The UK government Department For the Environment, Food and Rural Affairs reported 37,000 farmed deer in 2020 in England. No data were collected from Wales. Scottish farmed deer numbers are provided in Fig. 1.

Reference

DEFRA (1989) Department of Environment, Food and Rural Affairs (DEFRA) guidelines for the transport of farmed deer BL5922. DEFRA Publications, London

Managing a Traditional English Deer Park: A First-Hand Account

Callum Thomson

Abstract

A first-hand account of the complex issues to be prioritised in managing a large heritage deer park in southern England with public access and containing several different species of deer.

Keywords

Carrying capacity · Sex ratios · Feeding · Culling policies · Pasture management · Public access

Woburn Abbey Deer Park is one of the most famous Parks in the world; some say that it was once the greatest collection of deer ever brought together, with over 40 different deer species at one time (Whitehead 1950).

I managed the herds at Woburn from 1992 until 2012, but it is important that the reader understands my background before I took on the role as deer manager, in order to fully appreciate some of the management systems that I developed.

My background was in general livestock farming: cattle, sheep and pigs, until 1977 when I took on the role as a stockman on a deer farm. I was very fortunate to be trained by some very exceptional and experienced livestock managers in the cattle and sheep industries prior to working with deer.

These experiences moulded me more to the intensive side of managing livestock, controlling breeding, feeding, intensive pasture management and health issues.

When I was appointed Deer Manager at Woburn, I took all that I had learned from deer, cattle and sheep farming, and tried to implement different systems of livestock management within the Deer Park. For generations, at Woburn, the people that were

C. Thomson (✉)
Bedford Estates, Milton Keynes, UK

© The Author(s), under exclusive license to Springer Nature Switzerland AG 2022   105
J. Fletcher (ed.), *The Management of Enclosed and Domesticated Deer*,
https://doi.org/10.1007/978-3-031-05386-3_5

employed had come from a gamekeeping background, so it was a radical change of direction to appoint me, bringing a more detailed stock management approach.

As time went on I soon discovered that each Park is quite unique and varies depending on how you manage it. It was influenced by your instincts, basic stockmanship, your understanding of the park environment and how so many things outwith your control can affect the performance of the deer.

1   The Definition of a Deer Park

A deer park is an enclosed area of land where deer can roam freely and exhibit natural behaviour, where management is kept to a minimum in relation to feeding, pasture management, veterinary intervention, with no ear-tagging, where livestock movements in and out of the park are infrequent, and, in England where the area of land usually surrounds a large stately home. These are often ancient and may be called heritage parks.

2   Deer Parks with Many Deer Species

At Woburn, we had nine different deer species but for the purpose of this account we will concentrate on four species: red, fallow, sika and Père David deer. Ideally, each of these requires different management strategies but this is impossible given that they are all enclosed in the same area.

All planning in relation to culling, nutritional requirements at different times of year, feeding for different age groups within each species, mating dates, birth dates, should be treated separately, but how do you do that? There is no way you can actually achieve independent treatment for each species but given the habits and nature of the different species some progress can be made.

It is all about identifying the areas of the park that each species prefers. I am absolutely convinced that all the different species did not like each other, which was a great help with the virtual separation. So, given my background, all my thinking was directed at working out how I can micromanage the herds so as to improve their individual performance and give them what they need when they need it.

3   Stocking Densities and Carrying Capacity

There are a number of factors involved in deciding how many deer you can have in a park, the most important are the winter carrying capacity, the type and quality of the vegetation, the shelter, the inputs, i.e. fertiliser, feeding, and the quality of the people looking after the deer, as well as the amount of time they can spend managing the park. The stocking density will also be influenced by the level of public access as well as the economic objectives. As a rule of thumb 1 adult deer per acre (or 2.4 adult deer to the hectare), that is adults with the young born every year above this. It is vitally important that you get this right, or it will create so many problems from overgrazing to social problems within the species, disease, and what they call winter death syndrome—basically deer starving to death.

4 Culling

Every year a proportion of the total stock must be culled, but first of all, you need to determine how many deer you have, what is the carrying capacity of the park, and what are your objectives: is it venison production, trophies or is the park maintained just for aesthetic reasons and has the park to be financially viable?

Let us take a mixed approach with the objectives being venison, antler trophies, visitor attraction and financial viability. I will take red deer as an example. So say you have a total of 300 red deer comprising 200 hinds and 100 stags, all being 2 years old and above. There used to be a train of thought that a sex ratio of 1:1, stags to hinds, was the ideal, but if you want to produce venison you need more hinds, and if you want to produce trophies you need a high number of stags and although 100 stags is too many for the 200 hinds, you must also take into account that visitors like to see antlers.

An 80% survival of calves to October in a park is not unreasonable, that gives you 160 calves, and for ease 80 males and 80 females. These will be kept until the following autumn when they are approximately 18 months old. From the 80 males 50 will be culled, selected on body and antler quality, leaving 30 to go through to 2 years old, and at this age you have a better idea of what their antler quality will be so that you can then reduce the 30 down to 15. They will again be further selected over the next 2 years for antler quality, reducing your original 80 to 5–8 stags which will be breeding stock. When culling spikers (yearling stags), it is a good idea to leave a selection of antler styles giving the different genetics a chance to influence the herd. With the females I recommend keeping some for replacements every year, creating a good age structure throughout the herd. From the original 80 at least 20 should be left. Culling should be based on body size, but this is not easy from a distance with a rifle, and the skill of the managers is now very critical, it also means that 20 adult hinds have to be culled to keep the numbers at 200, but this may need to be altered to allow for deaths from natural causes. This is a basic cull plan for red deer and can be used for sika, fallow and Père David's deer, but I must stress that you have to be flexible, evaluating the situation all the time. A word of warning for the above culling strategy: it is like managing a herd of cattle, but with cattle, you have so much more control, they are all tagged so that you know their breeding and you can select replacements based on genetics, but in a park, you have no such idea and by reducing your recruitment to such low levels it may have an effect on genetic diversity.

As a footnote to all of this the people managing the park, their influence in decision-making and their skills, will determine what style and quality of antlers will develop over a period of time because they are shooting what they believe is good or bad. But, unlike the conventional farm situation, you are working with limited data and information, and it is a good idea to have more than one person making the selection decisions.

5   Feeding

This subject could fill a book by itself, and again, as I have referred to previously, you are never fully in control. The amount you feed is dictated by the deer numbers, the types of deer, your objectives, and the financial returns you think you can achieve.

Feeding in a mixed-species park takes careful planning and in a word a *ROU-TINE*. You also have to try and match the seasonal requirements in relation to ME (Metabolizable Energy), protein, and trace element requirements. The problem for the manager is that within each species there are different categories of stock each requiring different nutritional levels of feed at different times of the year. I have seen so many parks where the deer keepers put out hay or silage bales and let the deer get on with it. This is totally inadequate, with the dominant animals taking more than their share, leaving the young, old and less confident deer not getting what they need. The idea is to try and target the feeding to the species and this can only be achieved by using concentrates or other types of hard food.

Again, I will use red deer as my example, in the herd you will have adult hinds and stags, yearling hinds and stags, and calves, all requiring different amounts of feeding. The adult hinds just need a maintenance ration and in some circumstances, silage or hay will suffice through the winter. Adult stag's nutritional requirements are much higher coming out of the rut into the winter when their body condition is at its lowest level, and yearling stags, hinds, and calves born that year all require a higher plane of nutrition. Stags will naturally separate after the rut, which is your opportunity to feed them what they require. A stag weighing 200 kg requires 32 Megajoules (Mj) of ME per kg of Dry Matter (DM) per day, as maintenance. This is equivalent to approximately 4 kg of feed per day and how you make this up depends on what feed is available: silage, hay, root crops, concentrates. I used to feed 2 kg/head/day 16% protein concentrate and ad lib silage to the red stags, and at times they also were fed root crops. Concentrates are an expensive option, but I favoured this option for ease of use, and consistency, to meet the nutritional requirements for the deer.

The adult hinds, yearling hinds, calves, and probably young immature stags, will feed together, and here lies your problem: a deer weighing 50 kg requires 8.9Mj/ME/kg of DM per day, a deer weighing 120 kg requires 17.1Mj/ME/kg of DM per day, both of these just for maintenance. That is equivalent to 1.1 and 2.1 kg as wet food, and of course, to compound the problem there will be a whole range of weights in between. Financial constraints will play a part in how much you can feed this group. To try and cover all the different deer I fed 0.8 kg per head per day of 16% protein concentrate and almost ad lib silage, and again root crops. This was generally not enough for some but too much for others.

What is equally important as the amounts you feed, but also how you feed it. The following are some guidelines.

Concentrates Put this out in long parallel lines, 20 m apart, dispensed in piles, this is best achieved from a designated feeder, we called it a 'snacker', trailed behind a quad bike, allowing the feed to go out very quickly ensuring all the young animals get their share. It is also a good idea to gently go round the herd on the bike ensuring shy deer and especially the calves are moved towards the feed.

Silage, Hay, Roots These are best put out in one continuous long line and the quality of all these products must be first class so that the uptake is to a maximum. It is also a good idea to put out some bales whole in feeders which will give you a more ad lib approach.

Routine If you stick rigidly to timings the deer will always be waiting for you in the designated area, allowing you to split the different species. At Woburn, we had five different feed areas, and every day we would be at each site within a 10 min period to feed their concentrate. This keeps the herds within their species groups for feeding. As I said at the beginning, feeding can be related to your objectives in a park, but from my experiences deer in a park will perform better with a targeted approach to feeding, rather than a broad brush policy. This will allow you to exert some level of control, endeavouring to give them what they need when they need it.

6   Pasture

In most cases, the grassland in ancient parks has been there for hundreds of years, which has its good and bad points. The good is that it is tough, drought-resistant, deep-rooted and can supply year round grazing. The bad is that it goes to seed very quickly; it does not respond well to fertiliser like modern grasses whose production you can turn on or off like a tap with fertiliser. I used to spend a lot of money trying to get more productivity at certain times of the year with the likes of super phosphates, nitrogen, seaweed products and trying to correct the acidity of the soil using magnesian limestone. We also reseeded areas using direct drilling, broadcasting, and putting clover into the feed so as it was passed out from the deer to germinate. My idea was to improve the pasture moderately not change what had been there for a long time. We could have ploughed up some areas and reseeded properly, but I believe this would have been a mistake.

The results from years of fertilising and reseeding were very mixed, and I am not convinced that it was money well spent, I used to be asked by the accountants, when spending this amount of money, what returns could we expect in the productivity of the pasture and the deer; I could not say with confidence that I could improve either. My conclusions after 20 years were the inputs had to match the outputs, and if I had my time again, I would only use lime and small amounts of phosphate with no reseeding. Circumstances in different parks may allow for pasture improvements and may be the correct thing to do, but will you see the production and financial benefits, and how do you quantify them?

7 Public Access and How It Can Affect Performance

Disturbance is one of the main factors that affect the Deer and their performance. Most parks will have public access and some like Woburn will have regular events like pop concerts, festivals, and a whole host of others. All of these will have some impact on how the deer perform and behave. There is no question that if you had a totally private park with no events the productivity of the herd would increase, with less mis-mothering, the deer not so stressed, and with no young calves being picked up by well-meaning public, but having said all that, the deer can to some extent adapt their behaviour to the different situations that they face.

Woburn is one of the most public parks in England, we had regular events, pop concerts with 30,000 people, 10,000 cars all converging on the deer park and more than a third of the park was taken up with the infrastructure of the event. What do you do with 1400 deer given the damage to the pasture, the litter, etc.? I could go on! All of these have an effect on performance and financial returns from the deer at the end of the year.

The whole park had a series of footpaths going right through popular calving areas. Deer like the Père David would be easily disturbed often leaving newborn calves, and in some cases not returning. As a deer park manager at Woburn it was something you had to get used to and deal with as best you could. The bigger picture was that the income generated from these events brought in large amounts of money certainly outweighing anything you could earn from the deer.

8 Conclusions

Trying to bring a structured, more detailed approach to managing a deer park like Woburn does work to a certain extent. You can target feeding, improve pasture, and influence the quality of the antlers by selective culling. General farm management practices work to a certain extent, but this is limited given the nature of the park setup, and the many influences that are outwith your control. There is no question in my mind now that in regards to culling we overmanaged the deer, especially the Père David's deer, we did not have enough knowledge of the species to make decisions at certain ages. With the sika, fallow and red deer we had a greater understanding allowing my influence to dictate their style and quality. One of the main drivers was the correct density of deer for the given area and the correct balance of numbers in the different species. I would have kept more deer per year to give greater genetic diversity, but we were constantly trying to reduce numbers.

Targeting feeding with high-quality feeds does work, the performance of the deer will improve, giving you greater returns, and this can also be said for pasture improvements, but it all works only to a limited extent. How much you can run a deer park as a business is debatable. After all, at one time they were only there for the pleasure of the lords and ladies, and for hunting and were never expected to make money.

Bibliography

On Woburn Abbey Deer Park

Whitehead GK (1950) Woburn Abbey Park. In: Deer and their management in the deer parks of Great Britain and Ireland. Country Life Limited, Farnborough

Management, Diseases and Treatment of Deer in UK Deer Parks

Peter Green

Abstract

Free-living deer in deer parks are, in effect, wild deer confined within a boundary. They cannot be routinely gathered or handled for normal husbandry like farmed deer or domesticated livestock. Their welfare, however, is the responsibility of the owner or keeper of the park. This poses particular challenges in respect of welfare monitoring and disease control. This chapter sets out basic principles whereby animal welfare in a deer park can be safeguarded, and disease challenges can be met. The key to health and welfare is correct stock density, adequate supplementary feeding and appropriate responses to pathological problems in the deer. Tuberculosis is a particular challenge to the deer park and is discussed in detail.

Keywords

Park deer · Stock density · Trace elements · Winter die-off · Tuberculosis

1   What Is a Deer Park?

The difference between a deer farm and a deer park is not always clear. Both keep deer within deer proof boundaries at population densities that exceed those of wild deer in the open landscape. Both are subject to management, such as supplementary feeding, pasture management and the harvesting of venison. The range of premises that may be either a deer farm or a deer park can be considered as a spectrum of enterprises with characteristics that merge from end to the other.

P. Green (✉)
Peter Green Veterinary Consultancy, Barnstaple, UK
e-mail: mail@peter-green.org

At one end of the spectrum is the well-equipped commercial deer farm, where deer are farmed in ways similar to beef cattle: there are gathering and handling facilities, deer can be vaccinated, medicated, de-antlered and pregnancy tested, young stock is weaned into batches for finishing and may be weighed and housed, single sires are selected to cover groups of females. Farmed deer are processed through an abattoir for venison production, where full inspection of the whole carcass and viscera can be undertaken. Deer farms are usually divided into separate paddocks with a central set of yards, sheds and handling equipment.

At the other end of the spectrum is the large landscape heritage deer park, where deer are enclosed, but considered to be 'wild': they are never routinely gathered or handled while they are alive, there is no active weaning or separation of young stock from their dams, multiple males run with the herds and compete in the rut for females, deer are killed in the park by shooting and carcasses are treated in exactly the same way as wild deer hunted in the open countryside. Evisceration and dressing out occur in situ; only the empty skin-on carcass is consigned to the Approved Game Meat Handling Establishment (AGHE). Such deer parks are rarely sub-divided and have no yards or sheds for handling the deer, which are never vaccinated, weighed or pregnancy tested.

These two archetypal captive deer enterprises, the fully equipped commercial deer farm and the classical deer park, are clearly distinguishable, but there is a wide range of premises that fall between these two. There are self-styled 'farms' that have no facilities for gathering or handling the deer and therefore no prospect of testing them or sampling them for disease control purposes, yet sell 'farmed' venison. There are large deer 'parks' with catching pens and handling sheds, from which live sales are offered. There are 'visitor attractions' on tourist or holiday theme parks. There are deer 'parks' in conjunction with private houses in which deer are confined to a single paddock and fed throughout the year.

This range of premises, with a classical deer park at one end and a bona fide commercial deer farm at the other, presents difficulties for the statutory authorities, which adopt different labels and categories for enclosed deer. The processing and food hygiene of farmed game meat and wild game meat are regulated by separate statutes in the European Union, which means that the UK Food Standards Agency classifies deer as either 'farmed' or 'wild'. The UK Government Department responsible for animal health, Defra, makes no such distinction and insists that all enclosed deer are considered to be 'captive' and must be capable of being tested for notifiable diseases. The Farm Animal Welfare Committee of Defra has ruled that all enclosed deer come under the terms of the 2006 Animal Welfare Act (FAWC 2013), yet this Act specifically excludes wild animals. However, the non-departmental public body Natural England, sponsored by Defra, insists that deer in English parks are wild deer and subject to the legal constraints of closed seasons and the requirement of specific licences for any capture or gathering. The various statutory authorities are therefore unable to agree criteria or definitions that distinguish a deer farm from a deer park.

For the purposes of this chapter, a deer park is a land holding where:

- Deer are enclosed within a deer proof boundary.
- There are no permanent gathering or handling facilities, and the deer are not routinely caught for purposes of medicating, vaccinating, testing, weighing or de-antlering.
- There is no imposed weaning and separation of young stock.
- Deer are never housed.
- Multiple males run freely with the females and compete in the rut.
- Deer numbers are controlled by shooting deer in the park.
- All venison from the culled animals is sold as wild game meat.
- Artificial feeding is limited to supplementary feeding in the park during the winter.

2 How Many Deer Parks Are There in the UK?

Deer parks were formerly much more plentiful across Northern Europe than is now the case and sources agree that between the fifteenth and nineteenth centuries, a disproportionate number of these were in England (Whitehead 1950; Fletcher 2011). Christopher Saxton's county maps of England, published in 1579, show over 700 deer parks. Shirley (1867) states that Elizabeth first alone inherited some 200 deer parks from her father. Whitehead (1950) and Shirley (1867) both quote from Moryson's Itinerary of 1617, which states that at that time, there were more fallow deer in a single English county than in all Europe besides. This may be correct: Greswell (1905) lists records of 49 former and extant deer parks in the county of Somerset alone.

By 1867 Shirley lists 334 parks holding deer in England; Whitaker (1892) lists 395. After the second world war, Whitehead (1950) reports that this number had declined to 143 deer parks in England with deer still present.

The British Deer Society has recently established a register of enclosed deer in the UK, based upon information submitted by owners and by Society members. In June 2020, this register included 183 properties that were called 'deer parks' and a further 67 that kept deer enclosed as a visitor attraction but were not operating as a licenced zoological collection or as a deer farm. There were 74 properties identified as commercial deer farms.

3 The Character of Deer Parks

In the UK, most enclosed deer parks consist of a mosaic of open permanent pasture and wood pasture. There may be stands or copses of trees within the park, but these are unlikely to have significant woodland understorey because of the impacts of the deer. There are very few deer parks situated above the 400 m contour in the UK; most are lowland enclosures historically or currently associated with grand houses, or they are small remnants of royal or monastic hunting forests. The number and the persistence of so many deer parks in England is the reason why England has many more ancient and veteran trees than any other European country (Farjon 2017). Parks more recently established may simply be open grassland surrounded by a deer fence.

Deer parks may be bounded by walls, heritage palings, contemporary deer stock fencing or a combination of these. Occasionally, a deep ditch with one vertical side (a ha ha) persists to offer an open vista from the house. Long established deer parks may have a deer leap or 'salter' to enable wild or escaped deer to enter the park but not to leave.

The topography and landscape features of a given park will affect the welfare of the deer. Flat, open parks with little shelter will be more likely to experience the adverse effects of inclement winter weather on the deer than parks with slopes and hollows and banks and woodland. Deer cope much less well with adverse weather conditions than cattle or sheep of comparable body mass; they experience a greater intensity and duration of cold stress than cattle or sheep in cold, wet and windy conditions (Simpson et al. 1978).

Many urban and suburban deer parks are open to the public. Some have public roads running through them with high traffic volumes, some permit dogs to run loose. There may be golf courses, football pitches, jogging and cycling tracks, horseback bridleways, cafés and restaurants and other leisure and lifestyle offerings in the deer park.

4   Deer Species in Deer Parks

The majority of UK deer parks hold fallow deer (*Dama dama*). Some also hold red deer (*Cervus elaphus*) in conjunction with the fallow deer or alone. Many fewer parks have other species such as sika deer (*Cervus nippon*), Axis deer (*Axis axis*), Père David's deer (*Elaphurus davidianus*) or other more exotic species.

5   Stock Densities of Deer in Deer Parks

If supplementary winter feeding is to be limited to the coldest and wettest months of winter, the pasture of the deer park must support the deer for the rest of the year. Using the scheme proposed by Putman and Langbein (1992, 2003), whereby one fallow deer is counted as one deer stock unit (DSU), and one red deer counted as 2 DSU, experience has proven that most lowland deer parks can support between 2 and 6 DSU per hectare of good grazing. In calculating the available good pasture, areas of woodland, scrub, bracken (*Pteridium aquilinum*) and marginal wet ground should be excluded from the calculation. In practice, the optimum stocking level is best determined by monitoring the condition of the deer herd and reducing numbers if the combination of grazing and winter feeding does not maintain the animals in acceptable condition.

6   Routine Health and Welfare Monitoring

The welfare of wild animals has received increased attention in recent years, and current concepts offer a reasonable model for free-living wild deer in parks. These concepts suggest that welfare is best assessed by the ability of an animal to respond to an adverse stressor in ways that mitigate the stress. Welfare is, therefore, in a constant state of flux, not in a fixed state of 'good' or 'poor' (Ekesbo 2011; Ohl and van der Staay 2012; Sandoe and Jenson 2012; Appleby et al. 2014).

By these concepts, a deer park buck may be injured in the rut, which immediately depresses his welfare, but if he is able to find a quiet place in the park to recover, his welfare becomes more positive. If deer are suffering hypothermia because of bad winter weather their welfare becomes more negative, but if they can seek and find shelter where their core temperature rises, their welfare becomes more positive. Such a paradigm may be applied to hunger, thirst, dog chasing, road traffic disturbance and all other stressors that may bear down upon the welfare of deer in deer parks.

This model of welfare also illustrates why a deer sanctuary area is so vital in parks with high visitor numbers and intensive leisure and recreational usage.

Park deer are only routinely handled when they are dead. Health and welfare monitoring of the living deer is therefore best based upon three sets of observation, inspection or data:

- What is the body condition of the deer?
- How are the deer behaving?
- How is the herd performing?

Added to these assessments of the living deer, the careful examination of deer carcasses from the routine cull will provide valuable further information about the health and welfare status of the herd.

6.1   Body Condition Scoring of Living Deer

In temperate climates deer get fat in the summer and become lean in the winter. Mature fallow bucks and red stags eat little during the rut. They lose considerable amounts of weight. Before the onset of winter, males naturally are able to make up their lost weight because of the abundant autumn crop of seeds, nuts, berries, fruits, cereals and the autumn flush of grass.

The physiology of red and fallow deer predisposes them to severe metabolic pressures when poor winter weather is combined with depleted bodily reserves and reduced availability of forage. Unlike domestic ruminants, deer enter a period of significant metabolic slow-down in late winter, during which their reduced metabolic rate has three important consequences. It:

- Depresses their appetite
- Depresses their food conversion efficiency
- Reduces their ability to maintain their core temperature

Simpson et al. (1978) and Semiadi et al. (1996) established that red deer tolerate poor winter weather less well than domestic ruminants and have poor thermoregulatory capacity. From mid-winter to early spring, red and fallow deer are almost unable to gain weight or lay down fat reserves, irrespective of the amount of food and forage they consume.

The body condition of the deer (see also Green 2017) after the rut and before the winter is therefore critical: if they go into the post-mid-winter period in poor condition, they will be at risk of dying if the weather is poor and will have reduced ability to cope with parasites and other challenges. The phenomenon of 'winter die-off' in deer parks is well-recognised and is considered by FAWC (2013) to be the greatest welfare threat to deer in parks. Deer are unusual amongst ruminants in their seasonal reduction of both core temperature and heart rate as part of their winter metabolic depression. Once fat reserves are used up, the deer will start to catabolise muscle tissue; they risk becoming ketotic, and if this occurs, they will become depressed and completely inappetent and will pass the point where feeding can assist them (Ross 1994). Calves and fawns are likely to die first because they will, on average, have the poorest reserves at the end of the autumn. Stags and bucks will succumb before hinds or does; this is especially true of the males that were most active in the rut.

If a 'winter-die off' episode begins, it is too late to start supplementary feeding. The animals in poorest condition will be weak and inappetent. Food will be monopolised by the more vigorous deer. Deer will continue to die from metabolic collapse and exposure.

The condition of the herd should be assessed after the rut and before the onset of winter when there is still time to adjust supplementary feeding, but not all the deer are useful for this assessment. The body condition of mature males in late autumn after the rut will be influenced by their rutting activity. The dominant master bucks or stags may be in the worst condition. Similarly, the body condition of the mature red hinds and fallow is influenced by their social position in the herd, by their breeding performance, the age of their offspring and their continuing lactation.

The best way of assessing the condition of the whole herd in late autumn is by reference to the yearlings.

The yearlings provide a consistent cohort by which the condition of the herd may be measured and which is comparable year-on-year and between different herds. Yearlings are easily identified, and there are usually sufficient numbers of yearlings in a park herd to provide the necessary information.

Body condition scoring (BCS) of other livestock relies upon palpation, and similar BCS systems have been developed for farmed deer (Audige et al. 1998; Muller and Flesch 2001). The British Deer Farms and Parks Association has published a guide to body condition scoring of farmed deer in the UK, but hands-on palpation is part of the assessment. In the case of free-living park deer, a visual

Fig. 1 Body condition
scoring. BCS 1. Pelvic soft
tissues are markedly concave

pelvic condition scoring scale of 1–5 has proved to be extremely helpful (Figs. 1, 2, 3, 4, 5, and 6). The deer should be viewed at an angle, obliquely, not either from a lateral perspective or from the front or back. The contour from the tuber sacrum to the tuber coxa and tuber ischium should be assessed. This is most easily achieved by means of photographs, but can be undertaken in the field. It is important to score as many yearlings in the herd as possible and not to rely upon one or two animals. The median score of 3 is represented by a straight flat contour of pelvic soft tissues. Scores 4 and 5 have convex profiles, scores 1 and 2 have concave profiles.

As general rule, if 50% or more of the yearlings have a pelvic condition score (PCS) of less than 3 in the autumn after the rut, the herd as whole is in poor condition. In other words, the yearlings should have pelvic contours that are at least flat (PCS3) or better, and there should be few, if any, with any degree of concavity of the pelvic soft tissues (Figs. 7, 8, 9, and 10).

Visual assessment should also include careful inspection of the deer for signs of diarrhoea, skin disease or coughing, all of which may indicate unacceptable parasitism or bacterial infection. These may be linked to suboptimal nutrition, over population or to trace element deficiencies (see Sect. 8).

Fig. 2 Body condition
scoring. BCS 2. Pelvic soft
tissues are concave

Traumatic injuries in the rut are common and are not necessarily a sign of poor welfare (Fig. 11). By the welfare models outlined earlier, if a buck is injured in the rut and is able to adapt and respond to the injury by seeking isolation, shelter and food, his welfare is not unreasonably depressed, and intervention is not justified. Because of the remarkable osteogenic capacity of deer linked to seasonal antler growth, limb fractures heal speedily and can be monitored. Such trauma may be a reason to select an animal for culling when there are others that are injury free, but immediate euthanasia of deer with 'natural' injuries on welfare grounds is hard to justify unless they are clearly deteriorating or unless public scrutiny and concern warrant removal.

6.2   How Are the Living Deer Behaving?

Behaviour is a critical part of the welfare assessment of park deer. Deer in a good welfare state will be relaxed; there will be evidence of playfulness and mutual

Fig. 3 Body condition scoring. BCS 3. Pelvic soft tissue profile is flat

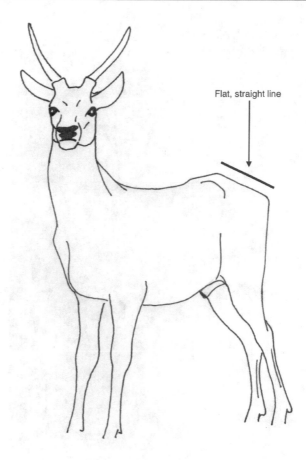

Flat, straight line

grooming within natal bands. Foraging for food will be obvious but leisurely. Deer will enjoy the sunshine and bask. They will settle in groups to ruminate (Fig. 12). Outside the rutting period park deer do not normally run about speedily. The great attraction of deer parks to the aristocracy of former times was the sense of calmness and tranquillity (Fletcher 2011).

If the deer are constantly unsettled or persistently trotting from one area of the park to another, this is an indicator of unacceptable disturbance and stress. Their welfare is being depressed because they are unable to find sanctuary and peace, despite their efforts to do so (Naylor et al. 2009).

Other indicators of adverse or poor welfare state include excessive bullying of subordinate animals, fighting amongst males or females outside the rut (to be distinguished from healthy, playful sparring, often seen in males in the spring), frantic foraging for food or obligate consumption of poor-quality fibre such as tree bark, dead stinging nettles or coarse rushes.

Calmness should not be confused with sluggishness or tolerance of extremely close human approach, which is not natural. Even deer that are very well habituated

Fig. 4 Body condition
scoring. BCS 4. Pelvic soft
tissue profile is convex

to human presence and will approach to beg for food will not normally tolerate a
bold human approach to them. Such tolerance suggests negative welfare for one
reason or another.

The presence of visitor activities can distress park deer, especially if dogs are
allowed to harass the deer or if they are subject to fireworks displays, low flying hot
air balloons or other similar stressors. Park deer will become habituated to many
activities like picnics, children playing, cyclists, golfers and other low-key activities.
Putman and Langbein (2003) report that disturbance reduces reproductive performance, with fewer calves and fawns reaching weaning in parks where levels of
disturbance are high. They suggest that this is because of neonatal mortality caused
by disruption of the dam-offspring bonding post-partum rather than because of
reduced conception rates.

If disturbance is perceived to be unacceptable, the park should provide a 'sanctuary' area into which no one is allowed so that the deer can always escape the stress of
a busy visitor day (Fig. 13). Dogs in deer parks should always be kept on leads,
although in public parks, this may be difficult to enforce.

Fig. 5 Body condition
scoring. BCS 5. Pelvic soft
tissue profile is rounded and
markedly convex

6.3 How Is the Herd Performing?

In a deer park herd that is performing well, in terms of recruitment into the herd, the park manager should expect at least 80% of adult does and hinds to raise a fawn or a calf. Females are considered adult when they are 2 years old at the rut. In many parks, more than 90% of adult females rear a calf or fawn. In many lowland parks, even yearlings achieve similar success rates, although in some more upland parks, yearlings may perform more poorly than this.

The size of the annual cull necessary to maintain herds at a constant number will depend upon the number of breeding females in the herd and the reproductive success of those females. Park populations and management plans are therefore best based upon the core herd of breeding females. Park population numbers are also best recorded in late winter, after the cull and before the spring and summer arrival of calves and fawns. A park with 300 fallow deer may have a core herd of 200 does, in which case an annual cull of 160–180 deer may be indicated. But if a park has 300 fallow deer with only 150 breeding does, the cull will be correspondingly smaller.

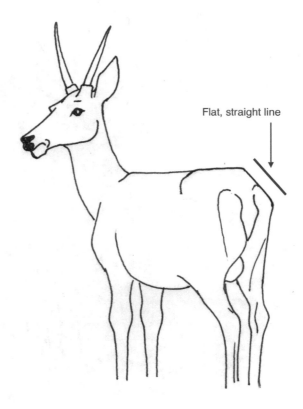

Fig. 6 Body condition scoring. BCS 3. Assessment can be made from either a frontal or rearward quartering angle

Flat, straight line

If annual recruitment in red deer or fallow deer herds falls below 60% of adult females, the park manager should investigate the possible causes, which might include overpopulation, insufficient males, male infertility, poor nutritional status of the females going in to the rut, neonatal losses and trace element deficiencies. Both male and female fertility declines in small, closed herds where in-breeding causes decline in vigour and in-breeding depression. This is especially problematic in closed red deer herds, but is also seen in fallow deer parks where no new blood has been introduced for many generations.

6.4   Deer Carcass Examination

Unless the herd is a non-breeding herd or has the facilities to sell surplus stock alive (in which case it is operating more as a farm than a park), there will be an annual cull of deer of both sexes and of varying ages. The carcasses from the routine cull can provide valuable information to the veterinary surgeon and the park manager.

The autumn period between the rut and mid-winter is the optimum time to gain information by post-mortem examination of deer culled and again, the yearlings provide the best consistent indicator of the condition of the whole herd.

Fig. 7 Body condition scoring. A fallow yearling buck in BCS 3. The pelvic soft tissue contour is a flat, straight line

Putman and Langbein (2003) suggested threshold carcass weights as indicators of when and what supplementary winter feeding should be provided. These absolute carcass weights have proved of limited value because of the variation in skeletal frame size between herds. Some herds of park fallow deer are large-framed animals, others are much smaller, irrespective of the habitat, climate or feeding. A very lean pricket (yearling fallow buck) carcass from one park may weigh 25 kg dressed out in the larder, while a fat pricket carcass from another herd may weigh only 21 kg.

Carcass condition should therefore be assessed by palpation of the pelvis, spine and ribs in the same way as one would BCS score a sheep and by reference to carcass fat deposits. In yearlings after the rut, the kidney fat index (the ratio of peri-renal fat weight to renal weight) should be at least 50%, measured in a number of animals. If most yearlings are found to have less kidney fat than this at this time, the management of the herd should be reviewed as a matter of urgency, before the onset of

Fig. 8 Body condition scoring. A fallow yearling doe in BCS 1–2. Note the concave soft tissues between the bony prominences of the pelvis

winter. Deer culled in late winter may have no peri-renal fat at all and may even have utilised all cardiac coronary groove fat. Deer in northern temperate climates differ from cattle and sheep in that catabolism of muscle as an energy reserve appears to be a normal part of their physiological annual cycle, especially in pregnant females, which are laying down protein in the embryo in utero over the winter at a time when their appetite reduces. They also excrete rather more protein than cattle and sheep in their urine and faeces (Haigh and Hudson 1993).

The dressed out carcass is of limited value for veterinary post-mortem examination. The veterinary surgeon should also examine all the internal organs and the heads of several deer in the post rut cull. For the assessment of herd bodily condition, yearlings are the animals of choice. For herd health and disease monitoring, animals of all ages and of both sexes are useful. This gross examination should include careful assessment of the retropharyngeal and sub-mandibular lymph nodes of the

Fig. 9 Body condition scoring. A yearling fallow buck (pricket) in BCS 2. Pelvic soft tissues are obviously concave

head, the bronchial and mediastinal lymph nodes of the thorax and the mesenteric, hepatic, rumenal and iliac lymph nodes of the abdomen. The viscera, pleurae and peritoneum should be examined in accordance with standard post-mortem protocols.

Faeces from the rectum should be examined in the laboratory for internal parasite eggs and oocysts. Fresh liver should be assayed for trace elements, particularly copper and selenium (see Sect 8).

7 General Considerations of Medicating Deer in Deer Parks

As deer park herds are not routinely gathered for medication and as the majority of deer parks have no facilities to gather the deer, the medication of deer by syringe-and-needle injection or by individual oral dosing is not possible. Fortunately outbreaks of infectious disease are not common in deer parks, but if they do occur, there may be a temptation to try to administer medicines by in-feed medication. This

Fig. 10 Body condition
scoring. A yearling fallow
buck (pricket) in BCS 3–4.
The soft tissues of the pelvis
are convex

is most likely to be contemplated in the face of an ostensible problem with nematode worms, when in-feed anthelmintics may be considered.

Unless the problem is affecting a small herd maintained almost entirely on artificial feeding, such medication must be resisted at all costs. In-feed medication in an extensive park will not deliver sufficient of the medicine to be effective. The deer worst affected by the problem will be least likely to consume sufficient to deliver a meaningful dose and the exposure of the pathogen to ineffectual low doses represents a high risk of inducing drug resistance. Most episodes of infectious disease in deer parks can and should be managed by changes to the numbers, feeding and husbandry of the deer. The exception is bovine tuberculosis (bTB), which is notifiable to the competent authority in most western jurisdictions. Although bTB can be eliminated from a park deer herd by changes to management, proof of elimination by testing is likely to be required, which requires the deer to be gathered and handled.

Fig. 11 Traumatic injuries in the rut are common in park deer but are rarely a reason for veterinary intervention

8   UK Park Deer Pathology

8.1   Winter Die-off Syndrome

Ruminants rely upon gluconeogenesis in the liver to a far greater extent than most other mammals. They do not produce fatty acids for milk fat in the udder from glucose, but rather from ketone bodies or acetate, and they do not synthesise fatty acids in the liver, but only in adipose tissue where it is synthesised from acetate. This system of perpetual reliance upon gluconeogenesis places particular metabolic stresses upon ruminants, especially when they are lactating and need glucose for conversion to lactose and in pregnancy when the demands of the placenta and foetus can only be met by glucose or amino acids (Cunningham and Klein 2007). Even amongst ruminants, there are differences. Cattle and sheep have greater reserves than red deer and cope with winter cold stress better (Simpson et al. 1978; Semiadi et al. 1996).

The maintenance of core temperature is a critical imperative for mammals. If the temperature of the abdominal and thoracic organs falls below about 38 °C the heart will suffer arrhythmia, and death will ensue. Heat is lost by mammals in a variety

Fig. 12 Deer in a positive state of welfare will be relaxed and at ease, as seen with these five fallow bucks

of ways: by convection into cold air or cold water in contact with the body, by conduction when the body is in contact with cold ground, by radiation into the surrounding environment and by evaporation when sweat, saliva and respiratory moisture is converted to water vapour. Heat to maintain core body temperature is produced by muscular activity, either by voluntary movement or by shivering, and by metabolism of carbohydrates, fats and proteins (Cunningham and Klein 2007). Deer are unusual amongst ruminants in their seasonal reduction of both core temperature and heart rate as part of their winter metabolic slow-down (Adam 1994; Clutton-Brock et al. 1982; Haigh and Hudson 1993; Andersen et al. 1998; Turbill et al. 2010).

The phenomenon of winter die-off is well known in deer parks subject to adverse weather conditions in winter and has been described in many other wild populations with large percentages of the population dying (Boyce 1989; Young 1994; Putman 2008; FAWC 2013). The cause of death is multifactorial, but at a group level, the herds of deer subject to catastrophic rates of mortality in single episodes appear always to have some common characteristics: the deer are present in high density in the park, and they are exposed to a combination of reduced nutritional intake and increased climatic stress. In temperate climates this combination is usually malnutrition and cold, wet weather (Young 1994).

The mortalities seen in deer parks subject to an episode of winter die-off therefore have an understandable pathogenesis. Stags lose considerable amounts of body weight in the rut, when they are extremely active and eat very little (Clutton-Brock et al. 1982). Unless they are able to make up this weight loss by access to good,

Fig. 13 A sanctuary area for the deer is an important part of welfare support in busy parks and should be clearly signed

nutritious food after the rut, they will enter the mid-winter with low fat reserves, while the hinds, which have been grazing throughout the rut, will almost certainly enter the coldest part of the winter in better condition. If severe weather closes in after mid-winter, all the deer will reduce their metabolic rates, have reduced appetites and will be unable to consume sufficient to maintain body weight and body temperature and will rely upon body tissue reserves. Exposure increases the effect of cold, wind and wet. Ketosis leading to depression and inappetence will soon affect the deer, because of their reliance on gluconeogenesis. The problem is compounded by the fact that foraging activity in winter is more costly in terms of energy and time expenditure, and this is exacerbated if the deer park has little cover or shelter from the wind. Fallow deer, which are essentially a Mediterranean species, are at the limit of their geographical range in northern England and Scotland. When deer are debilitated by this metabolic downward spiral and becoming terminal, other secondary pathologies may arise: *Yersinia pseudotuberculosis* or nematode bowel worms may give rise to loose faeces, but it is a mistake to make the diagnosis that yersiniosis or parasitic gastroenteritis is the reason for the decline and death of the deer.

In particularly severe weather another mechanism is important: simple physiological exposure. Red and fallow deer, especially the young, have relatively poor coats in terms of thermal insulation (Semiadi et al. 1996; FAWC 2013), and even 'fat' animals have less subcutaneous fat than sheep, which also have better fleeces (Simpson et al. 1978). If the ambient conditions are so cold and so severe that heat is lost at a greater rate than it can be produced, the deer will die of hypothermia as their core temperature falls below the mid 30s°C, irrespective of reserves. Good bodily condition will not guarantee survival in extreme conditions.

The signs of a significant winter die-off are obvious: deer are found dead in periods of severe weather, especially in prolonged periods of cold and wet. The dead deer are likely to be in poor condition. There may be deer lying or wandering around with depressed responsiveness to normal stimuli. Supplementary feeding once the deaths have started will not prevent more. This represents a welfare issue at the group level, but the negative welfare state for the group did not begin when the deaths started; it began when the deer were unable to adapt to winter by finding shelter and by laying down reserves at a time when their metabolism would have enabled them to make use of extra food. The individual deer, dying in such an episode, are clearly in a severe negative welfare state, but unless they are succumbing to extreme acute exposure, this state will have pre-dated their recumbency, depression, collapse and death by several weeks.

This is why pre-winter condition scoring of the deer park herd is so important. If condition scoring suggests that yearlings in late autumn are in BCS of less than 3 based on their pelvic contours (see Sect. 6.1), supplementary feeding should begin immediately, even if the weather is not currently poor. If financial constraints prohibit the feeding of the whole herd, a pre-winter cull should be undertaken to reduce the herd to numbers for which food can be provided. An episode of winter die-off is an indicator of poor management in the autumn and early winter preceding the death of the deer.

8.2 Trauma and Capture Myopathy Associated with Gathering

The definition of a deer park adopted for this chapter makes clear that the deer are not routinely caught up for management and husbandry purposes. The deer behave as wild deer, especially if attempts are made to capture them. It may, however, become necessary to make attempts to gather up the deer herd if the only alternative is to cull them all. This may be imposed upon a deer park if suspicion or confirmation of a notifiable disease such as tuberculosis arises, or if a deer park is sold and the deer need to be relocated. Experienced park keepers and veterinary surgeons working with deer will be cautious about such exercises. Fallow deer are especially challenging to gather because they panic and leap haphazardly to a greater extent than red deer and because bucks become extremely aggressive when they are stressed or crowded.

It is almost always essential to construct a deer-fenced catching paddock into which the deer are fed and habituated before the gates are closed for the actual gathering and handling. Systems of temporary yards and races can be erected on the side of the catching pen, but whatever system is used, quiet, slow and methodical patterns of human activity must be observed. (Fletcher 1994; Matthews 2007). Even in the best-designed systems, casualty rates may be high, and losses are to be expected, particularly with fallow deer.

Trauma from running or jumping blindly into fences is likely to include facial lacerations and fractures of limbs and mandibles. Unless bucks can be separated from the herd or culled before gathering, crowds of stressed park fallow deer will probably result in severe wounding and even bowel prolapse amongst juveniles and does.

If gathering and processing are successful, the syndrome of capture myopathy may appear later. This is a complex syndrome characterised by ataxia, weakness, paralysis, hyperventilation, myoglobinuria and death. It is not a simple matter of overexertion, myopathy and lactic acidosis, but has a complicated pathogenesis that includes gross metabolic disruption and a wide range of clinical pathological abnormalities. It is clear that stress, restraint and hyperthermia from excessive exercise and struggling all play a part in precipitating the clinical signs. Affected deer may die within hours or may survive for several days (Munro 1994; Kreeger and Arnemo 2012).

Attempts at treatment of park deer with capture myopathy are unrewarding. Reversal of acidosis by slow administration of sodium bicarbonate (5 meq/kg) together with sedation with diazepam (0.1–0.3 mg/kg) may be effective in mild cases, but park deer are usually so stressed by close human contact and presence that the act of administering the medication only worsens the syndrome.

Capture myopathy should not be confused with 'radial paralysis' sometimes seen after heavy stags and bucks have been in lateral recumbency for prolonged periods after darting. This is a pressure-induced musculo/neural compartmental syndrome, giving rise to forelimb lameness or disability of the dependent limb. It usually wears off after a few hours.

8.3　Rumen Acidosis and Feed Station Death

Supplementary winter feed must be provided in ways that allow all the deer in the park to access and consume the ration. This usually means putting the feed out in long lines. If feed is offered in piles or in a limited number of troughs, the dominant animals will monopolise the ration, and the subordinate animals will not be allowed access to it. In extreme cases, this can have two separate consequences leading to mortality.

The first is that the dominant animals may gorge themselves on the feed and suffer rumen acidosis, which may be fatal. Cereal concentrate rolls or nuts are especially problematic, but it is also seen with whole maize (corn) kernels and straight rolled barley. Master fallow bucks are most often affected by fatal rumen acidosis. The author has seen several episodes in which volunteers or other estate staff were engaged to feed the park deer when the regular deer manager was on holiday or indisposed and, rather than putting the feed out in long lines, simply tipped a large pile in one place in the park. The result was the death of three or four of the best bucks in the park in the vicinity of the feed pile within 24 h.

Post-mortem signs of rumen acidosis are typical: rumens gorged with fermented cereal gruel with a typical brewery-like smell, sloughing of the rumenal mucosa and a pH <5.

The second mortality associated with feed stations is correctly considered as part of the winter die-off syndrome. If feed station access is limited and supplementary feed is not sufficiently widely distributed, subordinate animals may loiter in the vicinity of the feed in the hope of getting some food, but are prevented from doing so by the more dominant animals. They are attracted to the prospect of food and do not forage for it elsewhere in the park, but are not permitted to access sufficient food to maintain themselves. Feed station death is the resulting phenomenon: emaciated young and subordinate deer are found dead within 50 m of the feed station, occasionally with several such carcasses in a ring around the feeding site or several collapsed and dying deer within sight and smell of the feed.

8.4   Trace Elements

Although a wide range of trace elements is essential to ruminant health and well-being, only two trace element deficiencies are regularly encountered in deer park deer. These are copper and selenium. Cobalt deficiency is well-recognised in beef cattle and sheep on marginal grazing but is not confirmed in park deer; trials with vitamin B12 supplementation of young deer have been equivocal (Jones 1994).

Both copper and selenium deficiencies, either alone or combined, give rise to ill-thrift, apparent immunosuppression and reduced fertility (Fig. 14). In addition, copper-deficient herds may have a higher incidence of broken antlers, and youngstock may develop swollen limb joints, which seem to be a manifestation of osteochondrosis. Both copper and selenium deficiency are commonly associated with higher than normal enteric parasite levels, sometimes manifest by loose faeces and high worm egg counts (see Sect. 8.5). The syndrome of enzootic ataxia or copper deficiency swayback (Fig. 15) is not common, even in herds with liver copper levels below the limit of detection of the laboratory assay. The reason for this is unclear, but it means that while parks herds in which enzootic ataxia is encountered will almost certainly be copper deficient, the absence of enzootic ataxia does not indicate that a given park herd is not severely copper deficient.

Liver tissue should be assayed for copper, selenium. Based upon more than 180 samples submitted from 35 deer parks over a 10-year period, Table 1 gives references ranges used by this author for these minerals. Note that ranges vary between red/sika deer, which are closely related and fallow deer. The current reference ranges used by commercial and APHA laboratories are based upon data

Fig. 14 Both copper and selenium deficiencies can lead to ill-thrift and poor condition. This fallow buck is in very poor condition, is dull and has broken antlers

from farmed red deer and derive principally from New Zealand (Grace and Wilson 2002; Haigh and Hudson 1993).

Supplementation of copper and selenium by means of salt or molasses-based mineral blocks will boost levels of these trace elements and enteric parasite burdens in the deer herd will usually reduce. The palatability and attractiveness of different mineral blocks to different deer herds is remarkably variable. Some herds will voraciously take a salt-based block and ignore a molasses-based block; other herds will enthusiastically consume a molasses block and shun a salt block. Park managers are best advised to trial a range of high-copper, high-selenium blocks to determine which their deer prefer. If supplementary winter feeding with commercial cereal-based rations is practised, these should have copper and selenium at levels in excess of those offered to cattle. High-copper mineral blocks and high-copper rations may be toxic to sheep; this is a problem in heritage deer parks where deer may run with rare-breed sheep.

Park deer appear to be at risk of accumulating both lead and iron (Vengust and Vengust 2004). Liver tissue levels in excess of 89 μmol/kg dry matter (DM) of lead have been encountered by the author in fallow deer herds with no clinical evidence of lead toxicity, although such levels exceed the thresholds permissible for human consumption. For all deer species when liver lead values exceed 10 μmol/kg DM (which is approximately equivalent to 2.4 μmol/kg wet weight (WW) or 0.5 ppm WW), there may be a need for further scrutiny in relation to the risk of lead exposure

Fig. 15 Enzootic ataxia is linked with hypocupraemia, but is not invariably seen in deer with low copper levels. Poor condition, low fertility and increase risk of winter die-off are more commonly signs of hypocupraemia

if parts of the carcase were to be consumed. Such values may also potentially have an impact on the health of such deer, although none have been detected. Iron accumulation in the liver, up to 380,000 μmol/kg DM in UK park deer, is evidently associated with heritable haemochromastosis (Olias et al. 2011). This accumulation appears not to be toxic to the deer but does result in swollen and darkly discoloured livers.

Table 1 Suggested reference ranges for liver minerals in park deer

Liver mineral	Species	Reference range µmol/kg dry matter
Copper	Fallow	200–6000
	Red and sika	200–3000
Selenium	Fallow	4–50
	Red and sika	4–20
Iron	Fallow	3000–200,000
	Red and sika	5500–33,000
Lead	Fallow	0–40
	Red and sika	0–10

8.5   Parasites

8.5.1   Helminths

Helminth parasite levels in deer park deer are usually lower than the levels seen in cattle and sheep, even when deer and domestic livestock are grazing together.

Liver fluke (*Fasciola hepatica*) is common in lowland parks, especially where historically very large parks have been reduced in size, because the ground to which the deer have been confined is often the poorer and more marginal part of the former park. Both red deer and fallow deer show livers with patchy fibrosis and adhesions obvious on the capsule, which extends into the parenchyma. The 'pipe-stem' bile ducts seen in cattle are uncommon. Sika deer infected with fluke may have little or no evidence on the capsular surface, and the liver needs to be incised to reveal the parasitised areas within. In all species, adult flukes can be squeezed out of the bile ducts.

Even in deer where fluke infestation is high, fluke eggs in the faeces are scarce. For this reason, if herds are being screened by faecal examination for the presence of liver fluke, pooled samples from at least five adult deer should be examined by the laboratory. Even then, no eggs may be seen when significant infestation is present.

Deer appear to tolerate liver fluke infestation without ill effect, and carcasses with extensive fluke damage to the liver are often found to be replete with fat in the late autumn. If there are roe deer (*Capreolus capreolus*) in the park, these may be found dead, as this species can suffer fatal infestations. If mixed grazing with cattle or sheep is undertaken in the park, regular flukicide medication of the domestic livestock will reduce prevalence of fluke in the deer.

Gastro-intestinal nematode parasites of park deer are similar to those of cattle and sheep, but are far less likely to be the primary cause of gastro-intestinal disease. The exception is acute disease of the abomasum caused by Osertagian parasites, which is sometimes seen in red deer in overstocked parks. The taxonomy of these worms is complicated and rather inconsistent, but the syndrome appears to be similar to the Type II ostertagiasis of cattle, with large numbers of encysted early L4 larvae erupting in late winter. Affected deer have intermittent loose faeces, usually dark in colour, and lose condition rapidly. Some may die, and the disease may form part of the winter die-off syndrome encountered in poorly managed parks. Typical nodular or red-spotted abomasal mucosa is seen at post-mortem examination. The Type I ostertagian disease of young cattle is not seen in park deer.

Nematode worms of the small intestine are usually insignificant in park deer. Numbers of strongyle-type eggs in the faeces are much lower than in cattle or sheep, but may rise if there are deficiencies of copper or selenium in the herd. In such cases, ill-thrift and poor late winter condition may be the only clinical signs.

The threshold of concern for deer park deer is 200 eggs per gram, much lower than that for other domestic livestock. At levels higher than this, even in the absence of patent disease, there may be some effect on body condition (Irvine et al. 2006). If counts above this level are encountered in a number of deer, the park should review its trace element status, stocking density and feeding protocols.

A number of lungworm nematodes are regularly encountered in park deer. Fallow deer are very commonly infected with metastrongyle lungworms of the genera *Protostrongylus* and *Meullerius*. These form small nodules in the lung tissue, often discoloured. Worms are difficult to discern by gross incision of the nodules because they are coiled and embedded in lung tissue. *Protostrongylus* worms may occasionally be discernible in the terminal bronchioles and may give rise to larger caseous or even calcified lesions that can raise alarm in respect of tuberculosis. Apart from this possible confusion, they cause no ill effect.

Dictyocaulus lungworms (both *viviparous* and *eckerti*) are seen more frequently in red deer. Worms are easily seen in the bronchi and bronchioles. Heavy infestations are associated with poor condition in yearlings and with death in adults as part of the winter die-off syndrome. Overstocking of the deer park and insufficient supplementary food leading up to mid-winter are usually contributing factors.

Treatment for nematode worms is rarely justified; changes in stocking density, trace element supplementation and increased feeding will almost always resolve any possible problem with these parasites. In-feed medication with anthelmintics should only be considered in small herds that are perpetually maintained on cereal-based concentrates or straight cereals with forage such as silage or haylage. These deer will be used to consuming most, if not all of their daily nutritional requirements from this diet, which may then be medicated with a 10% mebendazole premix in the diet, which is fed for 3–5 days. In larger herds and in more extensive parks where food is only provided as a supplement such medication should not be offered, for the reasons given earlier.

Deer parks with high public visitor numbers and high dog presence may have regular examples of liver white spotting in adult deer, in which the capsule of the liver shows multiple small discrete pale spots, often with tiny fibrous tags. The cause of this is not absolutely certain, but the canine roundworm *Toxocara canis* is almost certainly responsible. This parasite is known to infect many species other than canines, including birds, rodents, pigs and sheep and to cause liver white spotting by transhepatic larval migration (Lloyd 1998; Strube et al. 2013). The worm does not complete its life cycle, but viable larvae may remain the in the liver of these paratenic hosts for prolonged periods. For this reason, offal from affected deer should not be consigned to the human food chain to prevent the risk of human *larval visceral migrans*, a well-recognised zoonosis (Lloyd 1998). White-spotted livers are rare in deer in parks with no dog presence.

Cysticercus tenuicollis tapeworm cysts are regularly found on the liver and in the mesentery. This is the intermediate stage of the tape worm *Taenia hydatigena*, a parasite of foxes and dogs. It causes no ill effect to the deer and is considered an incidental finding.

8.5.2   Arthropods

Ticks are common on park deer, especially in parks where there are large tracts of bracken (*Pteridium aquilinum*). *Ixodes ricinus* is by far the most common species, with some deer carrying many thousands of individuals of varying life cycle stages.

The deer ked (*Lipoptena cervi*) is almost ubiquitous in UK deer parks and on wild deer. Lice of the genera *Bovicola* and *Solenopotes* may be encountered but are not common.

None of these external parasites cause primary disease but may be found in increased numbers when deer are in poor condition and as part of the winter die-off syndrome. In such circumstances, they are opportunistic indicators of malnutrition and of poor condition, not the cause of it.

The deer nasal bot (*Cephenemyia auribarbis*) may be discovered in the pharyngeal or nasal cavities of both red deer and fallow deer at routine post-mortem. The third stage larvae may be up to 40 mm in length and can be alarming to inexperienced larder workers. The head shaking and sneezing often seen in deer in deer parks in mid-summer may be an indication of nasal bot infestation. Apart from this irritation, the bots are not significantly pathogenic.

Deer warbles, the larvae of the fly *Hypoderma diana,* form nodules along the spinal skin of red deer and fallow deer. They may be more common in deer parks than is apparent, because park deer are almost all culled in the winter. This is when the resting phases of the migrating larvae are within the internal tissues, especially the epidural fat of the lumbar and thoracic spine, where they are unlikely to be detected. Patent warbles beneath and within the skin become apparent in late spring and summer, when few park deer are culled and examined. Deer hides from park deer are rarely used commercially. Warbles are not considered to be a cause of disease.

Fly strike with larvae of dipteran flies (myiasis) is regularly encountered after wounds are sustained by stags and bucks fighting in the rut, especially if the autumn weather is warm. Flies of the genera *Lucilia* and *Calliphora* are usually responsible. Limited lesions of fly strike around the head, especially in the frontal and coronet areas, are usually self-limiting. Affected males may shake their head and a hard, black eschar may be visible on the head or face. These animals should be monitored and culled only if they become depressed and lethargic. Wounds on the body that become infested with maggots carry a poorer prognosis and can quickly become very extensive (Fig. 16). Culling is the only option for these deer.

Fig. 16 Myiasis (fly strike) of the body is a serious condition. Note the extensive black eschar on the lumbar region of this stag and his poor bodily condition. The whole eschar was undermined by fly larvae

8.5.3 Protozoa

The protozoal pathogen *Cryptosporidium sp.* is a significant problem on red deer farms, where it is a common cause of neonatal diarrhoea and may lead to neonatal deaths. In deer parks, it occurs sporadically and is more likely to be seen in open grassland parks with little cover or shelter. In these parks, the female deer tend to calve and fawn in the same places each year, perhaps a stand of nettles or an area of coarse rushes or a single small stand of trees where contamination with the sporozoites builds up. Affected calves and fawns may be found weak and with profuse diarrhoea. They are more frequently found dead with soiled perineums and hind limbs. Diagnosis is by faecal examination, using modified Ziehl-Nielsen staining or PCR probing. As the infection is zoonotic, calf and fawn carcasses should be handled with care.

Park managers should be encouraged to provide different areas for calving and fawning each year, by mowing or cutting or fencing off the overused areas and encouraging cover growth elsewhere. Forestry brash and branches or scattered straw bales can be distributed over an area to provide artificial cover for the females to deliver their young if the park lacks sufficient natural cover.

8.6   Bacterial Diseases

8.6.1   Yersiniosis

Yersinia psuedotuberculosis and *Y. enterocolitica* may be cultured from deer with loose faeces in terminal decline during a metabolic winter die-off episode, but they are not the primary pathogen in such circumstances. The yersiniosis of farmed red deer, characterised by diarrhoea, depression and death of young stock, is rarely seen in park deer. It may occur in grossly overstocked parks, in which case stocking rates should be revised downwards. Treatment of individual animals is not usually practicable.

 Yersinia sp. Occasionally gives rise to a non-purulent but proliferative septic arthritis of the limb joints. This is more common in fallow deer than red deer and more common in the distal interphalangeal joint(s) than other joints. Affected deer become lame and the coronary area of the foot becomes swollen. The growth of the hoof of the affected digit may be distorted; changes are obvious radiographically (Fig. 17). No treatment is possible and affected deer should be culled if the swelling of the distal limb becomes obvious. *Yersinia sp.* can be cultured from the degenerate synovial fluid and joint capsule.

Fig. 17 Non-purulent proliferative septic arthritis of the distal interphalangeal joint of a fallow deer caused by *Yersinia sp.*

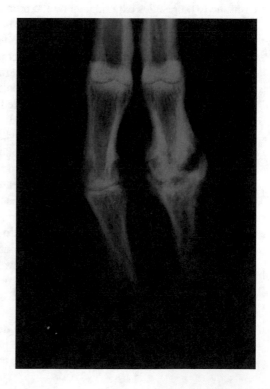

Figs. 18 and 19 External evidence of abscessation is uncommon in bTB, but is occasionally seen

8.6.2 Pasteurellosis

Acute pasteurellosis has caused outbreaks of sudden death in fallow deer (Carrigan et al. 1991; Eriksen et al. 1999) and has been encountered by the author on two occasions. In the peer-reviewed reports and in the author's experience, the outbreaks occurred in late winter when bodily reserves were depleted, but the syndrome was not typical of winter die-off because large numbers of deer died over a short period, and all showed haemorrhagic nasal and oral discharges, with swelling and inflammation of the pharynx and pneumonia. *P multocida* was cultured from several tissues in deer subject to post-mortem examination.

In the two episodes experienced by the author, the consumption of poor-quality forage appeared to predispose the fallow deer to the infection. In one case, the deer were being offered very poor hay with an abundance of dried thistles (*Cirsium arvense*) and in the other the deer were not offered any supplementary forage despite the park being overgrazed and the deer had resorted to eating gorse (*Ulex europaeus*) and blackthorn (*Prunus spinosa*). In both outbreaks, deaths ceased when good quality haylage was provided. It was concluded that abrasions and penetrations of the oral and pharyngeal muscosae had allowed *P multocida* to enter the soft tissues of the mouth and throat and become established, giving rise to septicaemia and pneumonia of deer that were already in poor condition. In the outbreaks described by Eriksen et al. (1999), in which over 300 deer succumbed in four outbreaks over four winters, the comment is made that the deer park was heavily grazed, and all vegetation up to a height of 2 m had been consumed.

8.6.3 Mycobacterial Disease: Tuberculosis

Bovine tuberculosis (bTB) has become problematic in deer parks in the west of England and parts of Wales, where the prevalence of the disease in cattle and indigenous badgers (*Meles meles*) is high. In deer parks with no contiguous cattle presence and no cattle grazing in the park, it is assumed that the badgers are the source of infection.

To date, the author has experienced eleven deer parks with bTB infection.

Typically, the first evidence of bTB in the park is the incidental discovery of a carcass with internal abscesses during routine culling. Clinical signs in living deer

Fig. 20 This fallow pricket has severe and extensive bTB. Other differential diagnoses of weight loss and diarrhoea include Yersiniosis, MAP and MCF

are unusual until the affected animal is severely diseased, at which point the deer declines in condition rapidly and dies (Figs. 18 and 19). Such evidence is usually only apparent several years after the first detection of the disease in the herd, by which time the prevalence in the herd may be high. Occasionally evidence of subcutaneous abscessation may be seen, but this is rare (Figs. 18 and 19). Deer in the terminal stages of the disease may have diarrhoea (Fig. 20). The prevalence of infection in an affected herd may remain low for several years and then suddenly increase.

Any internal abscess in a park deer should arouse suspicion of bTB because the character of the lesions is very variable. There may be purulent abscessation of the retropharyngeal and sub-mandibular lymph nodes, caseous focal nodules in the lungs and associated lymph nodes or abscesses on the mesenteric lymph node chain. Some carcasses, however, have large single abscesses, sometimes in the deep muscles of the limbs, others have miliary abscesses throughout the thorax and abdomen and yet others discrete abscesses on the parietal pleurae or peritoneum with no lesions elsewhere (Figs. 21, 22, 23, 24, 25 and 26).

By post-mortem examination of 219 red deer and 398 fallow deer from five infected parks between 2005 and 2020, bTB was confirmed in 31% ($n = 68$) of the red deer and 39% ($n = 155$) of the fallow deer by the author. Figures 27 and 28 show the distribution of lesions in red deer and fallow deer by sex in the 223 infected deer. Red deer had many fewer carcasses with lesions predominantly in the thoracic viscera (bronchial, mediastinal lymph nodes and lungs) compared with fallow deer

Fig. 21 Lesions of
Mycobacterium bovis in
park deer: caseous pulmonary
abscesses

Fig. 22 Lesions of
Mycobacterium bovis in
park deer: miliary lesions
throughout abdomen and
thorax

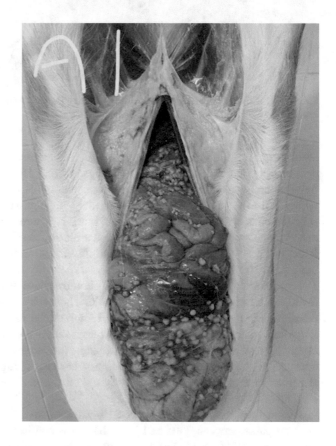

and many more deer with lesions predominantly in the mesenteric lymph node chain
than fallow deer. In both species, only between 16% and 25% of infected deer had
lesions in the lymph nodes of the head and throat with few or none elsewhere. This

Fig. 23 Lesions of
Mycobacterium bovis in
park deer: abscesses beneath
parietal pleurae

means that screening for bTB by examination of the head, heart and lungs alone may
not detect the majority of infected carcasses in red deer and may only detect 50% of
infected fallow deer. These findings concur with published data on the distribution of
lesions (Martín-Hernando et al. 2010).

Diagnosis on the basis of gross post-mortem is reliable in the hands of an
experienced clinician or pathologist. Laboratory confirmation of bTB infection
should not rely upon Ziehl Neilsen staining, since many bTB isolates from deer
are not acid fast (Gavier-Widén et al. 2012). Culture or genomic probing is therefore
essential.

It seems that the primary routes of transmission within an infected deer herd is via
nasal and oral secretions, rather than from faeces and urine (Santos et al. 2015). This
is in contrast to the disease in badgers and cattle, in which faecal and urinary
excretion is significant. It is also the reason why a critical element of bTB control
in an infected deer park is the cessation of supplementary feeding. The mycobacterial organism has been shown to persist on feed partially consumed by infected deer
and to be a viable source of infection to other deer (Palmer et al. 2004) and may
survive on partially consumed roots and fruit for many weeks (Palmer and Whipple

Fig. 24 Lesions of
Mycobacterium bovis in
park deer: caseous abscesses
in liver and diaphragm (red
arrows). Note the high kidney
fat index (blue arrows)

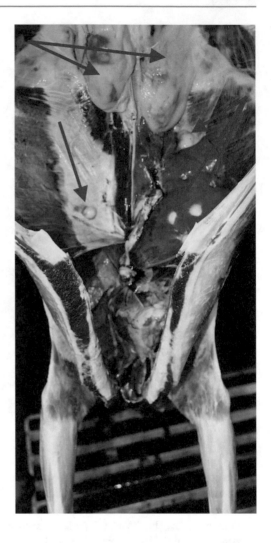

2006). It can survive in soil for over a year (Duffield and Young 1985; Ghodbane et al. 2014).

Deer parks faced with confirmed bTB infection have few options for control of the disease. After the initial index case, the park may choose to undertake a surveillance cull of an increased number of deer compared with the usual annual cull and to examine all carcasses carefully. This cull will also provide the opportunity to reduce the herd to a level that does not require supplementary feeding. This may be in the range of 2-3DSU per hectare of good grazing (See 5 above). Trace element status should be reviewed. Of the eleven deer parks with bTB infection in the author's veterinary practice, eight had marginal or deficient levels of either copper or selenium or both.

Fig. 25 Lesions of
Mycobacterium bovis in
park deer: cluster of small
caseous/calcified abcesses in
omentum

Fig. 26 Lesions of
Mycobacterium bovis in
park deer: large purulent
abscesses in mesenteric lymph
node chain. Note abundant
mesenteric fat

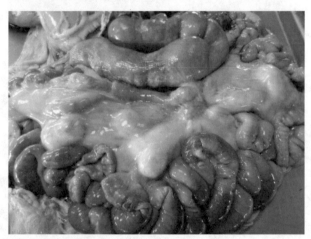

If the prevalence of infection is similar or lower than the prevalence in wild deer in the locality, the park managers may choose to monitor the park herd for a further year in the hope that the index case was a spill-over from a wildlife reservoir. As an example, the prevalence of bTB in wild deer is believed to be between 2% and 4% in the west of England (Delahay et al. 2007; Ward et al. 2008). This course of action may result in the park becoming epidemiologically free of bTB, with no further cases over several years, but if the disease is notifiable, this will not satisfy the statutory authorities that infection is eliminated.

Prevalence above expected local wild deer levels will suggest that the bTB is being maintained within the deer herd as a circulating reservoir of infection, and

a

Distribution of visible bTB lesions in male red deer
Lesions only or predominantly in these sites

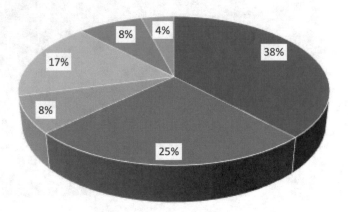

■ Mesenteric   ■ Throat   ■ Lung   ■ Bronchial/mediastinal   ■ Hepatic   ■ Miliary / other

b

Distribution of visible bTB lesions in female red deer
Lesions only or predominantly in these sites

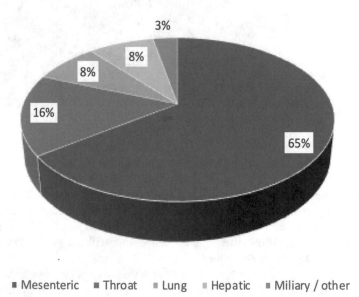

■ Mesenteric   ■ Throat   ■ Lung   ■ Hepatic   ■ Miliary / other

Fig. 27 Distribution of *Mycobacterium bovis* gross lesions in park red deer by sex ($n = 69$ deer with gross lesions)

a

Distribution of visible bTB lesions in male fallow deer
Lesions only or predominantly in these sites

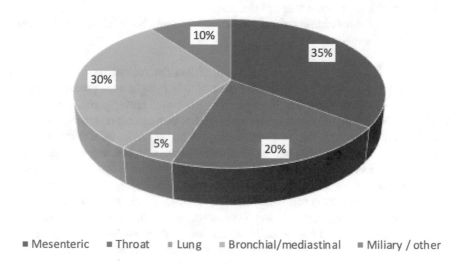

■ Mesenteric ■ Throat ■ Lung ■ Bronchial/mediastinal ■ Miliary / other

b

Distribution of visible bTB lesions in female fallow deer
Lesions only or predominantly in these sites

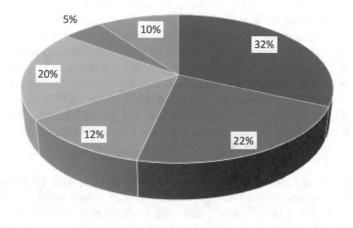

■ Mesenteric ■ Throat ■ Lung ■ Bronchial/mediastinal ■ Hepatic ■ Miliary / other

Fig. 28 Distribution of *Mycobacterium bovis* gross lesions in park fallow deer by sex ($n = 155$ deer with gross lesions)

even if there is only one index case for several years, the authorities may insist that freedom from disease is proven. The park then has only two options:

1. Cull all the deer, leave the park empty of deer for a prolonged period and then re-stock with deer tested free of infection.
2. Gather the deer for testing and achieve a sufficient number of clear tests to satisfy the authorities.

If option 2 is contemplated, considerable difficulties should be anticipated. As detailed above (see Sect. 8.2) gathering park deer is fraught with problems. In the absence of full deer farm-type facilities of paddocks, races, yards, pens and a handling shed, getting all the deer into a temporary testing unit may prove impossible. Deer will be injured. The deer will learn quickly, and an initial rate of success is unlikely to be repeatable. The comparative intradermal skin test requires the deer to be gathered twice with a 72 h interval—this has proved impossible on a number of occasions, and the prospect of repeating this double gathering for successive tests over a period of months is practically insurmountable. A scheme to gather and test the deer should only be contemplated if the prevalence in the routine or surveillance cull is less than 10%. If the percentage of carcasses with gross lesions is higher than this, there will be many more undetected deer in the early stages of infection and the likelihood of achieving bTB free status by successive rounds of testing is very small.

In the majority of deer parks, only one option remains if the owners wish to retain bloodlines: cull the majority of the deer and retain only a small number of yearling females, or better, female fawns or calves. These are less likely to be infected than older animals and more likely to be tractable to handling. Gather these and keep them in a yard or shed as on a deer farm and install adjacent workable facilities through which they can be regularly processed for testing.

The accuracy and reliability of various blood tests for the detection of bTB in deer are currently being evaluated by a number of statutory authorities (Busch et al. 2017), which may reduce the frequency with which the deer need to be processed through the handling facility. Irrespective of this, the cohort of young females will need to be yarded for many months before the herd is declared bTB free. It can then be released back into the park with new males tested free of infection. It is essential that the conserved cohort is not released into the park between tests because of the high risk of contracting infection from the contaminated pasture, soil and any wildlife reservoir.

However, a depressing caveat to the success of any such project must be emphasised. If the deer herd initially became infected from a wildlife reservoir of bTB infection such as local badgers, unless the infection in the wildlife reservoir has been eliminated, eventually putting deer back into the deer park will risk them contracting the disease once again.

8.6.4 Mycobacterial Disease: Johne's Disease (MAP, Paratuberculosis)

Infection with *Mycobacterium avium paratuberculosis* is well-recognised in farmed deer (Reid 1994; Haigh and Hudson 1993; De Lisle et al. 2003) and has been confirmed in wild deer in the UK (Williams 2001; Lindén 2012). Infection is almost certainly widespread in deer parks; park deer infected suffering Johne's disease were reported as long ago as 1916 by M'Faddean and Sheather (1916), however, clinical disease is not commonly apparent. This is in contrast to the disease on deer farms, where sub-adult deer lose condition, develop diarrhoea and die.

MAP may form part of a winter die-off episode, and when park deer are becoming emaciated and develop loose faeces despite good stocking levels, good feeding and good trace element profiles, MAP should be high on the differential diagnosis list. PCR probing of faeces and post-mortem examination of affected deer will confirm the diagnosis. Prominently swollen but not purulent mesenteric lymph nodes with thickening of the lower jejunum, ileum and ileocaecal junction are typical post-mortem findings (Lindén 2012).

8.7 Viral Diseases

The most significant viral disease of park deer is ovine herpesvirus 2, the agent of malignant catarrhal fever (MVF) in cattle (Reid 2012). Affected deer show a variety of clinical signs, including depression, pyrexia, diarrhoea, nasal and occular discharges. Pére David deer are especially susceptible to MCF, which has a very high fatality rate in this species; fallow deer are most tolerant of MCF and red deer somewhere between the two. There is no treatment.

Sheep, the reservoir host of the virus, are very commonly grazed with deer in deer parks, and yet outbreaks of MCF are uncommon. Where sheep are used to co-graze deer parks, there should be no lambing in the deer park, because of the increased shedding of ovine herpesvirus 2 in the periparturient period.

8.8 Foreign Bodies

Deer in parks with high visitor numbers, cafés and leisure offerings deer will frequently ingest litter and rubbish. The reason for this is not clear; fallow deer are particularly guilty of such scavenging. Plastic bags, rope, items of clothing, banners and flags are all regularly encountered in the rumen of the deer in public parks, sometimes in very large quantities (Fig. 29). Individual deer seem to acquire a habit for such behaviour, which is more prevalent in males than females.

The offending material, usually knotted and tangled into a tight mass, is often discovered during dressing out of animals routinely culled from the park. Most are in normal condition, but occasionally a deer will become emaciated, cease to thrive and will be culled on welfare grounds only for the deer manager to find the rumen contains a large bezoar of foreign material.

Litter bins in public deer parks should be deer proof and visitors encouraged to be vigilant in their litter disposal. Plastic banners on low fences, ropes and tapes to screen off areas from the visitors, plastic wrapping from silage and haylage bales should all be avoided if possible.

Fig. 29 Ingestion of rubbish and foreign bodies is a problem in park deer. John Bartram, retired deer keeper, Richmond Park, London shows the rumenal contents from two fallow bucks from the park

References

Adam C (1994) Husbandry: feeding. In: Alexander TL, Buxton D (eds) Management and diseases of deer. Veterinary Deer Society, London, pp 170–175

Andersen R, Duncan P, Linnell JDP (1998) The European roe deer: the biology of success. Scandinavian University Press, Oslo

Appleby MC, Weary DM, Sandoe P (2014) Dilemmas in animal welfare. CABI Publishing, Oxfordshire

Audige L, Wilson PR, Morris RS (1998) A body condition scoring system and its use for farmed red deer hinds. N Z J Agric Res 41:545–553

Boyce W (1989) Winter feeding. In: The Jackson Elk Herd. Intensive wildlife management in North America. Cambridge University Press, Cambridge, pp 129–163

Busch F, Bannerman F, Liggett S, Griffin F, Clarke J, Lyashchenko KP, Rhodes S (2017) Control of bovine tuberculosis in a farmed red deer herd in England. Vet Rec. https://doi.org/10.1136/vr.103930

Carrigan MJ, Dawkins HJS, Cockram FA, Hansen AT (1991) *Pasteurella multocida* septicaemia in fallow deer (*Dama dama*) Australian. Vet J 68:201–203

Clutton-Brock TH, Guinness FE, Albon SD (1982) Red deer. Chicago University Press, Behaviour and ecology of two sexes

Cunningham JG, Klein BG (2007) Textbook of veterinary physiology. Saunders Elsevier, St Louis

De Lisle DW, Yates DF, Montgomery RH (2003) The emergence of mycobacterium paratuberculosis in farmed red deer in New Zealand – a review of 619 cases. N Z Vet J 51(2): 58–62. https://doi.org/10.1080/00480169.2003.36341

Delahay RJ, Smith GC, Barlow AM, Walker N, Harris A, Clifton-Hadley RS, Cheeseman CL (2007) Bovine tuberculosis infection in wild mammals in the south-west region of England: a survey of prevalence and a semiquantitative assessment of relative risk to cattle. Vet J 173:287–301

Duffield DJ, Young DA (1985) Survival of *Mycobacterium bovis* in defined environmental conditions. Vet Microbiol 10(2):193–197

Ekesbo I (2011) Farm animal behaviour: characteristics for assessment of health and welfare. CABI Publishing, Oxfordshire

Eriksen L, Aalbaek B, Liefssen PS, Basse B, Christiansen T, Eriksen E, Rimler RB (1999) Haemorrhagic septicaemia in fallow deer (*Dama dama*) caused by *Pasteurella multocida multocida*. J Zoo Wildl Med 30:285–292

Farjon A (2017) Ancient oaks in the English landscape. Kew Publishing, Royal Botanic Gardens, London

FAWC (2013) Opinion on the welfare of farmed and park deer. Report by the Farm Animal Welfare Committee. DEFRA, London

Fletcher J (1994) Handling systems. In: Alexander TL, Buxton D (eds) Management and diseases of deer. Veterinary Deer Society, London, pp 30–35

Fletcher J (2011) Gardens of earthly delight. The history of deer parks. Oxbow Books, Oxford

Gavier-Widén D, Chambers M, Gortazar C, Delahay R, Cromie R, Linden A (2012) Mycobacteria infections. In: Gavier-Widén D, Duff JP, Meredith A (eds) Infectious diseases in wild mammals and birds in Europe. Wiley Blackwell, Oxford, pp 265–292

Ghodbane R, Mba-Medei F, Lepide H, Nappez C, Drancourt N (2014) Long-term survival of tuberculosis complex mycobacteria in soil. Microbiology 160:496–501

Grace ND, Wilson PR (2002) Trace elements metabolism, dietary requirements, diagnosis and prevention of deficiencies in deer. N Z Vet J 50:252–259

Green P (2017) Monitoring the health and welfare of free-living deer in deer parks. In Pract 39(1): 34–40

Greswell WHP (1905) Forests and deer parks of the county of Somerset. Barnicott and Pearce Athenaeum Press, Taunton

Haigh JC, Hudson RJ (1993) Farming wapiti and red deer. Mosby, St Louis

Irvine RJ, Corbishley H, Pilkington JG, Albon SD (2006) Low level parasitic worm burdens may reduce body condition in free ranging red deer (*Cervus elaphus*). Parasitology 133:465–475

Jones DJ (1994) Trace element deficiencies. In: Alexander TL, Buxton D (eds) Management and diseases of deer. Veterinary Deer Society, London, pp 189–191

Kreeger TJ, Arnemo JM (2012) Capture myopathy. In: Handbook of wildlife chemical immobilization. Wyoming Game and Fish Department, pp 135–137

Lindén A (2012) Paratuberculosis or Johne's disease. In: Gavier-Widén D, Duff JP, Meredith A (eds) Infectious diseases in wild mammals and birds in Europe. Wiley Blackwell, Oxford, pp 281–288

Lloyd S (1998) Toxocarosis. In: Palmer SR, Soulsby EJL, Simpson DIH (eds) Zoonoses. Oxford University Press, pp 841–854

M'Faddean J, Sheather AL (1916) Johne's disease; the experimental transmission of the disease to cattle, sheep and goats. J Comp Pathol 29:62–94

Martín-Hernando MP, Torres MJ, Aznar J, Negro JJ, Gandía A, Gortázar C (2010) Distribution of lesions in red deer and fallow deer naturally infected with *Mycobacterium bovis*. J Comp Pathol 142:43–50

Matthews (2007) Deer handling and transport. In: Grandin T (ed) Livestock Handling and Transport. CAB International, Oxon, pp 271–294

Muller RC, Flesch JS (2001) Nutritional requirements for pregnant and lactating red and fallow deer. A report for the Rural Industries Research and Development Corporation. RIRDC report 01/0095. University of Sydney, Sydney

Munro R (1994) Capture myopathy. In: Alexander TL, Buxton D (eds) Management and diseases of deer. Veterinary Deer Society, London, pp 165–167

Naylor LM, Wisdom MJ, Anthony RG (2009) Behavioural responses of North American elk to recreational activity. J Wildl Manag 73(3):328–338

Ohl F, van der Staay FJ (2012) Animal welfare: at the interface between science and society. Vet J 192(1):13–19

Olias P, Weiss ATA, Gruber AD, Klopfleisch R (2011) Iron storage disease in red deer (*Cervus elaphus*) is not associated with mutations in the HFE gene. J Comp Pathol 145:207–213

Palmer MV, Whipple DL (2006) Survival of *Mycobacteroum bovis* on feedstuffs commonly used as supplemental feed for white-tailed deer (*Ococoileus virginianus*). J Wildl Dis 42:853–858

Palmer MV, Water WR, Whipple DL (2004) Shared feed as a means of deer-to-deer transmission of *Mycobacterium bovis*. J Wildl Dis 40:87–91

Putman RJ (2008) A review of available data on natural mortality of red and roe deer populations. Report for the Deer Commission for Scotland, Inverness

Putman RJ, Langbein J (1992) Effects of stocking density and feeding practice on body weights, reproduction and mortality in park deer. In: Bullock DJ, Goldspink CR (eds) Management, welfare and conservation of park deer. Proceedings of the second deer park symposium. Leicester 1992. UFAW, Potters Bar, pp 55–63

Putman RJ, Langbein J (2003) Factors influencing levels of mortality in park deer. In: Putman RJ (ed) The deer managers companion. Swan Hill Press, Shrewsbury, pp 117–122

Reid HW (1994) Mycobacterial infections. In: Alexander TL, Buxton D (eds) Management and diseases of deer. Veterinary Deer Society, London, pp 93–106

Reid HW (2012) Malignant catarrhal fever. In: Gavier-Widén D, Duff JP, Meredith A (eds) Infectious diseases in wild mammals and birds in Europe. Wiley Blackwell, Oxford, pp 10–13

Ross HR (1994) Winter death syndrome. In: Alexander TL, Buxton D (eds) Management and diseases of deer. Veterinary Deer Society, pp 130–132

Sandoe P, Jenson KK (2012) The idea of animal welfare – developments and tensions. In: Wathes CM, Corr SA, May SA, McCulloch SP, Whiting MC (eds) Proceedings of the first international conference on veterinary and animal ethics. Wiley, pp 19–31

Santos N, Almeida V, Gortázar C, Correia-Neves M (2015) Patterns of *Mycobacterium tuberculosis complex* excretion and characterization of super-shedders in naturally infected wild boar and red deer. Vet Res 46:129–140

Semiadi G, Holmes CW, Barry TN, Muir PD (1996) Effects of cold conditions on heat production by young sambar (Cervus unicolor) and red deer (Cervus elaphus). J Agric Sci 126:221–226

Shirley EP (1867) Some account of English deer parks. Reprinted 2007. Grimsay Press, Glasgow

Simpson AM, Webster AJF, Simpson CA (1978) Energy and nitrogen metabolism of red deer (*Cervus elaphus*) in cold environments; a comparison with cattle and sheep. Comp Biochem Physiol A 60:251–256

Strube C, Heuer L, Janecek E (2013) Toxocar spp. infections in paratenic hosts. Vet Parasitol 193:375–389

Turbill C, Ruf T, Mang T, Arnold W (2010) Regulation of temperature and heart rate in red deer; effects of season and food intake. J Exp Biol 214:963–970

Vengust G, Vengust A (2004) Some minerals as well as trace and toxic elements in livers of fallow deer (*Dama dama*) in Slovenia. Eur J Wildl Res 50:59–61

Ward AI, Etherington TR, Smith GC (2008) Exposure of cattle to Mycobacterium bovis excreted by deer in southwest England; a quantitative risk assessment. Report SE3036 to TB programme DEFRA London

Whitaker J (1892) A descriptive list of the deer-parks and paddocks of England. Ballantyne Hanson, London

Whitehead GK (1950) Deer and their management in the deer parks of Great Britain and Ireland. Country Life London

Williams E (2001) Paratuberculosis. In: Williams E, Barker IK (eds) Infectious diseases of wild mammals. Blackwell, Oxford, pp 361–366

Young TP (1994) Natural die-offs of large mammals: implications for conservation. Conserv Biol 8:410–418

Alternatives to Culling in Deer Parks

Peter Green

Abstract

Deer in parks provide an attractive and popular spectacle to visitors, who may be upset to learn that numbers are usually controlled by shooting. Proposals to limit populations by alternative, non-lethal methods are therefore popular and are increasingly recommended by those who are opposed to killing healthy deer. Immunocontraceptives stimulate the production of endogenous antibodies to specific components of the natural process of fertility. These have been proven to be effective in a wide variety of animals, including several deer species. Unlike contraceptives based on reproductive steroids, immunocontraceptives pose no risk to the environment, scavengers or predators. However, current immunocontraceptives must be administered by hypodermic injection, which is not practicable for wild deer in deer parks. They also have adverse effects upon male deer. In small parks vasectomy of males may be possible, and females can be surgically sterilised. Contraception, vasectomy and sterilisation alter breeding behaviour and may disrupt social organisation, giving rise to compromise of welfare and to a cohort of senile animals that eventually need to be removed, killed and replaced. There is currently no realistic alternative to culling in most deer parks.

Keywords

Population control · Culling · Lethal control · Shooting · Contraception · Vasectomy · Sterilisation

P. Green (✉)
Peter Green Veterinary Consultancy, Barnstaple, UK
e-mail: mail@peter-green.org

© The Author(s), under exclusive license to Springer Nature Switzerland AG 2022
J. Fletcher (ed.), *The Management of Enclosed and Domesticated Deer*,
https://doi.org/10.1007/978-3-031-05386-3_7

There is regular opposition to the culling of deer in deer parks, especially in parks open to the public where the deer are maintained as an amenity spectacle, not primarily for venison production. Well-meaning and concerned users of the park and tax payers who contribute to the maintenance of public parks are increasingly suggesting that there are workable alternatives to culling. Many reliable opinion polls have shown that if effective, safe and species-specific alternatives could be implemented, both the general public and the wildlife biological and academic community would find such alternatives preferable to population management by culling. (Barr et al. 2002). In some circumstances and with some species, contraception is already employed to limit population growth of free-living large mammals, notably elephants, lions, bison feral horses white-tailed deer and North American elk (Kirkpatrick et al. 2009). Understandable public sentiment and the impracticality of the use of firearms in urban parks as a means of population control makes contraception a very attractive concept (Baker et al. 2002; Grandy and Rutberg 2002; Naugle et al. 2002; Kirkpatrick 2002; Kirkpatrick et al. 2009).

Since the 1970s, research has been directed at the development of contraceptive agents that safely limit the fertility of targeted wild animals. Research is also necessary in systems to deliver these agents to sufficient numbers of the target species to achieve population control (Harder and Peterle 1974; Plotka and Seal 1989; Tyler 1967; Kirkpatrick et al. 2009). Most recent research and effort has been concentrated on immunocontraception, which appears to offer efficiency, specificity and safety to the target species, without risk to other species or to the environment (Grandy and Rutberg 2002; Muller et al. 1997, Naz and Saver 2016).

The mammalian female sex hormones oestrogen and progesterone and their derivatives are the basis of long-established human female contraceptives. Such compounds have been used experimentally to control fertility in deer (Greer et al. 1968; White et al. 1994; Harder and Peterle 1974; Warren et al. 1997). Treated females do not exhibit normal breeding behaviour when steroidal hormones are used; they do not become fertile in the rut. Steroid hormones have significant undesirable effects on male animals, especially male deer. Antler growth is disrupted so that developing antlers fail to harden or established antlers fail to shed. Behaviour and secondary sexual characteristics are altered (Price et al. 2005).

Steroidal contraceptives present significant risks of pollution and persistence in both the environment and the food chain. They are not easily digested or metabolised and continue to have effects in both predators and scavengers. They are excreted in the urine and contaminate the habitat. (Harder and Peterle 1974; Warren et al. 1997; White et al. 1994). For these and other reasons, little current attention is directed at steroids as contraceptive agents in free-living populations such as deer parks, although they are successfully used in zoos and in limited situations where wildlife species are enclosed on reserves and not subject to predation (Nave et al. 2002; Hyndes et al. 2007). Animals subject to steroidal contraceptive treatment are either darted at close range or are captured, restrained and injected.

There are two types of vaccines that have been developed to reduce fertility in female wild animals by stimulating the production of antibodies that target key parts of the process of fertility.

The zona pellucida (ZP) is one of the layers surrounding the mammalian ovum after it is released from the ovary; it consists of a matrix of glycoprotein, with which the sperm must bind in the first stage of fertilisation (Dunbar et al. 2002). By vaccinating females with a form of ZP as the antigen, circulating antibodies are produced that attack the natural ZP proteins of the ova of treated female. Porcine ZP (PZP) has been found to be most efficient in many species and has been widely used because of availability. PZP may be natural (extracted from pig ova), genetically engineered or synthesised (Dunbar et al. 2002).

The effect of PZP vaccination is to interfere with sperm binding to the ovum after mating. Vaccinated females will therefore fail to conceive and will continue to cycle and to exhibit reproductive and mating behaviour (Fraker et al. 2002; Shideler et al. 2002; Muller et al. 1997; Killian et al. 2007). In some species, there are direct effects upon the tissues of the ovary and reproductive tract that may cause long-term infertility (Dunbar et al. 2002; Muller et al. 1997). Initial work with PZP vaccines in deer used two injections with a variable interval between the two to achieve contraception (Curtis et al. 2002; Naugle et al. 2002; Shideler et al. 2002). Single-injection commercial PZP products were then produced, and they have been used successfully in some deer species (Fraker et al. 2002; Fraker and Bechert 2007; Rutberg and Naugle 2007; Turner et al. 2007; Kirkpatrick et al. 2009). No PZP vaccine is currently available commercially in the UK.

Gonadotrophin-Releasing Hormone (GnRH) is a key regulator of the pituitary gland. By vaccinating animals with a GnRH analogue, antibodies are produced to the endogenous GnRH of the animal and the pituitary control of reproduction is disrupted (Baker et al. 2004; Curtis et al. 2002; Muller et al. 1997). Because GnRH function is essential for normal reproductive function in both sexes, blocking the effect of GnRH will affect both males and females (Botha et al. 2007; Killian et al. 2006). In females, the effect is to suppress ovarian cycling by blocking the surges of luteinising hormone (LH) and follicular stimulating hormone (FSH) from the pituitary (Baker et al. 2002); vaccinated females should therefore not come into oestrus.

Various GnRH analogue vaccines have been produced, including long-acting single-injection preparations, and these have been shown to be effective in a variety of mammals, including some deer species (Baker et al. 2002, 2004, 2005; Curtis et al. 2002; Naugle et al. 2002; Killian et al. 2006). Commercial long-acting GnRH vaccines are in production (Asa and Boutelle 2007; Fagerstone et al. 2007a, b; Miller et al. 2007a, b; Naz and Saver 2016), although none is available in the UK without a specific Home Office licence for research purposes.

Fallow deer have been successfully contracepted for more than 3 years by a single injection of long-acting PZP vaccine, with over 90% efficiency (Fraker et al. 2002). The fertility of white-tailed deer was reduced from 83% in untreated does to 11% after one single injection of an alternative PZP product, which only increased to 25% after 2 years, with no further injections (Turner et al. 2007). Long-acting GnRH vaccine has been shown to provide contraception for prolonged periods given as a single dose to white-tailed deer (Fagerstone et al. 2007a, b), and a similar product has achieved almost complete contraception in Rocky Mountain elk for one full year after a single dose given either by injection or by dart (Baker et al. 2004, 2005;

Conner et al. 2007). There is some evidence that GnRH vaccines are marginally more effective in young female deer and that PZP vaccines are better in older animals (Curtis et al. 2002), which accords with findings in domestic cattle and sheep (Brown et al. 1994; Evans and Rawlings 1994).

Effective contraception by injection or by darting has significantly reduced or even eliminated population growth in wild free-ranging African elephants, grey seals white-tailed deer, Rocky Mountain elk, feral horses, lions, cheetahs, koalas and kangaroos (Fraker and Bechert 2007; Turner et al. 2007; Herbert and Vogelnest 2007; Delsink et al. 2007; Bertschinger et al. 2007; Killian et al. 2007; Conner et al. 2007; Wilson et al. 2007). Captive populations of many ungulate species have been effectively controlled by immunocontraception injection of females. These animals include black-tail & mule deer, sika deer, Reeves' muntjac, wild boar, several antelope species, Javan banteng, thar and other ungulates, although the dose of antigen required, the interval of booster doses and the length of time to return to fertility has been very variable (Lamberski et al. 2007; Asa and Boutelle 2007; Penfold et al. 2007; Fraker and Bechert 2007; Frank et al. 2005).

There is no doubt that immunocontraceptives, administered by injection, could be effective in reducing fertility in some deer species, including fallow and red deer, which are the predominant species maintained in deer parks.

Such contraception is not without side effects. Female deer contracepted with PZP vaccines continue to come into oestrus and show normal reproductive behaviour. The effect of this is to prolong the rut in white-tailed deer, fallow deer and Tule elk, although researchers have noted that mature bucks and bulls cease to rut and contraceptive females are attended by young males in the later part of the prolonged rutting period (Curtis et al. 2002; Fraker et al. 2002; Shideler et al. 2002). In some studies, female deer injected with PZP vaccines were in poorer condition after the rut than the pregnant controls because they had been persistently harried and served by males without conceiving. However, contracepted females did not subsequently suffer the loss of condition associated with carrying and suckling a fawn (McShea et al. 1997; Fraker et al. 2002; Baker et al. 2004).

Female deer injected with GnRH vaccine should, in theory, fail to come into oestrus, but this prediction has been disproved: vaccinated female deer have continued to show normal breeding behaviour during the rut, despite blood samples proving that ovarian hormones and LH remain very low. The apparent oestrus behaviour is probably the result of very low background levels of oestrogen in the absence of progesterone (Baker et al. 2002). Treated females may actually show increased sexual activity as the rut progresses and as unvaccinated females in the population become pregnant (Baker et al. 2004). With both PZP and GnRH vaccines given in the short-acting form, treated females are effectively contracepted for 1 year and then may start rutting earlier the following year, although their fertility will be reduced. This may simply be a factor of them being in better condition because they have not been suckling a fawn (McShea et al. 1997; Curtis et al. 2002).

In feral horse herds, where immunocontraception has been used for many years, contracepted mares live significantly longer than mares that breed regularly, generating a completely new cohort of aged but healthy females. The same may be expected in deer (Kirkpatrick 2002; Turner and Kirkpatrick 2002; Cowan and Massei 2007). There is some concern that in long-lived species with strong social bonding, the presence of some neonates, juveniles and sub-adults is essential for normal social structure. In these species, high levels of contraception in a population may be undesirable (Delsink et al. 2007).

Breeding success or being pregnant may affect social status amongst females in a social group and contracepted females may therefore be condemned to perpetual low ranking. This has not been specifically studied in deer, but is seen in other species (Davis and Pech 2002).

Other behavioural effects have been reported in immunocontracepted feral mares and may be expected in park deer, since all are strongly herded, polygynous species. These include a degree of lethargy and reduced bonding to natal groups, with a consequent breakdown of social structure (Nunez et al. 2009). Such effects have caused ethologists to question whether immunocontraception of wild animals is humane (Hampton et al. 2015).

It is possible that immunocontraception may inadvertently select inferior stock over time. If the healthiest and most immunologically active females respond best to immunocontraception, it is predictable that weak, unhealthy females that fail to respond to the immunocontraception will be selected as the breeding stock (Nettles 1997; Muller et al. 1997). The most immunocompetent animals also have the best capacity to mount resistance to infection and disease; if these are prevented from breeding, whilst less immunocompetent animals remain fertile, this form of contraception may select against the fitter animals and eliminate their genes from the population over time. There is currently no evidence of this, but there has been no study of sufficient length or breath to answer the question (Hampton et al. 2015).

Local abscessation at the site of the injection and ovarian pathology are reported at low levels in deer; when it occurs, it is probably a reaction to the adjuvant, not the antigen of the vaccine (Dunbar et al. 2002; Naugle et al. 2002; Powers et al. 2007). When the vaccine is delivered by darting rather than by hand-held injection, local abscessation rates are far higher (Naz and Saver 2016).

GnRH vaccines delay puberty if given to pre-pubertal male animals, including male deer although the vaccinated male deer eventually go through puberty and develop normal antlers (Miller et al. 2007a, b). A commercial GnRH vaccine licenced for male pigs to delay puberty and reduce boar-taint of pork has been used successfully to control breeding in free-ranging mares in South Africa (Botha et al. 2007) and has been used on Dartmoor ponies. In zoo animals, GnRH vaccine has been successful in reducing aggressive behaviour in males, by means of the suppression of testicular activity. In some bovid species, aggression is reduced, but bodily condition, basal testosterone, testis size and sperm count are all increased (Asa and Boutelle 2007; Penfold et al. 2007).

Given to adult male deer, GnRH vaccines interfere with seasonal hormonal cycles, affecting rutting, antler growth and antler shedding; there is also some evidence that reduced testosterone caused by GnRH vaccines increases susceptibility to pulmonary infection in male deer (Killian et al. 2006). In populations subject to immunocontraception, male deer are in poorer condition after the rut compared with male deer in non-contracepted populations. This may be because the rut is prolonged, but may also be because contracepted females repeatedly cycle and are therefore repeatedly the focus of male-to-male competition (Hampton et al. 2015).

With both PZP and GnRH vaccines, the doses, responses and required booster intervals very considerably between deer species. Reeves' muntjac require very frequent boosting of PZP vaccines in comparison with other deer. Sambar deer failed to respond at all to PZP vaccination and were not contracepted (Lamberski et al. 2007; Asa and Boutelle 2007; Penfold et al. 2007; Frank et al. 2005).

PZP molecules are proteins, and GnRH molecules are smaller subunits of peptides. Both are combined with larger proteins and adjuvants to produce the active vaccine. There is no persistence of the active components in the environment and no risk of onward transmission through the food chain as the antigens are destroyed by normal digestion. Once they are administered to the target animal, immunocontraceptives present no risk to the environment or to non-target species. (Muller et al. 1997; Miller et al. 2007a, b). The adjuvant in the most widely used commercial GnRH vaccine (Gonacon©) is modified from a commercial Johnes disease mycobacterial vaccine (Fagerstone et al. 2007a, b), which may cross-react with tests for Johne's disease in ruminants and may therefore influence the licencing of the vaccine in some jurisdictions.

In long-lived species like deer, a reduction in fertility will not reduce the population until senile animals die or animals succumb to disease, road traffic accidents or culling (Davis and Pech 2002; Kirkpatrick and Tuner 2007). There have been few studies of long-term contraception in populations of deer. In a project where white-tailed deer in a suburban environment on an island were contracepted for 12 years, populations continued to increase by 11% per year for the first 5 years and then decreased by 16% annually as the effect of the contraception and management changes took effect. This required very intensive effort including annual vaccination by dart of as many females as possible and other management changes such as the prohibition of all artificial feeding (Rutberg and Naugle 2007). In another island situation with fallow deer, the females were rendered infertile by immunocontraception, the population was reduced by culling (Fraker et al. 2002). In both studies, the treated deer were either marked or individually identified to prevent repeat dosing of the same females. Deer are relatively poor candidates for population reduction by means of contraception because they are long-lived, with relatively low mortality. Widespread vaccination of female deer may halt population growth, but population reduction will take many years of continuing intensive effort (Kirkpatrick 2002).

On Assateague Island, USA zero population growth of horses was achieved in 2 years by means of contraceptive vaccines given annually to each mare, but it took a further 8 years to achieve a modest decline in population from 175 animals to 150 and a further 3 years to reduce the numbers to 135. Each mare was individually identified and immunised annually. The slow decline in population was caused by increasing condition in the mares, reduced mortality and significantly increased longevity (Kirkpatrick 1995; Kirkpatrick and Tuner 2007; Turner and Kirkpatrick

2002). Similar results have been reported with other feral horse projects (Turner et al. 2002).

The major obstacle to the use of immunocontraception in deer parks is the delivery of the vaccine to the deer. All current immunocontraception has to be administered to individual animals through a hypodermic needle, either from a hand-held syringe, a jab-stick or a dart. Almost all the published studies that demonstrate efficacy in deer report that the vaccines were administered by injection by hand to deer that had been captured or restrained (Baker et al. 2002, 2004, 2005; Curtis et al. 2002; Fraker et al. 2002; McShea et al. 1994, 1997; Powers et al. 2007; Shideler et al. 2002). Even in the very few studies of immunocontraception in free-ranging deer populations, the deer were either caught for their initial injections (Fraker et al. 2002; Shideler et al. 2002) or were darted at very close range. Darts were fired at distances of no more than 25 m (Naugle et al. 2002; Rutberg and Naugle 2007; Shideler et al. 2002). Slightly greater distances of 35 m by daylight or 45 m at night with a lamp were reported for deer confined within a pen (Curtis et al. 2002).

The deer in many deer parks can rarely be approached so closely. Darting wild deer is notoriously difficult, time consuming and inefficient (Nielsen 1999; Kreeger and Arnemo 2012). Fallow deer are especially cautious and flighty; experience has proven that only a handful of free-living fallow deer can be successfully darted from a population even with many hours of effort (Bergvall et al. 2015). Moreover, individual deer need to be identified or identifiable to prevent repeat dosing of the same individuals; this is impossible in deer park female deer without capturing them to mark them in some way.

Small numbers of individual deer can be darted or netted but catching deer in order to inject and mark them, is completely unrealistic at numbers sufficient to have any effect, even in a moderate-sized deer park (Hampton et al. 2019).

Injection by hand reduces the fertility rate more than injection by darting, probably because of the failure rate of darts to deliver the full dose: one-third of the darts administered by Baker et al. (2005) to sedated Rocky Mountain elk at close range failed to deliver fully. Large volume darts necessary to deliver depot combinations of long-acting contraceptives are less reliable than smaller, low volume darts (Thiele 1999 cited by Naugle et al. 2002, Shideler et al. 2002, Turner et al. 2002, Turner et al. 2007). Naz and Saver (2016) report that compared with hand-held injections, darts achieve only half the effective contraception and cause double the injection site abscesses.

In summary, immunocontraception in individual deer is effective, but the delivery of the immunocontraceptive to free-living park deer in sufficient numbers with accurate identification of individual animals is currently impossible.

A possible development in free-living wildlife contraception appears to be in the field of "ghost cell" technology, by which vaccines may be delivered orally by "hiding" the vaccine within the cell wall or envelope of dead bacterial cells or within empty pollen spore capsules (Miller et al. 2007a, b; Duckworth et al. 2007; Massei 2017). This will enable the peptide antigens to be absorbed without digestion by the target animal. The problems of target animal specificity remain, and although workers are developing promising species-specific systems, for example for wild boar (Massei et al. 2007) or for grey squirrels (Mayle 2007), the problem remains that in the case of deer, it is inadvisable to vaccinate males with GnRH vaccines.

Another alternative to culling deer in deer parks is to maintain female-only herds or to have only a small number of males that are vasectomised. Castration of male deer before puberty was widely practised in former times when these castrates, or 'haviers' were sometimes even kept in stalls and fattened for venison (Whitehead 1950). Such animals do not develop antlers at all. Males castrated after puberty develop disorganised soft antlers that are perpetually in velvet and prone to trauma and infection. In contrast, vasectomised males rut normally, grow and cast their antlers but cannot sire offspring. In small urban and metropolitan parks vasectomised males can be used to maintain the appearance of a 'natural' herd of deer, but without the generation of calves and fawns that leads to the need to catch or cull deer to limit population growth. If such an arrangement is undertaken, the herd must be established after fawning or calving and before the rut to avoid pregnancy in the females, which might deliver fertile male offspring. Because they rut normally, vasectomised males will fight and may injure each other or members of their female harems.

Park herds with vasectomised males are present in several small urban exhibition parks in the UK. Park managers and the public must be made aware that the herd will eventually become senile and will need to be replenished. This will almost certainly involve culling of very old animals that may have become favourites with the public.

Population control by surgical sterilisation of females has been undertaken in some urban deer populations in the USA (Merrill et al. 2006). Costs are very high, even with volunteer veterinary student input. Individual female deer must be caught, anaesthetised, sterilised, marked and released. It is doubtful whether such work would be licenced in the UK. The caveats in respect of catching up park deer apply once more, and it is difficult to contemplate a deer park of any significant size proposing to undertake such a project.

References

Asa CS, Boutelle S (2007) The AZA wildlife contraception Centre collaborative trial of the GnRH agonist Suprelorin® in North America. In: Proc 6th int. Symp on fertility control in wildlife. Wildlife Society, York

Baker DL, Wild MA, Conner MM, Ravivarapu HB, Dunn RL, Nett TM (2002) Effects of GnRH agonist (leuprolide) on reproduction and behaviour in female wapiti (*Cervus elaphus nelson*). In: Kirkpatrick JF, Lasley BL, Allen WR, Doberska C (eds) Proc 5th int. symp. on fertility control in wildlife. Reproduction supplement 60. Society for Reproduction & Fertility, pp 155–167

Baker DL, Wild MA, Connor MM, Ravivarapu HB, Dunn RL, Nett TM (2004) Gonadotrophin-releasing hormone agonist: a new approach to reversible contraception in female deer. J Wildl Dis 40:713–724

Baker DL, Wild MA, Hussain MD, Dunn RL, Nett TM (2005) Evaluation of remotely delivered leuprolide acetate as a contraceptive agent in female elk. J Wildl Dis 41:758–767

Barr JJF, Lurz PWW, Shirley MDF, Rushton SP (2002) Evaluation of immunocontraception as publicly acceptable form of vertebrate pest control. The introduced grey squirrel in Britain as an example. Environ Manag 30(3):342–251

Bergvall UA, Kjellander P, Ahlqvist P, Johansson IO, Sko K, Annemo JM (2015) Chemical immobilisation of free-ranging fallow deer (Dama dama): effect of needle length on induction time. J Wildl Dis. https://doi.org/10.7589/2013-11-290

Bertschinger HJ, deBarros VGMA, Trigg TE, Human A (2007) The use of deslorelin implants for the long term contraception of free ranging and captive lionesses. In: Proc 6th int. symp on fertility control in wildlife. Wildlife Society, York

Botha AE, Schulman ML, Bertschinger HJ, Annadale HC, Hughes SB (2007) The use of a GnRH vaccine to suppress mare ovarian activity under field conditions in South Africa. In: Proc 6th int. symp on fertility control in wildlife. Wildlife Society, York

Brown BW, Mattner PE, Carroll PA, Holland EJ, Paull DR, Hoskinson RM, Rigby RDG (1994) Immunisation of sheep against GnRH early in life: effects on reproductive function and hormones in rams. J Reprod Fert 101:15–21

Conner MM, Baker DL, Wild MA, Powers JG, Hussain MD, Dunn RL, Nett TM (2007) Fertility control in free-ranging elk using the GnRH agonist leuprolide. In: Proc 6th int. symp on fertility control in wildlife. Wildlife Society, York

Cowan DP, Massei G (2007) Predicting population levels effects of fertility control. In: Proc 6th int. symp on fertility control in wildlife. Wildlife Society, York

Curtis PD, Pooler RL, Richmond ME, Miller LA, Mattfield GF, Quimby FW (2002) Comparitive effects of GnRH and porcine zona pellucida [PZP] immunocontraceptive vaccines for controlling reproduction in white-tailed deer [Odocoileus virginianus]. In: Kirkpatrick JF, Lasley BL, Allen WR, Doberska C (eds) Proc 5th int. symp. on fertility control in wildlife. Reproduction supplement 60. Society for Reproduction & Fertility, pp 131–141

Davis SA, Pech RP (2002) Dependence of population response to fertility control on the survival of sterile animals and their role in regulation. In: Kirkpatrick JF, Lasley BL, Allen WR, Doberska C (eds) Proc 5th int. symp. on fertility control in wildlife. Reproduction supplement 60. Society for Reproduction & Fertility, pp 89–103

Delsink AK, Kirkpatrick JF, van Atena JJ, Grobler D, Beltschinger H, Slotow R (2007) Lack of social and behavioural consequences of immunocontraception in African elephants. In: Proc 6th int. symp on fertility control in wildlife. Wildlife Society, York

Duckworth JA, Cui X, Lubitz P, Molinia FC, Lubitz W, Cowan PE (2007) Bait-delivered fertility control for mammalian pest species in New Zealand. In: Proc 6th int. symp on fertility control in wildlife. Wildlife Society, York

Dunbar BS, Kaul G, Prasad M, Skinner SM (2002) Molecular approaches for the evaluation of immune responses to zona pellucida (ZP) and development f second-generation ZP vaccines. In: Kirkpatrick JF, Lasley BL, Allen WR, Doberska C (eds) Proc. 5th int. symp. on fertility control in wildlife. Reproduction supplement 60. Society for Reproduction & Fertility, pp 9–18

Evans ACO, Rawlings NC (1994) Effects of a long-acting gonadotrophin-releasing hormone agonist [leuprolide] on ovarian development in prepubertal heifer calves. Can J Anim Sci 74: 649–656

Fagerstone KA, Miller LA, Eisemann JD, Gionfriddo JP (2007a) Development and registration of GonaCon® immunocontraceptive vaccine in the United States. In: Proc 6th int. symp on fertility control in wildlife. Wildlife Society, York

Fagerstone KA, Miller LA, Gionfriddo JP, Eismar GS, Killian GJ, deNicola AJ, Sullivan KJ (2007b) Development of GonaCon® immunological vaccine for use in cervids. In: Poster presentation at 6th int. symp on fertility control in wildlife. Wildlife Society, York

Fraker MA, Bechert U (2007) SpayVac® - a long lasting, single dose PZP contraceptive vaccine for practical wildlife population control. In: Proc 6th int. symp on fertility control in wildlife. Wildlife Society, York

Fraker MA, Brown RG, Gaunt GE, Kerr JA, Pohajdak W (2002) Long-lasting single dose immunocontraception of feral fallow deer in British Columbia. J Wildl Manag 66:1141–1147

Frank KM, Lyda RO, Kirkpatrick JF (2005) Immunocontraception of captive exotic species. Species differences in response to the porcine zona pellucida vaccine, timing of booster inoculations, and procedural failures. Zoo Biol 24:349–358

Grandy JW, Rutberg AT (2002) An animal welfare view of wildlife contraception. In: Kiekpatrick JF, Lasley BL, Allen WR, Doberska C (eds) Proc 5th int Symposium on fertility control in wildlife. Reproduction, Supplement, vol 60, pp 1–7

Greer KR, Hawkins WH, Catlion JE (1968) Experimental studies of controlled reproduction in elk (wapiti). J Wildl Manag 32:368–376

Hampton JO, Hyndman TH, Barnes A, Collins T (2015) Is wildlife fertility control always humane? Animals 5:1047–1071

Hampton JO, Finch NA, Watter K, Amos M, Pople T, Moriarty A, Jacotine A, Panther DG, McGhie C, Davies C, Mitchell J, Forsyth DM (2019) A review of methods used to capture and restrain introduced wild deer in Australia. Aust Mammal 41:1–11

Harder JD, Peterle TJ (1974) Effect of diethylstilbestrol on reproductive performance of white-tailed deer. J Wildl Manag 38:183–196

Herbert CA, Vogelnest L (2007) Catching kangaroos on the hop: development of a remotely delivered contraceptive for marsupials. In: Proc 6th int. symp on fertility control in wildlife. Wildlife Society, York

Hyndes EF, Handasyde KA, Shaw G, Renfree MB (2007) Gestagen implants for fertility control of koalas in south-eastern Australia: effects on fertility, health and behaviour. In: Proc 6th int. symp on fertility control in wildlife. Wildlife Society, York

Killian GJ, Eisemann J, Wagner D, Werrner J, Engeman R, Miller LA (2006) Safety and toxicity evaluation of Gonacon immunocontraceptive vaccine in white-tailed deer. In: Proc. 22nd vertebrate pest conference. University of California Davis, pp 82–87

Killian G, Thain D, Diehl NK, Rhyan J, Miller LA (2007) Four-year contraceptive rates of mares treated with PZP and GnRH vaccines and an IUD. In: Proc 6th int. symp on fertility control in wildlife. Wildlife Society, York

Kirkpatrick JF (1995) Management of wild horses by fertility control: the Assateague experience. USA National Park Service Scientific Monograph. NRSM-95-26

Kirkpatrick JF (2002) Wildlife contraception – where have we been and where are we going? In: Kirkpatrick JF, Lasley BL, Allen WR, Doberska C (eds) Proc. 5th int. symp. on fertility control in wildlife. Reproduction supplement 60. Society for Reproduction & Fertility, pp 203–209

Kirkpatrick JF, Tuner AB (2007) Achieving population goals in long-lived wildlife with contraception. In: Proc 6th int. symp on fertility control in wildlife. Wildlife Society, York

Kirkpatrick JF, Rowan A, Lamberski N, Wallace R, Frank K, Lyda R (2009) The practical side of immunocontraception: zona proteins and wildlife. J Reprod Immunol 83:151–157

Kreeger TJ, Arnemo JM (2012) Approach. In: Kreeger T (ed) Handbook of wildlife chemical Immobilisation, pp 94–97

Lamberski N, Frank K, Lyda R, Liu IKM, Fayer-Hosken RA, Rieches R, Kirkpatrick JF (2007) Immunocontraception and taxon-specific antibody differences in captive nondomestic animals. In: Proc 6th int. Symp on fertility control in wildlife. Wildlife Society, York

Massei G (2017) Oral fertility control for grey squirrels. The Squirrel Accord. National Wildlife Management Centre APHA, York

Massei G, Coats J, Quy R (2007) The BOS [Boar-Operated-System]; a novel method to deliver contraceptives to wild boar. In: Poster presentation at 6th int. symp on fertility control in wildlife. Wildlife Society, York

Mayle BA (2007) Developing oral bait systems for fertility control in Grey squirrels (*Sciurus carolinensis*). In: Proc 6th int. Symp on fertility control in wildlife. Wildlife Society, York

McShea WJ, Monfort S, Hakim S (1994) Behavioural changes in white-tailed deer as a result of immunocontraception. In: Proc. 17th annual southeast deer study. Abstr 76. Southeast Deer Study Group, pp 41–42

McShea WJ, Montfort SL, Hakim S, Kirkpatrick LI, Turner JW, Chassy L, Munson L (1997) The effect of immunocontraception on the behaviour and reproduction of white-tailed deer. J Wildl Manag 61:560–569

Merrill JA, Cooch EG, Curtis PD (2006) Managing an over-abundant deer population by sterilisation: effects of immigration, stochasticity and the capture process. J Wildl Manag 70: 268–277

Miller LA, Fagerstone KA, Killian GJ (2007a) New contraceptive tools in the development phase at the National Wildlife Research Centre, USDA. In: Proc 6th int. symp on fertility control in wildlife. Wildlife Society, York

Miller LA, Fagerstone KA, Rhyan JC, Killian GJ (2007b) The effect of GnRH immunocontraception of male and female white-tail deer fawns. In: Poster presentation at 6th int. symp on fertility control in wildlife. Wildlife Society, York

Muller LI, Warren RJ, Evans DL (1997) The theory and practice of immunocontraception in wild mammals. Wildl Soc Bull 25:504–514

Naugle RE, Rutberg AT, Underwood HB, Turner JW, Liu KM (2002) Field testing of immunocontraception on white-tailed deer (Odocoileus virginianus) on Fire Island National Seashore, New York, USA. In: Kirkpatrick JF, Lasley BL, Allen WR, Doberska C (eds) Proc 5th int. symp. on fertility control in wildlife. Reproduction supplement 60. Society for Reproduction & Fertility, pp 143–153

Nave CD, Coulson G, Short RV, Poiani A, Shaw G, Renfree MB (2002) Long-term fertility control in the kangaroo and wallaby using levonorgestrel implants. In: Kirkpatrick JF, Lasley BL, Allen WR, Doberska C (eds) Proc. 5th int. symp. on fertility control in wildlife. Reproduction supplement 60. Society for Reproduction & Fertility, pp 71–80

Naz RK, Saver AE (2016) Immunocontraception for animals: current status and future perspective. Am J Reprod Immunol 75:426–439

Nettles VF (1997) Potential consequences and problems with wildlife contraceptives. Reprod Fertil Dev 9:137–143

Nielsen L (1999) Considerations and principles. In: Chemical Immobilisation of wild and exotic animals. Iowa State University Press, pp 17–29

Nunez CMV, Adelman JS, Mason CB, Daniel L, Rubenstein A (2009) Immunocontraception decreases group fidelity in a feral horse population. Appl Anim Behav Sci 117:74–83

Penfold LM, Jochle W, Trigg TE, Asa CS (2007) Inter and intra-species variability in response to GnRH agonists. In: Proc 6th int. symp on fertility control in wildlife. Wildlife Society, York

Plotka DE, Seal US (1989) Fertility control in female white-tailed deer. J Wildl Dis 25:643–646

Powers JG, Baker DL, Conner MM, Lothridge AH, Davis TL, Nett TM (2007) Effects of GnRH immunization on reproduction and behaviour in female Rocky Mountain Elk. In: Proc 6th int. symp on fertility control in wildlife. Wildlife Society, York

Price JS, Allen S, Faucheux C, Althnaian T, Mount JG (2005) Deer antlers: a zoological curiosity or the key to understanding organ regeneration in mammals. J Anat 203:607–168

Rutberg AT, Naugle RE (2007) Population effects of immunocontraception in White-tailed deer. In: Proc 6th int. symp on fertility control in wildlife. Wildlife Society, York

Shideler SE, Stoops MA, Gee NA, Howell JA, Lasley BL (2002) Use of porcine zona pellucida (PZP) vaccine as a contraceptive agent in free-ranging tule elk [Cervus elaphus nannodes]. In: Kirkpatrick JF, Lasley BL, Allen WR, Doberska C (eds) Proc. 5th int. symp. on fertility control in wildlife. Reproduction supplement 60. Society for Reproduction & Fertility, pp 169–176

Turner A, Kirkpatrick JF (2002) Effects of immunocontraception on population, longevity and body condition in wild mares [Equus callabus]. In: Kirkpatrick JF, Lasley BL, Allen WR, Doberska C (eds) Proc. 5th int. symp. on fertility control in wildlife. Reproduction supplement 60. Society for Reproduction & Fertility, pp 187–195

Turner JW, Liu KM, Flanagan DR, Bynum KS, Rutberg AT (2002) Porcine zona pellucida (PZP) immunocontraception of wild horses (Equus caballus) in Nevada: a ten year study. In: Kirkpatrick JF, Lasley BL, Allen WR, Doberska C (eds) Proc. 5th int. symp. on fertility control in wildlife. Reproduction supplement 60. Society for Reproduction & Fertility, pp 177–186

Turner JW, Rutberg AT, Naugle RE, Kaur MA, Flanagan DR, Bertschinger HJ, Liu KM (2007) Controlled release components of PZP contraceptive vaccine extend duration of infertility. In: Proc 6th int. symp on fertility control in wildlife. Wildlife Society, York

Tyler ET (1967) Antifertility agents. Annu Rev Pharmacol 7:381–398

Warren RJ, Fayrer-Hosken RA, White LM, Willis LP, Goodloe RB (1997) Research and application of contraceptives in white-tailed deer, feral horses and mountain goats. In: Kreeger TJ (ed) Contraception in wildlife. US Dept of AG Tech Bulletin 1853, Washington, DC, pp 133–145

White LM, Warren RJ, Fayrer-Hosken RA (1994) Levonorgestel implants as a contraceptive in captive white-tailed deer. J Wildl Dis 30:241–246

Wilson M, Coulson G, Fletcher T, Shaw G, Renfree MB (2007) Field applications of Suprelorin® implants in eastern grey kangaroos. In: Poster presentation at 6th int. symp on fertility control in wildlife. Wildlife Society, York

Whitehead GK (1950) The management of deer. In: Deer and their management in the deer parks of Great Britain and Ireland. Country Life London, pp 34–51

The Management of Enclosed Deer in France, Belgium and the Netherlands

Paul Audenaerde

Abstract

The extent and nature of deer enclosures and the different species enclosed in France, Belgium and the Netherlands are discussed, together with the historical and administrative background and the regulations in each country, including, for example identification and recording of the deer, the obligatory training for deer farmers, the breeders' associations, the slaughter and routes to market. The marketing objectives, including hunting and venison, the management of the deer and the principal diseases, are mentioned.

Keywords

Hunting parks · Hunting enclosures · Deer farms · Slaughter · 'Hertenkamp'

1   France

1.1   Background

Organisation of the country: France is a large country with approximately 65 million inhabitants, which is about the same as the population of the UK, but it has a surface area of 644,000 km^2 which is almost three times that of the whole of the UK.

Administratively the country is divided into 18 regions and 97 departments, which have a relatively strong devolved authority. Some departmental directives are very different from others.

The environmental and agricultural characteristics are very variable, ranging from a flat costal alluvial landscape (mostly along the Channel and Atlantic coasts and the

P. Audenaerde (✉)
Vétérinaire et Chef D'entreprise, Bruges, Belgium

© The Author(s), under exclusive license to Springer Nature Switzerland AG 2022   167
J. Fletcher (ed.), *The Management of Enclosed and Domesticated Deer*,
https://doi.org/10.1007/978-3-031-05386-3_8

Rhone estuary), to rolling hills all over central and Northeastern France and to young mountain chains (Alps and Pyrenees, both originating in the Tertiary period). Very many of these landscapes are home to the largest of the deer family in Europe: the red deer (categorised as an indigenous game species), which is present in 80 of the 97 French departments: there are 320 red deer populations in France, and half of these are descended from reintroductions which have taken place since 1950. The total number of wild red deer in France is estimated at 150,000 head (Klein 2019b).

Wild fallow deer, also categorised as game species, are less widely distributed. There are some 1000 heads, counted before the fawning season, in the Alsace, in Haut-Rhin and Bas-Rhin, and some 200 in the Seine-et-Marne. Elsewhere the so-called wild populations of fallow deer are much smaller and often originate from escapes, many after the winter storms around Christmas 1999 (Klein 2019a). The damage these 'wild' fallow populations do to agriculture and young forest plantations is in addition to the damage caused by the indigenous game species. Therefore, more and more departments are becoming concerned about their existence (Barboiron et al. 2020).

Sika are also categorised as game but a non-indigenous species. The wild populations originate from escaped animals. Sika can only legally be introduced into commercial hunting areas surrounded by deer proof fences. Because sika can crossbreed with red deer they risk defiling the genetics of the latter species, and therefore they are considered as an unwanted species in the wild in France.

France has 1,313,000 active hunters, which makes up 2.02% of the total population, this compares with the whole of Europe, which has seven million hunters in a total population of 747,400,000 people making only 0.94%. This demonstrates that hunting in France is of major importance to the rural population, mainly as a means of keeping game species at such a level that damage is manageable, but also as a food source and also as an important social activity that reinforces social bonds.

Besides hunting free-ranging game species, there is also the possibility of shooting game species in one of the hundreds of hunting enclosures or hunting parks. In 2012 France recorded 485 hunting enclosures and 587 hunting parks. The hunting parks differ from the hunting enclosures in not being surrounded by a fence conforming to the legal requirements, and the hunting enclosure can only be designated as such when it has a private house within it. Mostly kept for commercial purposes, these parks offer hunting facilities to those who have not the financial resources or the wish to maintain a hunting area the whole year round and only want to have a few game shooting opportunities per year—a bit like farmed trout fishing in a pond.

Some 100 enclosed hunting areas have red deer. About 300 have fallow deer.

France has a total of around 5 million licenced hunters in 2020. Numbers are declining because people in urban populations are often unfamiliar with the damage that game species may do to agriculture and a strong anti-hunting movement is very much alive. More specifically, many citizens are especially organised to fight against hunting in enclosures and have set up groups like the association for the protection of wild animals (Association pour la Protection des Animaux Sauvages—ASPAS).

ASPAS places films on the Internet which provide testimony as to the cruelty of hunting in fenced parks—see www.aspas-nature.org.

Summarising, it may be said that, in view of the importance of hunting and the legal possibilities, French deer farmers have potentially an extra market in the hunting and enclosed parks, for their surplus breeding stags and sometimes even for a part of their production.

Many *parcs animaliers et botaniques* have deer species in their animal collections which can hardly be compared with the English deer parks or Dutch *hertenkampen* but are rather zoos.

The park *'La Haute Touche'* in Obterre in the Indre department, is famous for providing accommodation to a major collection of deer species, several of them endangered. Artificial breeding techniques are studied in *'La Haute Touche'* to make them available for conservation purposes.

1.2   Deer Farming in France

Deer farming in France using the deer farming techniques which were developed in New Zealand and the UK, started in the early 1980s, mostly as a diversification of other red meat production or as an alternative land use for dairy farmers who had reached their milk quota limits (Theriez 1988).

The deer used on French deer farms are mostly red and fallow deer and the INRA (*Institut National de Recherche Agricole*) started a research project on red deer breeding in its research station in Clermont-Ferrand-Theix in 1988 (Theriez 1988).

Before being permitted to develop a deer farm, French deer farmers have to obtain a *certificate de capacité*. To obtain this certificate, they need to fulfil a period of practical training on one or more existing working deer farms and a written report on this work experience has to be provided. A relevant educational training or proven experience abroad is also acceptable as experience qualifying for this *certificate de capacité*.

After having obtained this certificate, a deer farmer can apply for permission to manage a deer farm if he has a particular farm in mind: the *permis d'exploitation*. A multidisciplinary jury including the hunting police, veterinary services, and others judge if the infrastructure of that farm suits the requirements: the height and quality of the fences, handling facility, raceway, etc. If all this is considered acceptable by the authorities, the new deer farmer must choose between a deer farm categorised as either:

- Category A: A farm that can produce deer, or other farmed game species, to be released into the wild, or, with the written consent of the hunting police, the OFB, 'elevages de gibier cat.A' can supply deer to hunting enclosures or hunting parks. They can also produce meat animals;
- Category B: A farm which can only produce deer for slaughter.

The *permis d'exploitation* is only valid if any person in possession of the *certificate de capacité* is permanently present.

French deer farmers work in a country where a relatively large proportion of the civilians is interested in hunting, and where enclosed hunting parks provide deer shooting facilities to people who appreciate a day in a hunting park and all the social events connected to this: an evening meal and enlarging their own network after sharing the common hunting experience.

Therefore some deer farmers aim at this market segment and produce animals to be sold to hunting parks or hunting enclosures.

Generally, only a limited number of deer produced on any deer farm can be sold to hunting parks or hunting enclosures. The rest will come onto the venison market which is very similar to that in other Western European countries: when the local market situation allows it, most French deer farmers aim at venison sales directly to the consumers (short chain). In rural areas, this may be more difficult as very often people living in the countryside have also access to venison derived from free-ranging deer. It is not an easy task to convince rural people that farm-produced venison is of a more standardised and better quality than that from free-ranging deer, especially as the latter is often much cheaper.

1.2.1 Identification and Registration of Deer

French deer farmers have some administrative obligations, mostly concerning traceability of their deer:

- They must be ear tagged to identify their farm of birth (Arr. 8.2.2010). This tag must stay in place for life, even when animals are released into the wild or, more commonly, into hunting parks or hunting enclosures.
- The farmer must keep a register up to date recording information about the following subjects (Arr. 8.6.2000):
  - A description of the characteristics of the breeding unit
  - Data on zoötechnical, sanitary and medical and veterinary control
  - Movements of deer on and off the farm
  - Data on maintenance of the deer including all handling procedures
  - Data on veterinary treatments
  - Slaughtering

1.2.2 Slaughtering

The French legislation is based on the EU-Directive 853/2004. Application of the legislation may differ slightly from one department to another, for example concerning the time when the ante-mortem inspection must be made before the animals are killed in the case of killing at the farm.

For French deer farmers, the lack of availability of an accredited local slaughterhouse for killing and processing may be a problem.

Trophy shooting, as well as the hunting of deer which do not carry a trophy of importance or even any at all, is common in the many enclosed parks. A few (one in the north west of France, and a second one in the south west) even aim at American customers who seek impressive trophies.

1.2.3   Association of Deer Breeders

France no longer has an association or union for deer breeders.. The former one (S.N. E.D. = Syndicat National des Éleveurs de Daims, which became later S.N.E. C. = Syndicat National des Éleveurs de Cervidés) stopped its activities through the lack of members.

1.2.4   The Marketplace

Most professional deer farmers sell their venison directly to the consumers and to restaurants, although a few have their own restaurant on the farm where they also serve venison from their own production. When selling directly to the consumer, the deer farmer often feels the competition from wild venison and from imports from New Zealand.

Redundant breeding stags often end up in hunting enclosures, and thus the deer farmer gets a premium on a deer that would otherwise be too old for sale as a prime meat animal.

1.2.5   Management on French Deer Farms

French deer farms vary in size from around 20 hinds to 300 hinds—the latter being a minority, and herds of some 40 hinds are more common. Except for the more fertile areas such as in Brittany, the numbers per hectare are often around 2–3 hinds plus calves, yearlings and breeding stags.

As many French deer farmers have their deer farm as a part of their agricultural enterprise, they have access to agricultural products or by-products to supplement their deer from late summer through the winter.

Some wean their calves for in-wintering, others leave them with the hinds.

Not all French deer farmers have handling systems—though the authorities advise them to have one (Theriez 1988). Consequently not all of them can load their animals without darting them with tranquillisers. Some de-antler their stags before the rut, others prefer the sight of the stag carrying its antlers and take the risk of injuries.

1.2.6   Current Disease Problems on French Deer Farms

Paratuberculosis (*Mycobacterium avium paratuberculosis*) is a problem on some French deer farms. The problem of paratuberculosis is particularly prevalent on deer farms which regularly take in deer from different sources. This is, by the way, not only a French problem though: it is a worldwide one. There have been cases of tuberculosis (*Mycobacterium bovis*) on others, and the presence of this disease in deer or other species (both farm and game species), within a department makes the authorities impose regulations for deer farms hoping to halt the spread of the disease.

The presence of Bluetongue (type 4 and type 8) in France makes the export of deer from deer farms complicated with stringent regulations requiring insecticides, vaccination, serology and PCR-testing.

Obviously French deer farmers also see those problems linked to intensive starch feeding: laminitis and bloody gastro-enteritis with hyperacute death.

As most French deer farms are managed extensively, gastrointestinal and respiratory tract parasites less commonly cause disease or important production losses. Nevertheless many French deer farmers feel comfortable when they treat their deer annually for internal parasites.

2 Belgium

2.1 Background

Organisation of the state: the surface of Belgium is almost 30,700 square km, with a population of 11 million people.

To better understand the specific Belgian situation concerning deer farming and hunting, we need to realise that Belgium is a federal country, consisting of three communities, corresponding to the three languages used in the country: Flemish (a variety of the Dutch language), French, and German; and three regions, which are linked to economic interests: Flanders, Wallonia and Brussels. They all have their own government (except for the Flemish community and the Flemish region, which have the same government), and each has a specific field of authority. Hunting and agriculture belong to the competence of the communities. For the whole of the country, a federal government retains part of some areas of activity, such as human health.

The French and German-speaking parts of Belgium are far less densely populated. It is there that wild fallow and red deer live. These species play an important role in the hunting activities of these communities, and they are intensively monitored by the authorities.

In Belgium, it is forbidden to release animals into the wild for any purpose, including for hunting purposes. Enclosed parks kept for hunting (sometimes known as *'canned hunting'* parks) are also forbidden.

2.2 Deer Farming in Belgium

The deer species kept in Belgium: In Belgium there is nothing that can be compared with the British deer parks or the similar Dutch *'hertenkampen'*—apart from a very few public deer parks near Antwerp and a few 'children's farms' which have a few tame fallow deer next to dwarf goats, sheep, donkeys, and similar domestic animals.

For many decades fallow deer and a few red deer have been kept by owners of smallholdings, mostly for amenity purposes. Often these small private deer parks were—and are—situated in the periphery of towns and villages where some people have relatively large gardens. They keep the deer partly to maintain some part of this available surface, and also because semi-rural people who do not need the income from the land because they have other sources of income, like the sight of these animals near their home. The deer are especially popular because they require hardly any assistance at calving time or at any time of the year. Obviously the surplus animals grown from birth are slaughtered and eaten by the owners, their family and friends. Some of these hobby deer owners also keep some sika or wapiti.

Almost all professional deer farmers in Belgium keep red deer, only a very few farm fallow deer. There are also four professional deer farms that use wapiti or F1 crosses as master stags on red hinds, as their progeny have a bigger carcass weight at slaughter than the pure reds (Sobry 2001).

In Belgium the first professional deer farm was conceived in 1987 and their first deer arrived in 1988 in the Ardennes—an old mountain chain with rolling hills where wild red and fallow deer are part of the wildlife and the traditional hunting of big game is incorporated into the local way of life. This first deer farm was managed on the principles developed in New Zealand and the UK, with the deer being mostly of Scottish origin.

In this part of Belgium, managing the big game species is taken very seriously by the authorities. They see it as their first priority that the management practices of deer farms, including the slaughtering of deer for venison and the transporting of venison or live animals, do not interfere with the official efforts to fight poaching. This is why, in the southeastern half of the country, lobbying for the legalisation of year round slaughtering of farmed deer has never been successful, and each movement of live deer continues to require prior permission from the Department of Nature and Forests (*Département de la Nature et Forêts*) (Regulation of the Government of Wallonia 25/4/1996). Please note that this regulation requiring permits for the transport of farmed deer dates from a decade after the creation of the very first deer farm: the authorities' regulations have been developed to take account of the evolution of this novel form of agriculture.

In 1989, a second deer farm was started, this time in the north west of Flanders, where no deer species live in the wild, apart from an occasional small herd of escaped fallow deer, and also some roe deer. Soon after the establishment of that second deer farm, several other entrepreneurs also started to keep deer—mostly reds. Most of these new entrants were landowners with occupations outside the agricultural sector. Whilst some of these continued to build a second source of income from their deer enterprises, others stopped when confronted with the administrative obligations connected to owning deer, and with the embarrassment and added complications of having to sell the progeny of their animals.

From the early years of Belgian deer farming, the legislation on how to commercialise venison has been created according to European directives (mostly Directive 853/2004).

2.2.1   Identification and Registration of Deer

There is also an obligation under Royal Decree 03.06.2007 concerning identification and registration of sheep, goats and deer, for every Belgian owner of deer to be registered as such, and they are then given a registration number.

All Belgian keepers of deer, whether hobbyists or professional farmers, fall under these identification and registration rules, as is the case for the farmed deer.

All enclosed Belgian deer, whether kept on hobby farms or professional farms, have to follow the national rules for identification and registration: they have to be identified by two ear tags, which must only be put in place whenever the animal leaves the premises of birth, whether it be dead or alive. Until then, the deer owner has to keep the ear tags ready and available. Each owner of deer has to be registered and to keep two registers: one for the deer coming in: bought in or born on the premises, and one to register the animals leaving the herd: deaths, sales and slaughtering.

A new EU-Directive 2019/2035 requires that countries will have to apply more stringent legislation on identification and registration of enclosed deer: they will all have to be identified by means of two ear tags or a subcutaneous transponder when they leave the premises of birth.

DGZ-ARSIA, the organisation which keeps records of all deer owners has reported that only 0.5% of all 1673 known Belgian deer owners have more than 100 animals (Van Mael 2019).

2.2.2 Slaughtering

Belgian deer farmers must follow the Belgian legislation about slaughtering deer which is based on the EU-Directive 853/2004. This requires that:

- Venison of a deer, that is to be consumed by more than one household must undergo both ante- and post-mortem inspection.
- As an exception (although it is actually used frequently), and only for reasons of human and/or animal safety, deer may be killed at the place of origin after live inspection. Thereafter, the carcass must be transported to an accredited slaughterhouse.

  The killing of deer at the farm may legally be done by free bullet. Deer farmers can base their demand for a firearms licence on the fact that they are legally allowed to kill deer on the farm (for animal welfare and/or safety reasons). Very few Belgian deer farmers have a killing crate where the use of a captive bolt pistol would be possible.

  Some Belgian deer farmers bring their animals inside a farm building in order to be able to kill them safely by free bullet.
- Deer can be legally slaughtered and butchered at the place of origin without any obligatory ante- or post-mortem meat inspection, provided that its venison will exclusively be consumed by the owner and his family members. However, the deer owner must declare such slaughter to the city administration beforehand.
- In the case of slaughterhouse killing, the necessary infrastructure to move deer into the slaughterhouse premises is lacking. Therefore, after arrival at the slaughterhouse the deer transport vehicle is considered as being part of the slaughterhouse, and the deer are killed in this vehicle by means of a captive bolt pistol.

A major problem for Belgian deer farmers is that very few slaughterhouses (which are private enterprises) have applied for permission to slaughter deer: their managers often do not see any economical benefits to justify applying for this licence because deer are a minor species. Also, the legal requirement that deer slaughter must be separated from the slaughter of other species either by time or space, does not encourage slaughterhouse managers to ask for permission to slaughter deer in their premises.

Some Belgian deer farmers transport their meat animals alive to a slaughterhouse because the ante-mortem inspection can be done at the slaughterhouse, which is less costly than the alternative system where deer are ante-mortem inspected and killed at home and the carcass is brought to the slaughterhouse for further processing to a clean carcass. However slaughtering deer on the farm creates a carcase that is perceived to have a better meat quality since the animals do not have to undergo the stresses of loading, transport and handling at the slaughterhouse.

Trophy shooting of stags in enclosed areas is forbidden in Belgium, as well as the release of deer into the wild.

2.2.3 Association of Deer Breeders

ABEC, the Belgian association of deer breeders, was founded in 1992. Since then, it has had a varying number of members, ranging from 30 to 65, from a total of about 1200 registered deer owners. However, ABEC reaches all professional and semi-professional deer farmers who are all members. These number about 10, and each has between 50 and 100 hinds. They farm their deer according to the deer farming methods developed in New Zealand and the UK, but most have another principal income outside agriculture and do not have to rely exclusively on the income they have from the deer.

ABEC has always insisted that deer be kept in animal friendly conditions: not too many deer per hectare, and with some permanent shelter, trees, hedges, or even buildings, available for harsh winter and hot summer conditions. In fact most (semi-) professional Belgian deer farmers in-winter their animals—at the least the calves, and some also house the adults, because outwintering can damage the pasture.

For 30 years, ABEC has been an active member of FEDFA, the European Confederation of deer Farmers Associations.

2.2.4 The Marketplace

Almost all professional Belgian deer farmers sell their deer through what is called the short chain: they sell directly to the consumers. The latter are thus in direct contact with the origin of the meat they buy, and the producers enjoy the feedback from their customers and receive the final price. Often this results in a good relationship between the producers and their clients. Obviously, this short chain selling is strongly facilitated when a city is nearby, which is almost always the case in relatively small and densely populated regions such as Flanders.

In the regime of direct sale ('short chain sale'), the farmer gets ± 14–15 € per kg carcass (autumn 2020) which must cover the slaughtering and butchering costs and VAT.

Several deer farmers create a second source of income from their animals by organising farm visits: people pay an income for the visit, and often they come back to buy venison.

No Belgian deer farmers are in the situation where old breeding stags or other trophy-carrying deer are shot on the farm and this would in any case not be legal.

2.2.5   Management on Belgian Deer Farms

Belgian deer farms vary in size from 5 to a maximum of 40 ha. and the stocking density depends on soil quality, ranging from 2–3 hinds per ha to 8 per ha.

Most Belgian professional deer farmers give additional fodder from August through the autumn and winter. Traditional farm products as used for other farm ruminants are provided: grass in any conserved form; corn (maize) silage; sugar beet pulp, small potatoes, cereals and their by-products, apple pulp when available, etc.

All Belgian professional and semi-professional deer farms are capable of bringing their animals into a farm building, and most are equipped with a deer crush: originally, these were all drop floor crushes, but these have almost all been replaced by hydraulic crushes or crushes invented by the farmer. Also they all have places for loading deer from that farm building into a vehicle. Having to dart deer before transport is only a necessity with hobby deer owners, as they are not sufficiently organised to be able to bring their animals into a pen.

Almost all professional deer farmers de-antler their master stags before the rut for safety reasons, although many miss the magnificent sight of a stag in hard antler in the rut.

Almost all Belgian professional deer farmers practice early weaning (before the rut, from the second half of August to the first half of September) and keep the weaned calves inside a barn until the next spring. Very few Belgian deer farmers wean after the rutting period.

2.2.6   Current Disease Problems on Belgian Deer Farms

Paratuberculosis (Godfroid et al. 2000) and the problems brought about by copper deficiency (enzootic ataxia in adults and arthritic joint disease in calves) are the most commonly seen diseases, as well as clostridial enterotoxaemia and hyperacute haemorrhagic enteritis followed by acute death caused by acute starch overfeeding (Haigh and Hudson 1993; Ciu 2014). Laminitis associated with chronic starch overfeeding is also common.

Gastrointestinal and respiratory tract parasites quite commonly cause disease on the more intensively managed Belgian deer farms, where the animal density on pastures is relatively high.

Tuberculosis has not been a problem on Belgian deer farms. In Belgium this disease would only be diagnosed when animals are tuberculin tested for international trade or in slaughterhouse findings. There are no reports of any cases of bovine tuberculosis in Belgian farmed deer.

A recent outbreak of confirmed Malignant Catarrhal Fever due to the presence of lambing Ouessant sheep in the same barn in winter time has decimated a deer herd by 77%.

The use of drugs on Belgian deer farms is minimal: hardly anything other than anthelmintics are administered to farmed deer in Belgium.

3 Netherlands

3.1 Background

The Netherlands has a surface of some 41.500 ha which, as the name of the country itself suggests, are flat and almost devoid of hills, except for some in the southeast, in the province of Limburg, where the river de Maas enters the country, as well as a very few others. The country is very much orientated towards agriculture but has a dense human population, which leaves hardly any space for wild deer. However, they do exist in a few natural reserves such as: the Veluwe (red and fallow deer), the Oostvaardersplassen (red deer), and the 'Amsterdamse Waterleidingduinen' (fallow deer). The Oostvaardersplassen was originally conceived as a nature reserve where human interference was to be avoided for a long time. Due to the lack of any large predators, the populations of introduced large herbivores (Konik horses, Heck cattle and red deer) became so large that hundreds died of starvation in each consecutive winter. The initial population of 40 red deer introduced from Scotland in 1992 had grown to become a herd of 3300 heads in 2011. The natural mortality of red deer in 2009 was 27%, and in later years, this became even worse. Finally, for animal welfare reasons, it was agreed (after several court cases) that a few hundred of the Konik horses should be transported to a nature reserve in Spain (where the same problems of overpopulation and starvation soon occurred), and also that 1745 red deer should be shot by free bullet in the winter 2018–2019. One thousand two hundred and thirty five of these carcases entered the human food chain; the rest were left in the nature reserve as food for scavengers.

A typically Dutch phenomenon are the municipal deer parks ('*hertenkamp*') distributed throughout the country and numbering around 200. A few are very old, about 10% of them date from the nineteenth century and the remainder were established in the twentieth century. These deer parks are almost always populated with fallow deer, although occasionally red deer and/or sika are kept.

The oldest Dutch deer park is the Koekamp in Den Hague, a 121 ha former bullfighting area where, since the seventeenth century, fallow deer, red deer and some sika have been enclosed. From the beginning, the Koekamp was a hunting area for the local Counts of Holland (City archives of Den Hague).

Another early *hertenkamp* is the one in Estate Elswout (85 ha), near Overveen in North Holland, which was established between 1812 and 1844. It contains only fallow deer (Tromp 1983).

Two *hertenkampen* dating from the first half of the nineteenth century are the Fogelsangh State, in Veenklooster, Friesland, and the Haarlemmerhout situated on the south side of Haarlem (Snijdelaar and van Klink 2002).

Today most of the Dutch *hertenkampen* are only intended as amenities for the citizens. The surplus animals are culled by the many local small dealers in deer who take the carcases to a game slaughterhouse or are sold to existing or new hobby deer owners as breeding stock.

3.2 Deer Farming in the Netherlands

The first deer farm to be managed according to New Zealand and UK deer farming principles was founded in Joppe in the early 1980s. For two decades, Dutch deer farming remained at a low level, until in the heart of the Netherlands, an entrepreneurial cattle farmer began to convince other farmers all over the country to breed deer for him, contracting to buy back the finished deer. As this enterprise had imported deer from England, it became involved in the government measures against foot and mouth disease when this disease struck England in 2001. To protect the export meat trade of the country all the deer on this farm and in all in-contact farms were killed and destroyed compulsorily for preventive reasons. This created a public outcry with significant emotional reactions from the public, especially since many deer in municipal parks, loved by many citizens as pets, were also compulsorily killed.

3.2.1 Identification and Registration

To date, identification of farmed deer in the Netherlands is only obligatory when the animals leave the premises of birth. This can be done using commercially obtainable, non-official ear tags (Gieskes 2016).

However, the recent EU-Directive 2019/2035 requires countries to apply more stringent regulations for the identification and registration of farmed deer. All farmed deer must be identified by means of two official ear tags or a subcutaneous transponder when they leave the premises of birth, or even once over 9 months if they remain on the premises where they were born. Thanks to lobbying by deer farmers' associations, the latter obligation has been taken out of the legislation. Like all EU countries, the Netherlands will have to apply this Directive in the near future.

3.2.2 Slaughtering

Most farmed deer are slaughtered in one specific game slaughterhouse in the province of Brabant, in the village of Mill, which is the main slaughterhouse authorised to slaughter farmed deer. Only about 10% are killed in the farm, after *ante-mortem* inspection, which is valid for 24 h, though often the authorities (the NVWA, the Dutch food and goods authority) extend this period to 3 days when a weekend is involved. Once killed on the deer farm, the carcases are brought to the slaughterhouse (Gieskes 2016). The remaining 90% are transported alive to the slaughterhouse, where ante- and post-mortem inspections take place (A. Timmerman, secretary of Vereniging van Nederlandse Hertenhouders, personal communication).

3.2.3 Market Situation

Dutch deer farmers have venison production as their almost exclusive goal—except for one in the province of Limburg, which also aims at the production of trophy stags for export to Germany and France. Occasionally the sale of breeding animals may also be a source of income.

The harvest of velvet antlers for the traditional Chinese medicine trade has always been rejected by Dutch deer farmers as a means of generating an alternative income, because it is felt that this would damage the perceived image of an animal friendly meat production system (Tom Krol, Friesian deer owner, personal communication).

Most professional or semi-professional Dutch deer farmers sell their venison through, what is called the short chain: the producers sell directly to the consumers via a farm shop or through the internet (Gieskes 2016). In this way, the producers can receive the final price and avoid competition with lower priced imports. This marketing system can create good relationships between the producers and their clients. It is more successful when the outlet is close to a city but this is almost always the case in a relatively small and densely populated region such as the Netherlands.

Several deer farmers have created a second source of income from their animals by organising farm visits: people pay for a visit, and sometimes they can buy food on the farm, which they can feed to the deer, and some will return to buy venison (Gieskes 2016).

Direct sales from farm to consumer are hardly influenced by New Zealand or Central European imports. The latter have however a major influence on the sales to restaurants (Gieskes 2016 and Rijksdienst voor Cultureel Erfgoed, Min. of Education, Culture and Science, *Historische hertenkampen*, 2016).

The price for boned out venison sold as direct retail was in 2019 about 15 €/kg meat.

3.2.4 Association of Deer Breeders

The Dutch deer breeders have organised themselves into an association, the Dutch Deer Breeders, the *Nederlandse Hertenhouders*. This association also attracts hobby deer farmers and numbers some 40 members.

3.2.5 Management on Dutch Deer Farms

The average area of Dutch deer farms is about 15 ha, and normally carries about 50–80 hinds (Gieskes 2016).

All Dutch professional and semi-professional deer farmers are able to bring their animals into a farm building, and most are equipped with a deer crush. Originally these were all drop floor crushes, but have now mostly been replaced by hydraulic crushes. Farmers also have facilities for loading deer from that farm building into a vehicle (van der Cruijssen, Personal communication n.d.).

Nearly all professional deer farmers de-antler their master stags before the rut for safety reasons, although many miss the magnificent sight of a stag in hard antler in the rut. And most professional Dutch deer farmers practice either early weaning (before the rut) or wean after the rut (late weaning) and then keep the weaned calves inside a barn until the next spring (van der Cruijssen, Personal communication n.d.).

Having to dart deer before transport is only a necessity with the thousand or so hobby deer owners, who are not sufficiently organised to be able to bring their animals into a building.

3.2.6 Current Disease Problems on Dutch Deer Farms

In the Netherlands, tuberculosis in farmed deer has not been an issue so far. Paratuberculosis and the problems brought about by copper deficiency (enzootic ataxia in adults and arthritic joint disease in calves) are the most commonly seen problems, as well as Clostridiosis followed by acute death following sudden exposure to high starch rations, and laminitis due to chronic overfeeding of high starch diets.

Gastrointestinal and respiratory tract parasitosis are quite common on the more intensively managed Dutch deer farms, where the animal density on pastures is relatively high.

An occasional outbreak of MCF in connection with sheep overwintering and lambing in the same barn has also been recorded.

The use of drugs on Dutch deer farms is minimal: hardly anything other than anthelmintics are administered to farmed deer in the Netherlands.

References

Barboiron C et al (June 2020) Faune sauvage n° 326, p 4

Ciu H (2014) J Wildl Dis 50(4):942–945

Gieskes JSH (2016) De koekamp en de toekomst

Godfroid J et al (2000) Vet Microbiol 77(3–4):283–290

Haigh J, Hudson R (1993) Farming wapiti and red deer. Mosby Yearbook Inc., p 222

Klein F (28 Mar 2019a) Office Francais De La Biodiversité, Le Daim

Klein F (28 Mar 2019b) Office Français De La Biodiversité, Le Cerf élaphe

Snijdelaar M, van Klink E (2002) Hertenhouderij, een verkenning. Expertisecentrum LNV

Sobry L (2001) *De hertenhouderij*, Fac. Landbouwkundige en toegepaste biolog. Wetenschappen, Univ. Ghent

Theriez M (1988) Elevage et alimentation du cerf (*Cervus elaphus*). INRA Prod Anim 1(5): 319–330

Tromp HMJ (1983) Elswout te Overveen. Zeist

van der Cruijssen E, Chairman of Vereniging van Nederlandse Hertenhouders (n.d.) Personal communication

Van Mael E (2019) DGZ Vlaanderen, report 2017

Management of Enclosed Deer in Austria, Switzerland, and Germany

Juergen Laban

Abstract

Historical backgrounds to the keeping of deer are provided for each country, as well as the numbers and areas of enclosures and the species of deer kept. Fallow deer predominate in all three countries. The regulations applying in each country are outlined together with the fencing employed, the feed used, and the means of handling and transporting. The objectives of the deer keeping and the major diseases are mentioned.

Keywords

History of deer keeping · Regulations · Fencing · Diseases · Venison · Numbers of deer enclosures

1   Austria

1.1   History

Wild animals, especially fallow deer, were kept by humans thousands of years ago, mainly to have fresh venison at their disposal. In Europe, the keeping of deer and hunting was reserved for the nobility. About 150 years ago, however, the right to hunt was extended to the entire population. The minimum area of land in Austria must be 115 ha in order to establish a hunting area. All smaller land holdings are grouped into community hunting areas and leased by the respective municipality to

J. Laban (✉)
Federation of European Deer Farmers Associations, Prague, Czech Republic

Obmann Bundesverband österreichischer Wildhalter, Semriach, Austria
e-mail: office@wildhaltung.at; gut.jaegerhof@laban.co.at

hunting communities for several years. In the large hunts, where red deer are the main quarry, it is common to create fenced areas where the game is caught over the winter in order to feed it more efficiently, to reduce the damage to the forest, and to count the animals and regulate the population. Such enclosures fenced in the forest are also called winter enclosures.

In the course of time, however, hunting enclosures were also formed, in which the game was kept all year round. Small hunting enclosures, however, were more and more frowned upon and finally banned. Currently, the minimum size of a hunting enclosure with 115 ha is now required by law. Animal rights activists, in particular, are trying to obtain a general legal ban on all types of hunting enclosures.

In the early 1970s, the first agricultural enclosures were formed with the aim of enabling alternative animal husbandry for agriculture and to produce venison. Hunting is prohibited in all these agricultural enclosures. The killing of the animals must be carried out by slaughter. A "stun shot" in the head is allowed and the slaughter is carried out by immediate blood withdrawal. The (hunting) shot in the shoulder is prohibited. These enclosures—deer farms—are subject to the Animal Welfare Act and there are many detailed requirements on the keeping of the deer.

1.2   Current Situation

- In particular, fallow deer has asserted itself as the main animal species for keeping in deer farms. Only in second place comes red deer. Other wild species such as sika, Père David's deer, moufflon, or wild boar are rather rare.
- There are currently 2025 game keepers/farms with a total of 48,410 animals in Austria.
- 65% hold fallow deer (and some other wild species) and about 35% keep red deer.
- About half of all deer farms are members of one of the federal land deer farmers associations.
- The average farm size is about 3.5 ha. The number of animals kept is increasing somewhat every year. Smaller hobby farms are slowly disappearing from the market. "Larger" farms where the deer are farmed are increasing. The prices of game meat are rising.
- The minimum size of an enclosure for fallow deer is 1 ha and for red deer 2 ha.
- A maximum of 20 adult fallow deer may be kept per hectare plus the offspring.
- In the case of red deer, a maximum of ten adult animals plus offspring may be kept, with two animals up to 18 months corresponding to one adult. In the first place, pastures are fenced, forest may only be fenced to a small extent.

Species	Minimum area	Maximum Stocking density adults/ha	Minimum area sq.m./Adult weather protection
Red deer, David's deer	2.00 ha	10	4.00 m²/adults
Fallow, sika	1.00 ha	20 adult	2.00 m²/adults

(continued)

Species	Minimum area	Maximum Stocking density adults/ha	Minimum area sq.m./Adult weather protection
Moufflon	1.00 ha	15 adult	1.50 m²/adults
Wild boar	2.00 ha	5 adult	5.00 m²/adults

For deer two animals up to 18 months correspond to one adult animal

1.3   Regulatory and Legal Background

The following are some of the Regulations which control the keeping of deer and other game animals:

* *Tierschutzgesetz*, Animal Welfare Act
* Bundesgesetz über den Schutz der Tiere (TSchG) BGBl. I Nr. 118/2004
* *Tierhaltungsverordnung*, BGBl. II Nr. 485/2004
* *Tierschutz-Schlachtverordnung* BGBl. II Nr. 312/2015
* *Forstgesetz* BGBl. Nr. 440/1975
* *Waffengesetz*, Arms Act BGBl 121/1967, BGBl. Nr. 121/1967

As·well as various hygiene regulations, such as the "Food Safety and Consumer Protection Act" *Lebensmittelsicherheits- und Verbraucherschutz-Gesetz* BGBl. I Nr. 13/2006.

1.4   Characteristics of Closed Systems, Technical Solutions

The fencing used for containing deer is now normally the high tensile deer nets such as TITAN from Tornado or similar. For fallow deer 1.80 m to 2 m, for red deer 2 m high.

Acacia wood is increasingly being used for the fence posts.

Forests may only be fenced to a small extent—a maximum of 5% of the enclosure area as a cover for the animals. Housing deer in buildings is not permitted.

1.5   Reproduction and Breeding

Some farms specialize in breeding deer for live sales. Bloodlines with particularly strong antlers such as those from Woburn in England are strictly rejected by animal rights activists and sometimes also by hunters.

Currently, no artificial insemination is carried out in Austria.

1.6   Feeding Technologies

In summer, only hay is fed next to the pasture. In winter mainly hay and grass silage (90%), as well as some corn (maize) silage, brewers' grains, apple pomace, barley, corn, turnips. Mineral feed is fed in small quantities. Ready-made game food from the trade is usually not fed for cost reasons.

1.7   Animal Health, Major Diseases

The most common diseases are parasites, which are combated with drugs via the food. TB and paratuberculosis are notifiable diseases which should be reported to the Authority. However, few cases have been diagnosed although there are bovine tuberculosis cases in Tyrol (Western Austria) in wild red deer.

1.8   Game Capture, Transportation

More than 95% of the animals are caught by immobilization using tranquillizer darts. Very few farms have their own handling facility and a crush. Deer are transported in common cattle trailers.

1.9   Utilization and Economy

The main objective is agricultural venison production (95%), the rest is for breeding for other holdings.

Breeding or keeping deer for hunting purposes is prohibited.

The degree of self-sufficiency with venison in Austria is about 50%—i.e. 50% of the venison is imported. The prices of venison directly from the farmer to the consumer are stable at a good level and are rising even further. e.g. saddles €35- to €45/kg, hind limb (haunch) € 25- to € 35/kg.

Keeping deer is only economic when the venison can be sold directly to the consumer. However, this requires a separate slaughterhouse, which involves high investment and work. The many small farms in Austria cannot be run for profit and are dependent on subsidies. Farmers all have a second occupation and mostly work in factories. The deer keeping only generates profits from holdings of more than about 10 ha. These are only some farms in Austria. In spite of everything, the deer keeping develops well here! The trend of consumers is increasingly towards healthy food and meat from animals that are kept in a species-appropriate and well-kept manner. Mass animal husbandry is rejected by most consumers. Here, game keeping offers many advantages. In addition, consumers want to shop more and more locally and directly from the farmer. The government is therefore seeking mandatory designations of origin. This helps Austrian deer keepers to differentiate their products from imported game meat, which partly also comes from New Zealand.

2 Deer Keeping in Switzerland

2.1 History

Agricultural deer farming in Switzerland started with a five-year project at AGRIDEA (a part-state agricultural advisory centre) to evaluate whether this species is suitable as an agricultural niche (1979–1984). In 1982, the *Swiss Association of Deer Farmers ("Schweizerische Vereinigung der Hirschhalter, SVH")* was founded, and, in 1999, the Deer Section of the *Advisory and Health Service for Small Ruminants* (BGK) (see www.kleinwiederkäuer.ch).

Over the past 40 years, game keeping has established itself as a popular niche and is still on the rise.

2.2 Current Situation

Official statistics (2019):

- 679 enclosures with approx. 26,000 animals, of which 333 are farms, the rest are classified as hobby enterprises.
- The average number of animals is just under 40 per farm and the average size of the enclosure is 7.3 ha.
- There are no hunting parks in Switzerland.
- There are also about 40 different wildlife parks/zoos with enclosures of several hectares up to micro-enclosures of <0.5 ha.
- Today, about 220 deer owners are connected to the SVH, about 170 to the BGK www.kleinwiederkäuer.ch. Every year, around 30 new interested farmers attend the specialist-specific vocational training, which is the prerequisite for obtaining a wild animal permit. Deer farming is still a niche animal husbandry, but enjoys a steady increase.

2.3 Regulatory and Legal Background

The following are some of the Regulations which control the keeping of deer:

- *Tierschutzgesetz* vom 16. Dezember 2005 (TSchG; SR 455)
- Tierschutzverordnung vom 23. April 2008 (TSchV; SR 455.1)
- Verordnung des BLV über den *Tierschutz beim Schlachten* vom 1. Juli 2020 (VTSchS; SR 455.110.2)
- *Tierseuchengesetz* vom 1. Juli 1966 (TSG, SR 916.40)
- *Tierseuchenverordnung* vom 27. Juni 1995 (TSV, SR 916.401)
- *Technische Weisungen des BLV über die Kennzeichnung von Klauentieren* vom 3. November 2021 (TW)

- *Bundesgesetz über Lebensmittel und Gebrauchsgegenstände* vom 20. Juni 2014 (LMG; SR 817.0)
- *Verordnung über das Schlachten und die Fleischkontrolle* vom 16. Dezember 2016 (VSFK; SR 817.190)
- *Heilmittelgesetz* vom 15. Dezember 2000 (HMG, SR 812.21)
- *Tierarzneimittelverordnung* vom 18. August 2004 (TAMV, SR 812.212.27)
- Verordnung des Regierungsrates zum *Gesetz über die Jagd und den Schutz wildlebender Säugetiere* und Vögel vom 20. Juni 1986 (RRV, RB 922.11)

2.4   The Main Features of Game Species Kept in Enclosures

The following wild species are farmed in Switzerland:

- (15% red deer, 80% fallow deer, 5% sika, as well as small numbers of axis deer, bison, reindeer, wapiti, moose, etc.).
- Average *enclosure size* 7.3 ha.
- Average *number of* animals per enclosure 40.
- *The minimum size* of enclosures according to the Animal Welfare Act is set at: 500 sq.m. with an additional 60 sq.m.per animal, and if the ground is unimproved then the minimum area is 1500 m^2.
- Stocking density recommendations for farmed deer are a maximum of 8–10 fallow deer, or 4–5 red deer plus their offspring per hectare.

2.5   Characteristics of Closed Systems, Technical Solutions

The most commonly used fences are high tensile deer nets or diagonal chain link at 2 m high for fallow deer, 2.5 m for red deer.

Deer can be housed but only one farm has made use of this and it has to ensure access to pasture at all times. Only juveniles are housed during the winter months. Deer farmers can in theory enclose forest but unfortunately due to the legal complexities this is only rarely done.

2.6   Reproduction and Breeding

There is no herdbook neither artificial insemination.

2.7   Feeding Technologies

Normally, hay is provided in hayracks even during the growing season.

In summer, minerals are mainly fed, usually mixed into a palatable feed such as apple pomace, sugar beet chips, or similar. In winter, in addition to the basic fodder (hay, aftermath, or grass silage), corn (maize) silage, corn cubes, grain mixtures, fallen fruit, crops such as potatoes or carrots are increasingly fed.

In some cases, special game food is fed, this is usually only a small percentage on farms, but in zoos certainly a larger proportion.

2.8   Animal Health, Major Diseases

The most common diseases are parasitic but necrobacillosis occurs as well as overmothering syndrome, acidosis, and neonatal mortality. (tuberculosis, paratuberculosis, brucellosis, etc., have never been diagnosed in Swiss deer enclosures.)

2.9   Game Capture, Transportation

Only about 1–2% of animals are caught in handling systems and crushes, usually the animals are immobilized using tranquillizer darts (98%).

The animals are transported in trailers or in special transport boxes. This may only be done once the deer have been awoken.

2.10   Utilization and Economy

About 85% of deer farms revenue is from venison production with a further 5% of income derived from sales of livestock. A further 10% keep deer solely as a hobby. Hunting is not permitted.

Deer farming can be profitable as yields are approximately the same as for other agricultural grazing animals such as sheep or cattle and direct marketing of venison is possible.

3   Deer Keeping in Germany

Jürgen Laban, Ionel Constantin, Gina Strampe, Wolfgang Eggers.

3.1   Short History of Game Keeping

In 1971, employees of the "Chamber of Agriculture of Rheinland", led by Prof. Dr. Günter Reinken, were already looking for alternatives to the question of whether livestock farming had become less profitable during the twentieth century and whether cattle, sheep, and goats were adequate for the use of pasture land and less favoured. Finally, the experts considered that an additional species should be identified which was more suitable for extreme conditions. For the selection of an alternative animal species, the following criteria had to be priorities:

1. Longevity
2. Resistance to disease
3. Winter resistance, to enable out wintering without housing
4. Good temperament
5. Good feed utilization
6. Low demand for food, especially in the winter
7. Precocious, i.e. early maturing
8. High fertility
9. Easy calving
10. High killing out percentages and excellent meat quality
11. High proportion of valuable meat parts

After a thorough examination of the results of the investigation and observations in wild animals, these characteristics could initially be met by several animal species. After further critical consideration red and fallow deer remained the short-term choice. In an accurate and critical comparison, the experts found that the meat quality of the fallow deer was better.

On the basis of the above criteria, an experimental plant to investigate the management of deer for meat production was established on an area of 4.4 ha in 1973 at the "Teaching and Experimental Institute for Animal Husbandry", Haus Riswick in Kleve of the Rheinland Chamber of Agriculture using 28 fallow does and one buck. In the autumn of 1974, another experiment was started on a research farm in the Bergisches Land (Marienheide) with 33 does and three bucks.

Due to the good results in the two experimental stations, 46 pilot farms were established in Nordrhein-Westfalen from 1976 to 1979 in different regions under various location and climatic conditions—from the Lower Rhine to the highlands of the Sauerland—and among different farmers. This enabled several years of experience to be gained in practical deer farming. In other federal states, fallow deer farming was also considerably expanded, so that by 1985, there were about 715 holdings in Nordrhein-Westfalen with approximately 14,162 animals and in the Federal Republic about 2050 holdings with approximately 37,606 animals.

In the former DDR, Sachsen-Anhalt, there were also early attempts to farm fallow deer. This privilege was granted to the LPG farm under the chairmanship of Dr. Hünsche. Only after the merger of the DDR and the Federal Republic of Germany BRD did deer-keeper associations emerge in all federal states and the number of farms grew steadily.

3.2   Current Situation

There are currently more than 7000 deer farms in Germany (see Table 1), of which about 25% are members of one of the federal state deer keeper associations. The

Table 1 Deer farms in Bundesrepublik Deutschland

State	Farms	Of which member of an association	% Share
Baden Württemberg	600	216	36%
Bayern	2300	694	30%
Brandenburg/Mecklenburg-Vorpommern	300	117	39%
Hessen/Rheinland-Pfalz/Saarland	750	132	17.6%
Niedersachsen	871	116	13.3%
Nordrhein-Westfalen	1132	252	22.3%
Sachsen	520	35	6.7%
Sachsen-Anhalt	405	43	10.6%
Schleswig-Holstein	120	37	30.8%
Thüringen	235	124	52.7%
Totals	7233	1766	25%

Source: Jahrbuch der Landwirtschaftlichen Wildhaltung in Deutschland 2020

majority are used for the production of high-quality venison. A few are involved in the sales of live deer for breeding and a small number are purely hobby farms.

About 80% of the farms keep fallow deer, the rest work with red deer, sika deer, moufflon, wild boar, and other species.

3.3   Regulatory and Legal Background

In addition to the laws in force in the EU, there are a wealth of national laws and regulations, as well as state-specific regulations controlling, for example, the number of animals, the fences, buildings, feeding, animal welfare, nature conservation, gun law, food hygiene, and slaughter while many other issues are also regulated very precisely.

These controls can be found in the following laws and regulations, among others:

- Leitlinie für eine Tiergerechte Haltung von Wild in Gehegen
- Arzneimittelrecht
- Tierhalter- Arzneimittelverordnung
- Betäubungsmittelrecht
- Bundesjagdgesetz
- Bundesnaturschutzgesetz
- Tiergesundheitsgesetz-Verordnung zur Bekämpfung von Seuchen
- Tierschutz-Schlacht-Verordnung
- Tierschutzgesetz
- Viehverkehrsverordnung
- Waffengesetz
- Tierische Lebensmittel-Überwachungs-Verordnung

3.4   Fences

The fences use knotted galvanized wire nets. In the case of fallow deer, the net must be 1.7–1.8 m high for the outer fence and 2.0 m high for the red deer. In the case of fallow deer, a line wire must be erected 20 cm above the net or two line wires at 10 cm intervals to secure the upper part of the fence. For internal fences, a height of 1.4–1.5 m is sufficient for fallow. For the outer fence, posts must be inserted at intervals of 15–20 m, depending on the ground and degree of frost or in the case of steep, rocky, and broken terrain at slightly closer intervals of 10–12 m. The posts must be 2.5–2.7 m long with an upper diameter of 8–10 cm and be driven into the ground 60–70 cm. Posts must be vacuum-impregnated, round, and peeled using oak, acacia, pine, larch, or Douglas fir and with no splits to ensure long life. Between the posts, the so-called herring posts of 1.25 m length and 5–7 cm top-diameter are introduced into the ground at a distance of 4–5 m, so that the fence is held close to the ground.

Farmed deer cannot be kept in buildings.

Forests may only be fenced with the consent of the local forestry authority and young trees must be fenced (protected).

3.5   Breeding

In BRD, no artificial insemination nor embryo transfer can be carried out on farms.

3.6   Feeding

Basically, the wild ruminants feed from mid-April to the beginning of October from the grass in the enclosure. In the case of reductions in grazing capacity, during droughts or in the case of high levels of animals, it may be necessary to feed hay or grass silage to meet the maintenance needs of the deer. To meet the energy requirements for maintenance during the spring and summer as well as for moulting, antler growth, pregnancy, and lactation and to compensate for the protein surplus in the young grass, energy-rich feed can be fed such as corn (maize) silage, carrots, potato, cereals, concentrated feed. The proportion of high-energy feed fed in terms of dry matter should be only 20–25%.

About 6 weeks after birth, the calves begin to require an additional intake of plant-based food. From this point on it is worthwhile feeding the calves in a calf creep which excludes adults. The feed used for this is cereal mixtures, mixes of grain and molasses, sugar beet chips, or concentrates up to 250–350 g/animal/day.

From mid-October and in winter, the deer are basically fed forage of hay or grass silage (for larger herds) as basic fodder. Up to temperatures of +5 ° C, the basic feed is sufficient for the maintenance of life functions. At outdoor temperatures from 5 °C to −10 °C, the animals need an additional energy boost for heat production. Energy compensation can be achieved by feeding maize silage, fodder beet, potatoes or cereals, dry sugar beet chips, wild fodder, or concentrated feed formulated for sheep.

If the outside temperatures drop below −10 °C and the enclosure is wild outdoors without shelter, the additional energy requirement for heat production via energy-rich feed can be compensated for with a higher digestibility ration using cereals, dried sugar beet chips, game fodder, or concentrated feed which can be fed as an organic coarse feed mix.

The proportion of the artificial feed in the total ration in terms of dry matter must be no more than 25–30%. In North Rhine-Westphalia, compound rations designed for wild animals are rarely fed in agricultural game farming but is generally used in zoos.

3.7 Animal Health

The susceptibility to disease in fallow or red deer is low. Diseases of the gastroin-testinal tract (parasites), feeding-related and deficiency disorders, bacterial (coliforms, necrobacillosis) and mycotic diseases were most commonly reported.

Exposure to foreign bodies (plastic, poisoning) can be a problem for fallow deer.

3.8 Catching and Transporting

Only about 10% of farms with wild game holdings have a handling facility for treatment, sorting, and live sales. In the other 90% of farms, immobilization is made using tranquillizing drugs administered by dart.

The transport of live animals is basically carried out in a trailer, which is easy to clean and disinfect (horse trailers, special game transporters, or game trailers). The transport of slaughtered animals to an EU-approved slaughterhouse is carried out hanging in the above—but is a costly means of transport.

3.9 Purpose of Game Keeping

In principle, the deer are kept for meat production and for the sale of breeding animals. Approximately 80% of the agricultural farmed game is used for meat production, about 10% for the sale of farmed animals, and 10% for hobby.

According to the local hunting authority, there are no hunting enclosures in NRW.

3.10 Economy

Demand for game meat and game meat products from farmed game varies widely from region to region. Deer keeping for venison production is a real alternative to grassland use, which can contribute to improving the income situation on the farm. The mastery of production technology and, above all, the marketing of animals are two factors which have a decisive influence on the economic viability of the enclosures. Profits can only be generated through careful quality control and direct marketing.

Further Reading

The BLW (Bundesverband für landwirtschaflicher Wildhaltung e.V.) publishes a yearbook, currently the Juni 2020 Jahrbuch BLW in Deutschland which provides an excellent analysis of game keeping in Germany www.wildhaltung.net

A very informative and clear summary of deer keeping can be found at www.wildhaltung.at

The Management of Enclosed Deer in Hungary

Janos Nagy and Julianna Bokor

Abstract

The Hungarian historical legendary world is intertwined in many places with the mythical creature of the red deer. There were few records of Hungarian game parks in the early Middle Ages. During the centuries, more and more parks established beside castles and later destroyed. Several of them reached the nineteenth century but broke down due to the World Wars.

The natural endowments of the Carpatian Basin are excellent for red and fallow deer, which greatly contributed to several world record trophies in both species.

At the end of twentieth century, after the land privatization, the hunting parks have been revived and developed continuously. The first establishment of the deer farms was in 1985, in which the Kaposvár University played a significant role. Although the environmental conditions are good for deer, the outdated pasture management, and the strong lobby of hunters and crop farmers do not promote the spread of deer farming.

Several researches and studies have been carried out at the university, which have greatly contributed to the achievement of the results so far and to the establishment of the foundations.

Keywords

Hungarian red deer · Deer farm · Game parks · World record · Hungarian fallow deer

J. Nagy (✉) · J. Bokor
Hungarian University of Agriculture and Life Sciences, Gödöllő, Hungary
e-mail: nagy.janos@uni-mate.hu

1 Historical Overview

Based on genetic research work, it has been established that three genetic lines have been involved in the development of Europe's red deer population: the Western European (A), Eastern European (B), and the Mediterranean (C). After the last glacial, Central Europe was populated by the Eastern European (B) subspecies from the Balkans and the Western European (A) subspecies from the west, where the contact and mixing zone of the two haplogroups was formed (Ludt et al. 2004; Skog et al. 2009; Niedzialkowska et al. 2011; Karaiskou et al. 2014; Krojerová-Prokešová et al. 2015). The red deer formed in this way awaited our ancestors in the eighth to nineth century.

The Hungarian historical legendary world is intertwined in many places with the mythical creature of the red deer. Thus, according to legend the sons of Nimrod, Hunor and Magor, chased a female red deer during their hunt and followed it to the swamp of Meotis, where they lost her traces. The area was found to be very favourable and they asked their father for permission to settle there. This is the territory of today's Carpathian Basin, Hungary.

1.1 History of Game Parks in Hungary

There were few records of Hungarian game parks in the early Middle Ages. Perhaps the first written record dates from 1055, appearing in the deed of foundation of Tihany. In addition, the name "Game park" can be found in several place names, which may have referred to this facility (Csőre 1997).

The first definite written record dates from 1238, in the form of a charter of Zsédenypuszta. We also find the type of game park or garden game park linked to castles in Hungary, similar to other European countries, with the park located directly beside the castle or in the vicinity of the castle (Uzhgorod, Gömör).

By the 1400s, several game parks were already appearing in the descriptions, but it was mainly the royal parks which gained mention, especially that of the kings Sigismund and Matthias. Matthias' game parks already reflected the taste of the Renaissance. The parks were decorated with colourful flowers, groves, sometimes fishponds and often non-domestic animal species. There was already a birdhouse in the Buda castle park with domestic and foreign species. Matthias also kept his lions in the Buda castle park. Csőre (2000) considers it probable that at this time the fallow deer was introduced to Hungary. For a long time they were kept only in game parks.

At this period it was popular among the rulers to travel by red deer-drawn carriage on special feast days and from this our King Matthias was no exception. Such domesticated red deer represented great value.

The wars, castle sieges, and the march of the military did not leave the game parks untouched. Most of them started to perish from the sixteenth century, and only a few survived in the western part of the country (Somlyó, Sárvár), in the Highlands (Füzér castle game park), and in Transylvania (Gyalu castle game park, Mondra).

In the 1700s, large estates began to form in Hungary. The centre of the estate was usually a castle, which also had a park and often also a game park. In the nineteenth century, the number of game parks increased again and their character also changed. In addition to admiration, hunting received increasing emphasis.

The inhabitants of these facilities were mainly red deer, fallow deer, wild boar, and roe deer, but in many cases also rabbits, pheasants, and guinea fowl. Rarely the parks incorporated an aviary too, which gave home to non-native birds.

In the nineteenth century, free-living red deer were rare in the wild and game parks were established to increase the number of wild red deer or to introduce new species.

One of the most notably successful introductions is the mouflon. These were brought by Count Charles Forgách in 1868–1869 to Ghimes (now Romania) from Frankfurt and Brussels. From the Ghimes animal population, 80–100 individuals were released in 1883 and later settled in other game parks (Nagyappony, Betlér). It is also worth mentioning the sika deer introduction in Fehérvárcsurgó in 1910, which also proved successful (Csőre 2000).

Sometimes the purpose of establishing a game park was to prevent damage to crops caused by wild game. The wild boar parks were the main ones for this purpose and Joseph II took such measures to mitigate the damage caused by the wild boar.

The laws regulating the hunting seasons were not applied to hunting in game parks.

The heyday of game parks in Hungary can be traced back to 1880–1920. During this period, according to Csőre (1997), there were 235 game parks and 97 pheasant parks. In 1887, the trade press (Vadászlap) described the establishment of game parks and provided detailed guidelines as to site selection, extent, accessibility, fencing, species composition, etc.

In 1880, Count Iván Draskovich established a 1300-ha deer park in Sellye, which was made to prosper by his son, Iván Draskovich Jr., who achieved outstanding results in the breeding and development of a superior strain of red deer in the game park. In his book entitled Deer Management, he summarizes his views and the results achieved in red deer breeding; he considered the red deer in the wild to be the noblest.

Thanks to careful farming and breeding, trophy sizes have improved, and stags with long antler beam (115–120 cm), large circumferences (18–20 cm), and with large burr size (28–30 cm) have characterized the bloodlines of the Sellye game park. The outstanding results are well illustrated by the fact that Draskovich's trophies easily won the 1910 World's Hunting Exhibition in Vienna and then the Berlin Exhibition in 1937. These successes led to the setting of further goals, and in order to increase the number of crown points, he used for breeding some newly born stag calves caught from the floodplains of the Danube and Drava which were raised and strictly selected. In just a decade, the occurrence of the multipointed crown increased significantly, and in addition, the good beam length and circumferences of the antlers were retained (Draskovich 2007).

Through the results of the Sellye game park, Iván Draskovich proved that the red deer stags in the Carpathian Basin could produce high-quality trophies in game park conditions using sensible game park management and professional selection. Also arising from his work was his great contribution to raising the standard of Hungarian game management for which he set an example.

During the World War II, the Sellye game park was destroyed, but the stags in the area still carry the antler characteristics of Draskovich stags.

The number of game parks decreased significantly after World War I, partly due to the events of the war and partly due to the Treaty of Trianon by which, in 1920, two thirds of Hungary's territory was lost. With a few exceptions, the remaining game parks on the territory of the mutilated country were swept away by World War II.

The world wars did not spare the free-living wildlife populations either, it is estimated that 75–80% of our red deer population disappeared.

In order to develop deer management and improve quality, in 1964, on the proposal of Dr János Tóth, Head of the Department, the National Trophy Judging Committee was established as the first in the world. Its task was to judge trophies according to standard principles, recording data and using photo documentation (Szidnai 2013).

Efforts to establish game parks re-emerged in the 1970s, mainly due to restrictions on the wild boar population. These parks were established primarily to reduce the damage to agriculture. The parks were stocked with wild boar caught from the free hunting area in summer and then profited by organizing large-scale driven hunts in the parks during the winter.

By 1987, there existed a technology developed in the field of capture and breeding (Ákoshegyi 2005).

In 1971, the World Hunting Exhibition was held in Hungary, where we presented three world record trophies (red deer, fallow deer, and roe deer) to the general public. This exhibition greatly contributed to the publicization of hunting opportunities and the launching of hunting tourism.

By 1998, the Game Management Department of the Ministry of Agriculture had already registered 68 parks, with a total area approx. of 38,200 ha. In 26 of these parks there were only wild boar, in 30 parks there was a mixed population, and in 12 parks there were only red deer (Ákoshegyi 2005).

The change of regime in 1990 was followed by the privatization of land, and then in 1998, the Hunting Law assigned the hunting right to land ownership. The re-emergence of large estates gave a new impetus to the establishment of game parks, so that their number has now reached 150.

1.2 The Intensive Deer Breeding

1.2.1 Formation

Péter Horn—the visionary leader of the deer project in Kaposvár—first encountered the dynamically developing New Zealand deer breeding in 1975.

In the early 1980s, the international reputation of the Hungarian red deer attracted deer breeders of this distant continent to our country. In order to increase body weight and antler size, breeding animals were imported to New Zealand from several Central European countries. In Hungary, the Southern Transdanubia region was visited in the hope of buying animals.

Len Crowsky and Don Craig, among others, visited Hungary as employees of Taupo Forestry, and Bernard Pinney, a private breeder, and Don Matson, representing the firm Wrightson, organized the capture of red deer and their export to New Zealand.

One feature of live game captures was the communication difficulty. Hunting guests came to Hungary from the German-speaking area at that time, so the professional hunters knew the German language, while the Kiwis spoke English, which led to a number of interesting situations. On one occasion, Len Crowsky arrived at the hunting lodge SEFAG in Zsitfa-Puszta for several days, where at that time Emil Sas, a well-known and highly knowledgeable professional hunter, was working. According to good Hungarian custom, at 5 o'clock every morning, the chief hunter knocked on Len's room with a bottle of Hungarian brandy in his hand. The guest stood in the doorway incomprehensibly, so Emil took a sip of brandy in his mouth, showing the guest what to do—he rinsed, gargled, and then swallowed the 50-degree alcohol. A few years later, during a trip to New Zealand, János N. learned that according to the Kiwis, the Hungarians brush their teeth with brandy every morning.

Apart from the misunderstandings, the live game captures ended with good results, after a quarantine in England, more than 100 individuals came to New Zealand, including also such red deer stags (Kapos, Laci, Emil), which provided outstanding performance in the land of Kiwis.

Regular visits by New Zealand professionals accelerated the development of deer husbandry, with preparations beginning at three locations between 1984 and 1986. Consultant Mike Harbord and Don Matson helped to get through the initial difficulties, and with their help, several professionals went on a study trip to New Zealand. Fencing and deer handling systems were built on the deer farms of Kaposvár College and SEFAG according to the plans of New Zealand specialists.

1.2.2 Artificial Calf Rearing

At the Kaposvár College—on the recommendation of the consultants, the settlement of the 40-ha farm in Gálosfa started with artificial calf rearing.

In 3 years, without a single death, more than 50 individual calves were successfully raised, with the carers selected mainly from college students.

The success of raising calves depended on early imprinting and the use of deer milk replacer. Two to three days old calves easily got used to their new environment and carers in a few days.

The researchers of the college, led by János Csapó, based on their own research results, compiled a low lactose milk substitute, which perfectly replaced deer milk. The hind suckles her calf 8–10 times a day, so the workers also fed them every 2 h, initially with only 40–50 mL of milk, and then gradually increased the dose to 800–1000 mL. The goal was for the calf to consume 4–5 litres a day by the age of 3 months.

Calves' milk consumption, weight gain, and changes in body size could be easily followed because they were so tame and easily handled. Their daily weight gain during the milk feeding period ranged from 320 to 630 grams. The date of weaning was timed for the end of August, by which time the live weight of each individual exceeded 50 kg, they were regularly consuming grain feed and were grazing.

The development of one of the stag calves was outstanding, its body weight was 80 kg at the end of August and the antler pedicles had already appeared on his forehead.

Although the New Zealand consultants helped a lot, the real breakthrough came from the New Zealand study tours. At the end of the 80s, Dr. László Sugár, in 1989 János Nagy, later many young colleagues visited this wonderful country. It was then that several colleagues became aware of what it meant to belong to the family of deer breeders. The farmers were characterized by openness, creativity, perseverance, and striving for professionalism—with a focus on the red deer.

These personal experiences greatly helped to develop the following technologies:

- construction of deer fences
- design and operation of handling systems
- pasture management, grazing of red deer
- live animal capture, transport
- velvet antler production
- artificial insemination
- performing veterinary tasks.

1.2.3 The Deer Farm Is Moving to Bőszénfa

In 1990, deer breeding was expanded with new areas, and the institution purchased a 1360-ha hilly farm. In 10 years, 55 km of fences were built in Bőszénfa, using its own construction, and 28 breeding paddocks and two hunting parks were built with New Zealand technology, a post driver machine and deer netting imported from England.

The properties of Bőszénfa are favourable for the design of deer enclosures, because the topography is varied and the forests are located in a varied and patchwork fashion, so that every paddock has a natural hiding place and the trees provide shade. It was important not to get too much forest or bush in the paddocks, because Hungarian red deer can easily revert to their wild behaviour.

Thanks to continuous fence building the area has expanded more and more and the number of the deer herd has increased due to net traps.

In addition to fencing systems, it was essential to build a handling system where up to 80–100 individuals could be handled at the same time.

The handling system in Bőszénfa is a simple board fence system where ear tagging of animals, taking their blood sample, preparation for delivery and their loading onto a trailer can be carried out.

After several minor modifications, the conditions for the safe handling of red deer were finally established (Fig. 1).

Fig. 1 Handling system in Bőszénfa (Archive of Game Management Centre)

Red deer are handled 3–5 times a year, depending on their age, sex, and animal health status.

1.2.4 Scientific Activities on the Deer Farm

In connection with cervids, within the field of game breeding in enclosed land, the Game Management Centre of the University of Kaposvár has carried out active researches in many research areas.

Between 1981 and 1983, the researchers of the Kaposvár Agricultural College carried out several surveys in the hunting areas of Somogy County. The most important of these were: the study of the density and chemical composition of red deer antlers, the assessment of drivers of body size, reproductive biological parameters, the mineral supply of red deer in Somogy County, the adaptation and development of red deer, fallow deer capture, and *per oral* immobilization procedures.

Later, the body size and growth of Hungarian red deer were studied, and it was found that this type (Hungarian) has larger body sizes and slower development than the western one. Early (postnatal) growth was also studied in the early period of the deer farm when calves were reared on artificial deer milk (Horn 1987). Increases in body weight and body sizes were measured and analysed from 2 months to 11 months of age (Bokor 2016; Bokor et al. 2010). Models have been tested for the mathematical characterization of juvenile growth (from birth to 8–9 months of age) (Bokor et al. 2015).

Zomborszky's name is associated with the introduction of biotechnological methods in the reproductive biology of red deer in Hungary. The most important of these are the technology of post-mortem semen extraction and freezing (Zomborszky et al. 1999) and its application, as well as the initiation of storage of extracted semen for gene conservation purposes. Also Zomborszky et al. made an attempt to collect oocytes in cervids.

Under Hungarian climatic conditions, János Nagy et al. investigated the effect of different grass mixtures on the weight gain, meat yield, and meat quality of young red deer, which was supported by CT examinations (Nagy et al. 2019) and blood parameters (Szabó et al. 2013).

The constantly developing field of molecular genetic research also reached the Hungarian red deer and it was natural to use the Bőszénfa deer farm, the unique and largest farm in Hungary. Viktor Stéger et al. (2010) studied the associations between antler development and human osteoporosis in terms of their molecular genetics, during which they dealt with deer genome mapping too.

2   Present Situation

2.1   The Rebirth of Hunting Parks

In the late 1990s, after land privatization, large estate systems were re-established, and the new landowners began to build game parks (mainly to contain wild boar) in increasing numbers.

By 2019, there were 107 hunting parks registered by the authorities, with a total area of 32,786 ha.

The number of wild animals kept in the parks or shot can be estimated most accurately by comparison with data from free range. According to the data of the National Game Management Database (OVA), the estimated total of red deer in the country was 114,500 in the spring of 2018, which demonstrates a steady increase compared to the estimates from previous years.

Every shot game trophy has to be shown to the hunting authority for official judgment by CIC. The International Council for Game and Wildlife Conservation (CIC) is a politically independent organization which aims to preserve wild game and hunting.

The *CIC Trophy Evaluation System* (TES) is an international system created for monitoring and comparing hunting trophies from various species with species-specific, unified measurement parameters (CIC Homepage 2020, www.cic-wildlife.org).

The growth in the number of individuals shot is a good illustration of the real dynamics of the population, which is shown in the Table below based on data of 20 years (Table 1).

Table 1 Estimated populations of each of the four large game species and the numbers killed in hunting parks as a proportion of the total. The total includes both free range animals and those in hunting parks (Csányi et al. 2020)

Year	Wild species	Estimated populations			Numbers killed		
		Total	H. Park	Proportion	Total	H. Park	Proportion
	red deer	69 294	4 515	6,5%	19 692	1 092	5,5%
1997	fallow deer	13 777	4 328	31,4%	4 722	1 060	22,4%
	mouflon	9 364	1 013	10,8%	1 464	70	4,8%
	wild boar	57 161	5 446	9,5%	38 126	3 020	7,9%
	red deer	79 941	4 531	5,7%	33 967	1 233	3,6%
2007	fallow deer	25 193	5 828	23,1%	9 301	1 533	16,5%
	mouflon	10 090	2 610	25,9%	2 564	318	12,4%
	wild boar	81 723	13 666	16,7%	94 015	10 239	10,9%
	red deer	101 464	5 169	5,1%	58 068	1 149	1,9%
2017	fallow deer	34 725	7 770	22,4%	13 730	1 686	12,3%
	mouflon	10 819	3 453	31,9%	3 623	373	10,2%
	wild boar	105 196	14 838	14,1%	148 044	12 647	8,5%

Among the game species in hunting parks, between 1997 and 2017, the wild boar population increased the most, and the bag of wild boar quadrupled in 20 years. The park population of mouflon and fallow deer can also be considered significant. Among our big game populations, the keeping of red deer demands the most time and money which may explain why the population of red deer in the hunting parks stagnated.

It is interesting to note that despite the significant increase in free-range area populations, the bag within the parks did not decrease significantly, in fact it increased for all species except red deer where the bag remained static. This demonstrates that the park operators found sufficient demand from hunters willing to shoot in the parks. Nevertheless the number of red deer shot in the free-range area has doubled in the last decade.

The data in Table 1 show that the bag is high compared to the estimated population. In 2018, compared to an estimated 114,500 individuals, 63,751 red deer were shot, representing a utilization rate of more than 55%. The reason for this is to be found in the fact that game managers, in order to protect their deer population, "carefully estimate": they report a smaller number so that they have to comply with a lower shooting quota.

The bag of red deer in the hunting park is small compared to the free-range area data, it is only 2% of the total bag (Csányi et al. 2020).

Figure 2 demonstrates that there is a higher rate of fluctuation in park shooting, ranging from 1000 to 2000 individuals per year. This is mainly due to the fact that for many hunting organizations, the hunting park serves as a "safety reserve", stags are only shot if the desired game has not been shot on the free-range area.

Figure 2 illustrates the evolution of park shooting data over the past two decades.

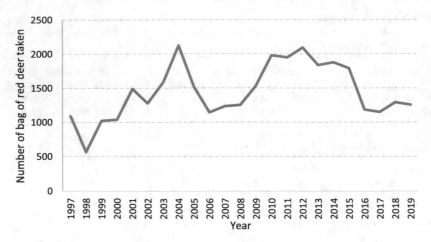

Fig. 2 The bag of red deer taken in hunting parks

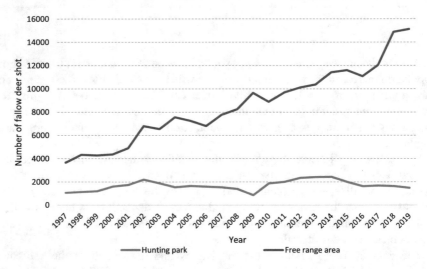

Fig. 3 Numbers of fallow deer shot in free-range areas and in hunting parks

Our other important and continuously growing wild species is the fallow deer, which has also doubled its estimated free-range population during the period presented.

Figure 3 shows the annual cull of fallow deer shot in free-range areas and in hunting parks. There is a very significant increase in the free-range bag and a stagnation in hunting parks.

There have also been changes in the composition of the guests with the proportion of domestic guests increasing. Based on OVA-s 2018 data, in the hunting parks 45% of red deer stags were shot by foreigners and 35% were shot by domestic hunters, 20% of the total number were shot by the staff. In case of hinds and calves, 27% of the individuals were shot by foreigners, 34% by Hungarian hunters, and 39% by the staff.

Today, the game population of hunting parks can also be enriched from game farms—thanks to a change in the hunting law—which could be a market opportunity for farms in the future, although the opportunities are limited for the time being due to the large number and high quality of deer populations in free-range area. In general, breeders can sell their trophy animals to the hunting parks, for 45–50% of the hunting price list.

2.2 Game Parks in the Service of the Public

The game parks differ completely from hunting parks in that their purpose is to provide education, carry out research, and meet tourism demands. They are primarily entitled to keep and present only those game species that can be hunted in Hungary. Their number shows an increasing trend in the recent period: in 2013, 24 were already operating, and, by 2019, there were 32 game parks in Hungary. The most popular animal species were fallow deer. In 2019, a total of 162 fallow deer, 98 red deer, 91 mouflons, 34 wild boars, and 32 roe deer were kept in game parks. These facilities usually farm on a smaller area of 3–5 ha, often keeping other popular pets such as goats, sheep, lamas, ponies, etc.

2.3 Game Farms Are Modern Establishments

The main goal of establishing game farms is to produce carefully controlled venison of high quality. In Hungary, due to the "abundant", relatively cheap free-range game production, farms mainly choose the sale of animals for breeding, and for trophies. Only individuals that do not meet breeding objectives will be sold as meat.

The game farm category provides an opportunity for people, even individuals, legally to keep game species for hobby purposes. This is partly due to the fact that the average farm size is only 4.3 ha. In 2019, there were 332 game farms operating on a total of 1450 ha of land, with the number and species listed in the table below (Table 2).

Table 2 The numbers of large game kept on farms by year and by species

Year	Red deer	Fallow deer	Roe deer	Mouflon	Wild boar	In total
2013	333	1006	131	205	5210	6885
2014	298	908	107	178	3233	4724
2015	404	1181	103	306	4348	6342
2016	408	1383	115	333	5175	7414
2017	451	1418	143	383	5002	7397
2018	448	1457	159	410	5531	8005
2019	428	1370	178	502	4891	7369

Since 2007, there has been the opportunity to provide young people, through a young farmers' scheme, support for the establishment of game farms. As a result, the number of farms is constantly increasing. This can provide a good basis for the expansion of quality-controlled game production by creating the necessary conditions (slaughter and processing plants).

3   Characteristics of Cervids Kept in Enclosures

In enclosed situations both red and fallow deer are the most important cervids in Hungary.

Other cervids can be found in zoos and game parks as exhibits.

3.1   Fallow Deer

Among the cervids the fallow deer are the most commonly enclosed species.

The main reasons for this are:

– They have been known and accepted as game park species since their introduction to Hungary (probably in the fourteenth to fifteenth centuries)
– They adapt well to being enclosed
– The intensive breeding of fallow does not encounter resistance from the hunting profession in the same way as does that of red deer
– Fallow can be bred with good results even on grasslands of medium quality
– They do not require a significant amount of grain during their feeding
– They achieve good weight gain even in dry summer droughts
– They tolerate being kept in large groups
– They are not susceptible to disease including parasitic infections
– They have an attractive appearance
– The hunting of fallow is exciting and their trophies are popular hunting
– Fallow meat is an excellent, tasty, and valuable cooking ingredient.

The fallow deer population of our country is of outstanding quality. Over the last three decades, a number of world record fallow bucks have been shot, which is why breeding animals have good sales opportunities. The trophy shown in Fig. 4 is the current world record that was shot in Guth in 2002.

As the number of farms increases, fallow breeding on farms is becoming more and more popular.

Fig. 4 World record fallow deer trophy (Guth 2002, CIC: 237,63) (Gábor Bognár)

Fig. 5 World record red deer trophy (Karapancsa 1986, CIC: 271) (János Nagy)

Fig. 6 Capital red stag from
Vörösalma (2002, CIC:
263,88) (Gábor Bognár)

3.2   Red Deer

The free-ranging red deer population is also of outstanding quality in Hungary and
enjoys a high international reputation due to its world-ranked trophies (Figs. 5 and
6).

Intensive breeding of red deer divides the hunting profession.The hunting authority does not consider this method of utilization to be desirable and, accordingly, does
not support the spread of agricultural red deer breeding.

The main reasons for this are:

- They fear for the quality of the free-range population. It is not illegal for farmers
  to import breeding animals or to import and use genetic material from other
  countries, so the protection of the Hungarian deer gene pool cannot be ensured.
- Over the past decade, there have been a number of abuses in connection with the
  release of imported stags for hunting into free-range areas and this is prohibited
  by hunting law. Such cases create distrust among hunters, causing significant
  damage to the reputation of Hungarian hunting.

It is in the long-term interest of breeders to comply with the legal requirements.
Quality breeding of purebred Hungarian red deer requires good farm conditions,
knowledge, time, money, and handling practice.

The main features of the subspecies:

– The Hungarian red deer has a high growth rate, but it needs abundant and easily digestible food. During summer droughts, it is often necessary to feed with supplements, which makes husbandry costs more expensive.
– A well-functioning handling system and experienced staff are needed to ensure regular deworming and handling.
– This subspecies is in the early stage of domestication, so farmed populations need to be further selected in order to increase stress tolerance.

Farmers require good production conditions and a large area of land (at least 30–50 ha) to be suitable for the farming of red deer.

4 Legal Background

In Hungary, hunting, game management, and game protection are regulated by Act LV (the Hunting Act) of 1996 which controls the protection of game, game management, and hunting. In addition, also the general protection rules for wildlife protection of Act LIII of 1996 legislate for game protection and nature protection.

The Hunting Act lays down the regulations for the exercise and application of hunting rights, for the protection of huntable species and their habitat, as well as rules for game management, and compensation for damages caused by game.

The Minister responsible for game management—in agreement with the Minister responsible for nature protection—states in the enforcement decree, the species of game that qualify as being native to Hungary, or occurring, or introduced with permission, or passing through, our country and which are qualified as large game not protected under nature protection, or small game species that can be hunted.

The Hunting Act defines three forms of game keeping in enclosed areas. These are hunting parks, game parks, and game farms. The hunting park primarily serves an economic purpose, so the reason for its establishment is to provide hunting, the production of breeding animals, and the reduction of damage caused by game. The game park serves scientific and educational purposes, therefore the reason for its establishment is the study, observation, and presentation of wild species. The purpose of establishing a game farm is to produce game meat as food. On a game farm, the game is essentially considered as "agricultural farm animals".

The establishment of a hunting park, game park, and game farm is permitted by the hunting authority having national competence.

4.1 Hunting Park

The hunting park is a fenced part of the hunting area for the keeping and breeding of red deer, fallow deer, roe deer, mouflon, and wild boar for hunting purposes. In the case of hunting in a hunting park, the requirements concerning hunting seasons and the game-keeping capacity of the hunting area do not have to be applied. In this case, however, the generally accepted ethical rules for the reproduction of game must be followed, and hunting must be reported in advance to the hunting authority (Metzger 2013).

The fencing of the hunting area for the purpose of establishing a hunting park is permitted by a decision of the hunting authority with the prior consent of the owner or user of the land, at the request of the organization authorized to hunt. The application must include the reasons and purpose for the establishment of the hunting park(s), as well as the most important data of the game park(s) to be established (area, external fence length, game species, number of animals planned, etc.).

The minimum size of a hunting park is 200 ha for wild boar and mouflon, and 500 ha for cervids.

If game is bred and hunted in the same hunting park, the size of the contiguous hunting area must meet the minimum requirements of 200 and 500 ha, respectively.

The utilization of game kept in the hunting park is authorized by the hunting authority in the annual game management plan, every hunting year.

Game from a free hunting area or from enclosures can be acclimatized and introduced into a hunting park if the phenotype of the individual bears the characteristics of the given game species. The introduction of game into a hunting park is licensed by the hunting authority. In the hunting park, natural feeding, cover for hiding and seclusion of the animals must be ensured.

4.2   Game Park

The game park is a fenced area for research, education, and exhibition purposes.

The establishment of a game park must be authorized by the hunting authority with the prior consent of the owner or user of the land. The hunting authority must specify in the license (similar to those conditions restricting the hunting park):

– the wild species that can be kept, as well as the number of species;
– the animal health conditions related to game husbandry;
– in the case of forests, the conditions related to the protection of the forest and to the exploitation of forest resources;
– the conditions relating to the protection of the area in the case of area being under nature protection;
– other conditions related to the keeping of the game.

Compliance with the above conditions shall be reviewed annually by the hunting authority and, if necessary, a decision shall be taken to withdraw the permit.

Hunting in a game park is only allowed with the special permission of the hunting authority. If the number of animals and the sex ratio specified in the decision authorizing the establishment of the game park cannot be restored otherwise, the hunting authority may order (upon request or ex officio) a hunting to regulate game numbers.

4.3 Game Farm

A game farm is an establishment used for the husbandry of huntable game species of Hungary, for the production of meat. However, game farms cannot be established in protected natural areas. Hunting is prohibited on game farms.

The establishment of a game farm is authorized by the hunting authority at the request of the owner or user of the land. The application contains the reasons and purpose for the establishment of the farm, as well as the most important data (area, external fence length, planned game population and species).

Appropriate farming conditions for the game species authorized to be kept must be ensured on the farm. There must be feed storage, a handling system, a supply of drinking water, and quarantine facilities. The operator of the game farm is obliged to keep a stock register and to report changes in the stock data annually up to and including the last day of February each year.

5 Technological Characteristics

The largest deer farm in Hungary is at Bőszénfa, which measures 250 ha and where more than 1000 red deer are kept. Average farm sizes are only 4 ha, although a deer farm needs to be around 25–50 ha to be sustainable.

5.1 Fence

In recent decades, almost every farmer uses "New Zealand type" fences, because they are long lasting and safe, and at the same time they provide animal friendly solutions. According to the regulations of the hunting authorities, a deer net of 240 cm height is required for red deer and 200 cm for fallow deer. Against wild boar, golden jackals, and dogs, 20 cm netting is often sunk into the ground. In many cases, in addition to the fence, one or two electric wires are also used to keep out the undesirable animals. Currently, the biggest headache for farmers are the golden jackals, against which the electric fence can be a solution.

Fence building businesses have been established recently to serve the demand of fencing for highways, to protect forested areas, and for farming purposes. These teams are using New Zealand technology with post driver machines, local acacia posts, and high tensile netting. The wooden posts are barked and burned, or treated with preservatives, to about 20 cm above the ground level then they are knocked into the ground to a depth of 80 or 100 cm. The durability of acacia posts is 15–20 years. The corner posts are placed deeper and they are also 20–30 cm longer and 5–8 cm thicker.

Although the fences have to be checked regularly, so far there has been little vandalism by humans.

5.2   Feeders, Waterers

In the case of deer farmers, winter feeding means hay, silage, and grain crops.

Round bale feeders have proven to be good for feeding the hay and grass silage.

On large farms, silage is fed from a feeding tray, sometimes using mixer wagon for making TMR (Total Mix Rations) for deer.

The grain is fed from the ground or from troughs, which are mostly made of wood.

Farms have different water supply conditions. On several farms, where water can be piped, frost-free waterers or plastic crates are used. Many farmers also dig pits for red deer, to allow them to wallow during the hot summer days; it is an important part of their welfare.

5.3   Handling System

The professional handling of red and fallow deer is still in its infancy on most farms. The first handling systems were developed with advice from New Zealand deer consultants, at Gálosfa on the farm of Kaposvár University, and at Rinyaszentkirály on the farm of SEFAG Zrt. Thereafter newly established farms drew in part from these experiences and in part from New Zealand experiences.

Only a few farms (3–4) have handling systems for fallow deer, and the handling techniques are still undeveloped focussing on the capturing and penning of a few individuals, mainly for sale. In some case, the owners capture the animals with a dart gun and narcotics.

For red deer, we also still find systems that were developed from basic capturing yards where the animals are squeezed into a funnel-shaped corridor. In the last section, the animals are then individually boxed. Fortunately, these methods are slowly being displaced by the more gentle handling methods. In these more modern handling units there are central revolving doors, and drop floor and mechanical crushes (Fig. 7). Most livestock exports are handled by these farms.

The handling of Hungarian red deer requires attention, expertise, and practice. These large, semi-wild animals sometimes attack each other in the yard, and sometimes even attack people, especially when they are carrying antlers.

6   Reproduction

The primary breeding goal of the farms is to increase the antler weight. The market is chiefly looking for good quality breeding deer and especially for trophy animals, hence the farmers select their breeding stock for trophy quality. The live animal prices are mainly determined by the hunting pricelists. The trophy fee to be paid by the hunter is determined on the basis of the weight of the trophy. The score measured using the C.I.C. system only serves to decide the ranking of the trophy. To select

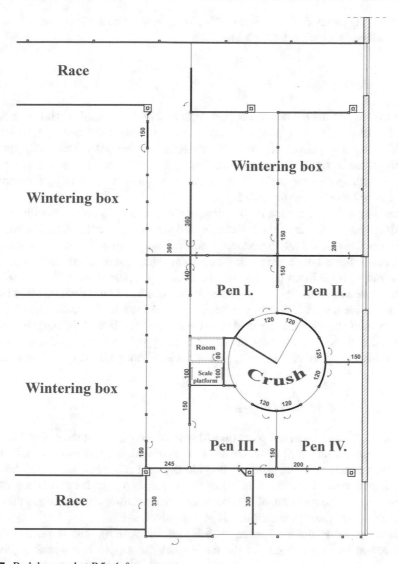

Fig. 7 Red deer yard at Bőszénfa

hinds, it is generally accepted in professional circles that large, long-headed individuals should be selected for as being more likely to achieve high antler quality.

During the last decade, there has been a growing dislike amongst hunters for the unnatural, many pointed, "bushy" antlers. Therefore, in selecting animals for breeding, preference should be given to Hungarian-type stags having antlers with long beam and relatively few points (max. 18–20 points).

Deer breeding is based on Hungarian blood lines, which are typically large-bodied (hinds weighing 130–150 kg, and stags 250–300 kg), and with the stags carrying long-beamed, few-pointed (12–16) antlers. Some smaller farms however also use foreign genetics for their breeding.

6.1   Mating

The peak time of the red deer rut is from 10th to 25th of September, and from 5th to 20th of October for the fallow deer. On most farms single sire mating (one stag to 30–40 hind) is applied, resulting in 80–85% calving rate. Does this include yearlings? Multi sire mating has unsatisfactory results in the case of red and fallow deer. In both species the males fight too much and lose a lot of weight, ending up with a poor calving rate (65–70%).

Breeding stags are chosen for breeding at the age of 3–4 years (sometimes older) by which time their antler quality can be judged by their antler size. On red deer farms, the stags are usually introduced into the hind group at the end of August and taken out at the end of October–early November although there are farms where the stag is run with the hinds all year round and is only replaced every few years.

Of assisted reproduction techniques artificial insemination is used on some farms, but mainly for research. Semen has been collected from live animals and also post mortem from stags shot in the wild, however due to the lack of interest, it was not marketed and the development of the sector stopped. Embryo production and/or implantation does not take place in game species although it may occur in zoos.

6.2   Calving and Weaning

The main calving season is around mid-May for the reds, and at the end of May and the beginning of June for fallow. On most farms the calving paddocks are in wooded areas with good cover and shelter. The golden jackal (*Canis aureus*)—has become more numerous and extended its range in the last two decades and is a growing problem in the South-West of Hungary, especially during calving when they can cause big losses through the killing of young calves.

There are many different strategies adopted for weaning with the calves being separated at various times from the early (mid-August) to late weaning (end of November). There are even farms which do not separate the calves from their mother. The system chosen depends on the farm conditions: its pasture quality and productivity.

6.3   Breeding Organization

The Association of Hungarian Deer Breeders was established in 1998, then in 2010, it was transformed into the Association of Hungarian Game Breeders. The primary goal of the organization is to represent the interests of farmers, to promote game meat, to preserve and pass on the culture of hunting and game management. It currently has 56 members, many of whom are engaged in wild boar breeding.

7 Feeding

The keeping of cervids is primarily based on pasture, although the amount and distribution of annual precipitation in Hungary cannot be called ideal for grassland management. Irrigation is not typical on deer farms, so in the summer months, the grasslands burn out due to the heat and the lack of rainfall. In addition to this, it is also true that in Hungary grassland management is "in its infancy". One of the main reasons for this is that intensive field crop production results in the greatest profit, so grassland management is restricted to those areas with poor productivity. There is no research on grassland management and so little knowledge reaches farmers.

We find planted grasslands on only a few farms, so the standard of farming lags far behind the possibilities. Ancient grassland associations also contain valuable plants that can be put to good use with regular care. Unfortunately, the high demands of the Hungarian red deer cannot be fully met on medium-quality pastures. In this respect, fallow deer can achieve much better results on domestic grasslands. In both species, it may be necessary to feed supplementary feeds, primarily grains of corn, barley, or oats.

On medium-quality grasslands, 3–4 red deer or 8–10 fallow deer can be kept per hectare, on planted grasslands, with good production site conditions, this number can be increased by 45–50%.

Several research programmes have been carried out and are ongoing in the Game Management Centre of the University of Kaposvár, the aim of which is to improve the quality of grassland management in hilly areas. Possibilities for improving pastures are also being explored in hunting park and farm conditions.

In Hungary, two trends can be observed in the field of red deer keeping and feeding. One is the primarily grassland-based husbandry, where a small amount of supplemental fodder is used, and the other is where large amounts of grain are fed with supplementary forage, primarily alfalfa hay.

The former management system based on former extensive farms suits the digestive physiological and ethological or behavioural characteristics of red deer much better. As a result, there is less incidence of metabolic diseases and injuries caused by stress and of course also the conditions are much more aesthetically pleasing. Obviously however, these farms also need more investment and land.

In the case of the intensive husbandry method, it can be said that the advantage is that it requires less area and investment and that stags with a high antler weight can be reared in a short time.

8   Animal Health

In Hungary, the official health procedures for game species are determined by the legislation created for the closest related domesticated species. The legislation applicable to cattle is therefore applied on red and fallow deer.

According to the current legislation, a tuberculosis and a brucellosis test must be performed within the EU or in Hungary within 30 days prior to transport and, if all are negative, may be shipped after the necessary accompanying documents (veterinary declaration, traces, etc.) have been issued. Of course, in addition to what is required by law, the buyer may also request other tests.

8.1   Viral Diseases

Bluetongue
In 2014, bluetongue appeared in cattle populations in Hungary. Although it does not cause disease in deer, it has significantly limited livestock sales opportunities.

Vaccination of susceptible species in the protection and surveillance zone around an outbreak of bluetongue is required by the authority of the national chief veterinarian at the expense of the keeper. In case of vaccination unique marking and traceability are very important. This can be difficult for cervids because there is no central identification register.

Immunization of animals in areas free from bluetongue is prohibited. The last outbreak of blue tongue disease occurred in 2018.

8.2   Bacterial Diseases

Brucellosis
Testing for brucellosis (*Brucella abortus*) is obligatory during animal transport, but there have been no cases of the disease occurring under enclosed conditions.

Mycobacteria
Tuberculosis (*Mycobacterium bovis*, *Mycobacterium avium*) occurs in the country and has also been detected in red deer and fallow deer in some infected regions, but populations of wild boar play a significant role in the maintenance and spread of the disease. At the University of Kaposvár, Csivincsik et al. (2015) examined the occurrence of *M. bovis* in the South Transdanubia region in 833 boars shot, where 30% infection was found.

The intradermal (*M. bovis*) skin test is required to be performed before each shipment. Blood test results are not accepted by the authority, therefore its application is not widespread.

Paratuberculosis (*Mycobacterium avium* paratuberculosis) infection occurs sporadically. However, the disease still remains at a low level. The reason for this may be that very intensive housing conditions are not typical in Hungary, and the domestication of the Hungarian red deer has not yet reached the level where the functioning of the immune system declines.

8.3 Cervid Parasites

Flukes

Deer are susceptible to the common liver fluke (*Fasciola hepatica*), but severe infections are rare in the wild, free-ranging population. Under farm conditions infection is very rare.

American liver fluke (*Fascioloides magna*) has also appeared in Hungary, according to the investigations of Králová-Hromadová et al. (2011) who found that there was a wave of infection in the floodplains along the Danube starting in the 1990s; it also reached Austria, Slovakia as well as Hungary. It often occurs in red deer, fallow deer, and roe deer population in floodplain areas and is still spreading. Medicinal treatment of infected populations is cumbersome, the parasite causes serious damage and occasionally results in significant quality deterioration.

Nematode Worms

Lung nematode parasites (Dictyocaulus species and members of the Protostrongylidae) as well as intestinal nematodes are the most common causes of clinical disease. In young adults, especially in the spring, if their immune systems are weak, they can cause problems. Deer farmers are using antiparasitic drugs, twice a year, in autumn and spring, and in many cases without any test—just to be on the safe side.

In Hungary, it is not possible to perform the Carla test, which indicates the degree of resistance to internal parasites.

8.4 Anomaly of Antler Pedicle

In the early 1990s, wild fallow deer bucks, living in free-range areas were found to have a unilateral distortion of the palmated antler with inflammation and purulence of the skin tissue around the antler pedicle. The anomaly was described in the hunting areas of Somogy and Tolna counties (in the southern part of Hungary) and affected a large population of fallow deer. The first research reported *Staphylococcus xylosus* infection (Gál et al. 2011).

It has appeared in other fallow deer populations over the past decades and we now find it also in hunting park conditions. Those two counties remain the worst affected but the infection is also present in another seven counties.

It causes very significant damage in some hunting areas, 45–50% of stags can be affected by the disease.

Researchers initially considered it species-specific, but they have already found such an anomaly in red deer stags and roe deer, although in much smaller numbers.

Fig. 8 Broken pedicle, abnormal casting plane (Julianna Bokor)

In the spring of 2016, when the cast fallow deer antlers were collected, five fallow antlers were found in the hunting park of Bőszénfa, in which there were pieces of pedicle under the burr (Fig. 8).

The fracture surface of the broken pedicle was very similar to the normal antler casting plane, indicating abnormal functioning of bone-eating cells, the osteoclasts. Because of this in the following hunting year all fallow bucks with abnormal antlers were shot in Bőszénfa (Fig. 9).

The proportion of stags with abnormal antlers has not changed so far and only 3–5 individuals are shot per year, which accounts for 10–15% of all fallow stags shot. The anomaly has not yet affected red deer stags and roe bucks in the park.

Due to the spread and the serious damage caused by the disease, a consortium was formed involving several universities and research institutes to explore the problem and the causes. The studies also cover the study of all sick and healthy individuals in the populations and consider environmental effects. In addition to these antler anomalies, reports also reported reproductive problems including prolonged oestrus and testicular inflammation but without any deterioration in condition. The study has shown that the bacterial infection is secondary and it is not the trigger. Today, it is thought to be a polyfactorial disease caused by a combination of several factors including the adverse effects of fungal toxins.

Fig. 9 Fallow stag with abnormal antler at Bőszénfa (Máté Nagy)

9   Game Capture

For farms, hunting parks, and game parks, animals have often to be captured for handling or animal transport. With the development of deer farms and the spread of hunting parks, there was a great demand in the 1990s to capture live red and fallow deer.

9.1   Live Game Captures

In 1990, a new method was developed at the University of Kaposvár. Previously wild deer had been captured by feeding them into yards but this caused a lot of injuries to the animals, it was expensive, and often ineffective as well. Some researchers then started to use the "Komarov bullet" 0.22 LR where the drug was put into the drilled hole of the ammunition core (Succinylcholine Chloride). Another solution was to mix sedatives such as diazepam (Seduxen) into the feed of the animals. These techniques were used during the live capture of deer for export to New Zealand, with relative low efficiency, and also with deaths. For these reasons a more gentle technology was developed by using 20 m × 20 m nets, which were hung on metal legs above the feeding place so that deer (and occasionally, mouflon and

Fig. 10 Game capture with net (Archive of Game Management Centre)

wild boar) could easily go under the net (Fig. 10). The stretched net was "shot" at the animals below with an explosive device, and the deer became entangled. After that it was easy to narcotize them with a hand syringe, than take them out from the net, and place them in a transporter.

After the initial difficulties, deer were captured in big state owned forestry companies like GEMENC and SEFAG. These hunting grounds are the best habitat for red deer, and the quality is also outstanding (three world record stags were taken in these hunting grounds in 1968, 1981, and 1986). At that time there were so many deer, that on a cold winter day, several groups of 150–200 head would follow a horse drawn cart while the daily corn portion was fed out. This capture system became the best method, around 25–30 animals were caught at a time, with the best result being 44 animals. In addition, because the more careless (and hungriest) youngsters went under the net most easily, the older deer stood outside looking upwards and spying at the unknown object. This created the most favourable solution because the farms required young individuals, they were much cheaper, and it was much easier to get them out from the net, and transport them to the farm. Thanks to the successful captures in GEMENC, these floodplain deer became the most important genetic component on the farm at Bőszénfa.

The effectiveness of the method is well demonstrated by the fact that between 1990 and 1995, more than 400 red deer were captured from different hunting areas without significant mortality. One of the disasters was at the very beginning, the first attempt in Kaszó, which belonged to the Hungarian Defence Forces, as a VIP hunting ground. With no experience but big enthusiasm the biggest stag of the area and some hinds were captured on the first day. At that time nobody knew that the antlered animals are much more complicated to work with, and the stag turned his antler under himself, and drowned before he could be taken out from the net. It required an awkward explanation to the leaders of the army. This painful lesson was very valuable over the following years. Capture with nets proved to be the most successful method, with very few injuries, being safe for the staff and also for the animals.

The same technique was used when introducing fallow deer and mouflon into the hunting park at Bőszénfa.

9.2   Transport

In the 1980s, hunters used individual boxes for transport, this required the members of the mob to be separated causing another stressful situation for the animals. Many of them jumped against the walls of the box and became seriously injured. New techniques were developed during the New Zealanders' visits and improved the live captures. For group transport, a boxed transport vehicle had to be developed, where in one box 2–3 stags, or 6–8 hinds and calves could be transported at the same time.

Nowadays, the rules and guidelines for animal transport in Hungary are defined by the Animal Protection Act. The 30 years of work of the experts in the Game Management Centre of the University of Kaposvár have greatly contributed to the development of the transport methods that are routinely operating today.

In the case of domestic animal movements, tuberculosis and brucellosis tests must be carried out before transport. For shipments abroad, the requirements of the country of destination must be met during the examinations.

The transport of large game requires a special vehicle (Fig. 11), which must be approved separately for this purpose. Few companies have such a permit, so deliveries are handled by only a few people/companies in the country.

10   Commercialization

10.1   Venison Sales

In Hungary, until now, farm-like venison production from enterprises run using the type of management developed in New Zealand has not been able to develop significantly, the main reasons for which are:

– Few livestock breeders keep deer, the numbers of individuals are low, and the amount of product coming from farms is small.
– Deer breeding does not receive state subsidies, so it is at a competitive disadvantage compared to beef cattle and sheep breeding, therefore efficient systems cannot be established.
– In Hungary, the consumption of venison per head is barely more than 200 grams per year, one of the main reasons for which is the lack of retail demand.

Fig. 11 Animal transport—the modern way (Foki Károly)

- The Hungarian red deer is not the most ideal subspecies for venison production, its selection in this direction is still in the initial phase. The animals are sensitive to stress and find it difficult to bear the handlings and paddock changes. Large population groups that can be grazed efficiently cannot be formed, high individual densities adversely affect the growth of animals, and the risk of developing diseases increases.
- The red deer is able to reach a high body weight (the adult live weight of red deer stags can reach 300–350 kg, that of hinds 150–160 kg), but it is almost impossible to meet the high nutrient demand without forage, which significantly increases the costs of keeping.
- Hungarian red deer are slow maturing reaching favourable carcase weights relatively late so that meat production indicators measured at the usual slaughter age of 14–16 months remain below the data published in New Zealand.

In the future, it is expedient to develop the conditions for venison production further, to examine the possibilities for fallow deer breeding for meat production, and to increase the selection of the red deer population. In order to increase the consumption of venison, marketing activities need to be improved.

The gastronomic revolution has brought significant change in recent decades. Thanks to televised cooking programmes and popular movements towards healthy eating, venison has become increasingly important in the offerings of restaurants. As a result, demand is rising, especially for products from individuals shot or slaughtered at a young age.

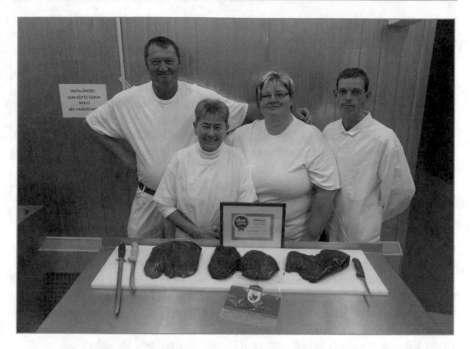

Fig. 12 Employees of the processing plant with the "Golden Ribbon" award (Archive of Game Management Centre)

To explore the potential of the venison trade, the University of Kaposvár established the Zselicvad venison processing plant in 2002, which significantly contributed to the promotion of venison.

Between 2003 and 2010, the turnover of the plant increased due to the product presentations, and its products are especially popular among demanding restaurants. As it is a small plant, excellent quality and reliability are the most important aspects from the very beginning. This was recognized by the Hungarian Gastronomic Society when it awarded the "Golden Ribbon" prize to the plant's products in 2013 (Fig. 12).

In connection with venison consumption, the Department of Agricultural Marketing of the University of Kaposvár conducted several studies in order to find out the characteristics of domestic consumption (Torma et al. 2014). In 2014, in a focus group study, the participants characterized venison as follows:

- "Venison is prestigious"
- "Venison consumption is a community experience"
- "Venison is usually eaten in a restaurant (at gatherings of friends and relatives)"
- "Special", "Unusual"
- "Game-based foods are superior compared to other types of meat"
- "More delicious, tastier, healthier"

- "Tradition, health, natural, success in the kitchen"
- "Expensive, inaccessible, unknown".

72% of adult men participated in the study are happy to eat venison, the consumption of which is especially popular among conscious eaters.

All this provides an excellent basis for a significant increase in domestic venison consumption.Direct marketing of venison from deer farms is used in many countries with good results. Unfortunately, this form of sale cannot be adopted in Hungary because strict animal health conditions prevent direct sales.

10.2   Livestock Sales

Typically, deer farms have the facilities and a suitable livestock population to conduct livestock sales. Also this market segment provides the greatest revenue opportunity, especially if the sale is for export purposes. The quality of the Hungarian red and fallow deer is world-famous, especially the excellent trophy quality of the antlered males shot during hunting. The most significant market is in Eastern European countries, where the establishment of hunting parks has become fashionable after the land privatization.

As the purchases are mainly made for hunting purposes, breeders can achieve high selling prices.

As a result, farmers have mainly specialized in the breeding and selling of animals for breeding.Between 1994 and 2004, more than 600 red deer were sold from Bőszénfa to 13 European countries. Hungarian red deer can be seen on Fig. 13.

However, it must be accepted that this market may easily become saturated with and buyers becoming sellers, and so reducing sales opportunities. Even today, there is already a significant decline, the hunting parks have been replenished, and today, instead of buying a large number of individuals, customers are only looking for individuals with outstanding qualities.

Rearing trophy game for hunting and domestic sales is relatively recent. An amendment to the Hunting Act in 2015 made it possible for breeders to sell live animals from farms to hunting parks. However, rearing trophy stags takes a lot of time, and small farms need to cut the antlers from the stags every year in order to prevent them injuring each other. This work requires expertise, experience, and a well-functioning handling system. The law continues to prohibit the release of game reared on farms into the wild.

10.3   Hunting in Park

Although hunting in enclosed areas is controversial and divides the hunting community because of ethical issues, the right of parks to exist is indisputable. The management of deer in hunting parks has many advantages: hunting seasons do not have to be kept, the trophy quality and age of the individuals that can be shot are

Fig. 13 Hungarian deer in Bőszénfa (Archive of Game Management Centre)

not regulated, and the shooting of the desired wild species can be ensured in a shorter time. In addition, game management of wild deer in free-range areas has been affected in recent years by significant adverse effects which create many problems for hunting—they "drive water to the hunting parks' mill". Thus the increase in damage caused by game, the negative effects of fencing systems and electric fences surrounding afforestation and agricultural land, the increase in the number of collisions between wild animals and vehicles, and the development of hiking tourism are leading to a steady decline in red deer living space. Today, hunting parks need to be set up to ensure adequate living conditions and tranquillity for the game population.

The most difficult challenge is to achieve success in hunting park conditions for red deer because the Hungarian climate cannot be considered as ideal for cheap feeding: for grassland management. Parks are usually established in forested areas, where there are minimal possibilities for the creation of the required amount of game cover crop and game pasture. In addition, park owners often want to keep wild boar in the park in addition to ruminant game species, and this further reduces the chances for rational pasture management. Without the regular use of fodder from external sources, the appropriate level of production, reproduction, and antler quality of the population cannot be ensured.

It is much more common to keep fallow deer in a hunting park, the benefits of which have been discussed earlier. Significant trophy weights can be achieved in parks where the conditions for grassland management are good and the proportion of pastures is around 20–30%.

The possibilities for deer farming in the hunting park are well illustrated by the hunting park Bőszénfa, where more than 700 big game live on 604 ha. The varied surface area and vegetation has always provided popular living conditions for Zselic's big game. Its woodland cover remains significantly below the average for hunting parks: the proportion of forests is only 45%, providing ample space for the establishment of wild pastures.

The dominant game species are red deer and fallow deer, both of which have a stock of more than 300 individuals in the park. Mouflon and wild boar can also be found in the hunting area, although the latter occurs rarely.

In order to keep the number of red deer stags at the same level, 20–25 red stags will be moved into the hunting park in June from the breeding area of the farm. This gives time for the stags to get used to their new living space so that they will behave naturally during the hunts. With this method, the appropriate age composition and size of the population can be ensured. Nowadays, around 40 red deer stags, 15–20 fallow deer stags, and 6–8 mouflon are shot every year, which can ensures the profitable operation of the park.

The hunting park Bőszénfa performs above the national average, the reason for which is to be found in the good genetic foundations, the continuous development of game cover management, and the reduction of the wild boar population.

Several national records have been shot in the last decade, with the highest-scoring stag shot in 2010 with a CIC score of 262, while the highest trophy weight in 2019 was 15.4 kg (Fig. 14). With regard to fallow deer, a buck shot in the 2020 season achieved the highest score and trophy weight till now, weighing 5.55 kg and scoring 208.64 CIC points.

10.4 The Tourist Attractiveness of Deer Breeding

The sight of wild animals at close quarters attracts many visitors to zoos and game parks. The red deer has a particularly important and mythical significance for the Hungarians, so from the very beginning there has always been a need for visitors to see and touch the animals (Fig. 15). Taking advantage of this, many people have set up demonstration parks, where the game species which can be hunted in Hungary can be seen. In many of these parks educational projects take place and liaison with forest schools is encouraged. Today, our most characteristic wild and domestic animals need to be shown not only to children who live in cities but also to generations growing up in the countryside who also only rarely encounter these animals.

Game park services are usually accompanied by other programmes encouraging traditional occupations, biology lectures, and also the sale of souvenirs connected with animals.

Fig. 14 Red deer with record score in Bőszénfa (János Nagy)

Fig. 15 Visitors in the Bőszénfa game park (Julianna Bokor)

10.5 Velvet Antler and Other By-Products

Every single particle of deer is valuable, but real market opportunities are not easy to exploit. Deer leather, for example, is an excellent raw material, but the leather industry in Hungary has shrunk, and large quantities of hides of uniform consistency are needed to take advantage of export opportunities.

One of the main barriers to the use of deerskin in the case of red deer is infection by the larvae of the warble fly. The vast majority of individuals shot in the wild, free-range areas, or in hunting parks are infected, making the skins unsuitable for processing.

The tendons, deer tails, and the genitals of stags are sold. They are collected by buyers and exported to Far Eastern markets.

The collection and trade in deer antlers has been surrounded by many abuses, especially since the increase in demand. In the last 5 years, purchase prices have doubled, so unauthorized people are also involved in the black trade.

In the case of red deer, the velvet antler represents a higher value than the cast or fallen hard antlers. During the 1990s, attempts were made to produce velvet in Bőszénfa. The velvet antler cutting is a high-precision operation that can only be performed under farm conditions and with veterinary intervention. The low number of domestically bred deer population did not allow the production of competitive quantities, and the largest batches were only around 2000–2500 kg per year. Initially, $ 100 per kilogram of frozen antlers was paid, which decreased to $ 50 by the turn of the millennium. Low prices, ever-increasing costs, and EU regulations eventually led to the abandonment of velvet antler production in 2004. In addition to the economic considerations, a number of animal welfare concerns have also been raised, which also accelerated the cessation of production.

References

Ákoshegyi I (2005) Nagyvadtenyésztés és farmi vadtartás. Jegyzet vadgazda mérnöki szakos hallgatók részére. Szent István Egyetem Vadbiológiai és Vadgazdálkodási Tanszék. Gödöllő. p 90

Bokor J (2016) Egyes fenotípusos tulajdonságok összefüggései gímszarvasban. Dissertation, p 165

Bokor J, Nagy J, Szabó J, Bokor Á, Szabari M, Horn P, Nagy I (2010) A magyar gímszarvasok (Cervus elaphus hippelaphus) növekedése választástól 9 hónapos korig. Agrár- és Vidékfejlesztési szemle 5. évj 2010(1):398–403

Bokor J, Nagy J, Szabó A, Nagy I, Szabari M, Bokor Á, Horn P (2015) A közép európai gímszarvas (Cervus elaphus hippelaphus) borjak növekedésének modellezése születéstől 7-8 hónapos korig. Magyar Állatorvosok Lapja 137(10):633–640

CIC Homepage (2020). www.cic-wildlife.org. Accessed 15 Dec 2020

Csányi S, Márton M, Kiss K, Köteles P, Schally G (2020) Vadgazdálkodási Adattár - 2019/2020. vadászati év. Országos Vadgazdálkodási Adattár, Gödöllő, p 66

Csivincsik Á, Rónai Z, Nagy G (2015) Post mortem examination of submandibular lymph node in wild boars (sus scrofa) as a beneficial part of bovine tuberculosis surveillance systems. Acta Agraria Kaposváriensis 19(1):8–12

Csőre P (1997) Vadaskertek a régi Magyarországon. Mezőgazda kiadó, p 140

Csőre P (2000) A magyar vadászat története. In: Sándor O (ed) A magyar vadászat ezer éve. Millenniumi Vadászati Bizottság, Budapest, pp 21–36

Draskovich I (2007) Szarvasgazdálkodás: tanulmányok a hibás gazdálkodásról és tévedésekről, javaslat az országokban alkalmazható állománykezelésre. Budapest, Vadászlap Kft, p 94

Gál J, Jánosi K, Marosán M, Sugár L (2011) Purulent inflammation of the basis of antler in fallow deer (Dama dama) caused by staphylococcus xylosus. Magyar Állatorvosok lapja 133(3): 161–164

Horn P (1987) Új hústermelő ágazat: gímszarvastenyésztés. Az Állattenyésztési Tudományos Napokon elhangzott előadás. Állattenyésztés és Takarmányozás TOM 37(2):106–112

Karaiskou N, Tsakogiannis A, Gkagkavouzis K, Papika S, Latsoudis P, Kavakiotis I, Pantis J, Abatzopoulos TJ, Triantaphyllidis C, Triantafyllidis A (2014) Greece: a Balkan subrefuge for a remnant red deer (Cervus elaphus) population. J Hered 105(3):334–344

Králová-Hromadová I, Bazsalovicsová E, Štefka J, Špakulová M, Vávrová S, Szemes T, Tkach V, Pybus M (2011) Multiple origins of European populations of the giant liver fluke Fascioloides magna (Trematoda: Fasciolidae), a liver parasite of ruminants. Int J Parasitol 41:373–383

Krojerová-Prokešová J, Barančeková M, Koubek P (2015) Admixture of eastern and western European red deer lineages as a result of postglacial recolonisation of the Czech Republic (Central Europe). J Hered 106(4):375–385

Ludt CJ, Schroeder W, Rottmann O, Kuehn R (2004) Mitochondrial DNA phylogeography of red deer (Cervus elaphus). Mol Phylogenet Evol 31(3):1064–1083

Metzger Sz (2013) Vadászati jog és igazgatás. „E-tananyag" a Vadgazda BSc szak hallgatói számára. p 125

Nagy J, Szabó A, Donkó T, Bokor J, Romvári R, Repa I, Horn P, Fébel H (2019) Body composition and venison quality of farmed red deer (Cervus elaphus) hinds reared on grass, papilionaceous or mixed pasture paddocks. Arch Anim Breed 62(1):227–239

Niedzialkowska M, Jedrzejewska B, Honnen A-C, Otto T, Sidrokovics VE, Perzanowski K, Skog A, Hartl GB, Borowik T, Bunevich AN, Lang J, Zachos FE (2011) Molecular biogeography of red deer (Cervus elaphus) from eastern Europe: insights from mitochondrial DNA sequences. Acta Theriol 56(1):1–12

Skog A, Zachos FE, Rueness EK, Feulner PGD, Mysterud A, Langvatn R, Hmwe S, Lehocky I, Hartl GB, Stenseth NC, Jakobsen KS (2009) Phylogeography of red deer (Cervus elaphus) in Europe. J Biogeogr 36(1):66–77

Stéger V, Molnár A, Borsy A, Gyurján I, Szabolcsi Z, Dancs G, Molnár J, Papp P, Nagy J, Puskás L, Barta E, Zomborszky Z, Horn P, Podani J, Semsey S, Lakatos P, Orosz L (2010) Antler development and coupled osteoporosis in the skeleton of red deer Cervus elaphus: expression dynamics for regulatory and effector genes. Mol Gen Genomics 84(4):273–287

Szabó A, Nagy J, Bokor J, Fébel H, Mezőszentgyörgyi D, Horn P (2013) Clinical chemistry of farmed red deer (Cervus elaphus) yearling hinds reared on grass or papilionaceous pasture paddocks in Hungary. Acrh Tierz Arch Anim Breed 56(1):443–454

Szidnai L (2013) Dr. Szederjei Ákos erdőmérnök. Erdészeti Lapok. CXLVIII. évf 3:86–88

Torma D, Böröndi-Fülöp N, Szente V (2014) Vadhúsfogyasztói magatartás a minőségi szarvashús imázsa és pozícionálása. évf 1(1–2):141–147

Zomborszky Z, Zubor T, Horn P (1999) Sperm collection from shot red deer stags (Cervus elaphus) and the utilisation of sperm frozen and subsequently thawed. Acta Vet Hung 47(2):263–270

The Management of Enclosed Deer in Latvia

Dainis Paeglitis

Abstract

The history of deer enclosures in Latvia is described and the current numbers of deer within deer farms. The regulations controlling the development and management of both intensive and extensive deer farms are explained and the creation and present status of associations representing deer breeders together with a national breeding programme and an annual antler competition. The various income streams: venison; live deer sales; trophy hunting; and tourism are mentioned. A scientific research programme is proposed.

Keywords

Deer farms · Trophy · Antler competitions · Regulations

1 History

The beginnings of deer farming in Latvian territory can be traced back to the seventeenth century in the Duchy of Courland (Lange 1970). These were hunting parks near the manors of noble rich people.

The best known of them are the families of Duke von Ketteler, Duke von Biron, Count von Medem, Count von Borh, Furst von Lieven. By the end of the nineteenth century, the number of parks exceeded 20 but during the first World War and the subsequent revolution most parks were destroyed and the rest ceased to exist during the second World War (Skriba 2011). During the Soviet period under a policy for population recovery programme a number of temporary enclosures were erected for deer before releasing them into the wild.

D. Paeglitis (✉)
BLSDAA Association, Rīga, Latvia

© The Author(s), under exclusive license to Springer Nature Switzerland AG 2022
J. Fletcher (ed.), *The Management of Enclosed and Domesticated Deer*,
https://doi.org/10.1007/978-3-031-05386-3_11

229

Deer breeding in Latvia resumed after the collapse of the USSR and the restoration of Latvia's independence. The first permit for the establishment of a deer farm issued by the Ministry of Forestry was received in 1994 by the Farm "Saulstari-1". To start this herd animals were imported from Poland, UK, Hungary, Sweden, Denmark.

Currently there are 119 deer farms in Latvia enclosing a total of 17,000 animals made up of 13,000 red deer and 4000 fallow deer (www.ldc.gov.lv). The total fenced area is about 12,000 ha. With individual farms having fenced areas ranging from 2 to 700 ha. Most of the larger enclosures (over about 30 ha) are game parks. Some ten of the enclosures, with territory of over 100 ha are sub-divided into park and more intensive breeding areas.

2   Regulations

In order to be granted a permit for the installation of a deer farm, the technical design of the fencing and a short business description must be approved by the Construction Board and the Nature Protection Board (Regulation of the Cabinet of Ministers No.4 from 04.01.2011).

They identify two types of enclosure:

Intensive breeding areas—in which woodland must represent no more than 30%, but the number of animals is not limited (however the recommendations of BLSDAA (see below) are for a maximum of 10 red deer or 20 fallow deer per 1 ha pastures).

Slaughtering with a free bullet is permitted.

Extensive breeding areas—these aim to provide animals with living conditions similar to the wild, in a minimum area of 10 ha with a maximum stocking density of one red deer or two fallow per hectare, or in mixed herds one red and one fallow require two hectares.

Hunting is permitted in accordance with legal hunting regulations.

Animals can be moved from intensive areas into extensive areas and vice versa.

For *organic farming* in intensive breeding areas the stocking density is restricted to seven red deer or ten fallow deer per ha.

In order to receive grassland subsidies the stocking density must be at least 0.3 cattle units per ha. one red deer corresponds to 0.6 cattle units and 1 fallow deer corresponds to 0.3 cattle units.

2.1   Association

In 2000, the number of deer breeders reached 14 and the Wild Animal Breeders Association (SDAA) was established. They elected as Chairman Dr. Dainis Paeglitis, the owner of Farm "Saulstari-1" and vice president of the International Deer and Ungulate Breeders Association (IDUBA). Today there are 40 members in the association and it has been renamed the Organic Farmers and Wild Animal

Breeders Association (known as the BLSDAA from its Latvian name). The Association is a member of the Federation of European Deer Farmers' Associations (FEDFA).

2.2   Breeding Programmes

In 2005, the BLSDAA developed red deer and fallow deer breeding programmes which were approved by the Ministry of Agriculture and the BLSDAA became an Approved Deer Breeding Organization. The programmes were revised in 2019 and will be valid until 2029.

The Association has 23 certified breeding farms under its supervision—17 red deer farms and 6 fallow deer farms with 4554 red and 670 fallow deer.

According to the breeding programmes all animals must carry identification numbers and undergo an annual genetic quality assessment based on a 10-point system. This assessment evaluates the exterior appearance, the level of aggression, susceptibility to stress, productivity, antler quality, and the quality of the progeny of each animal in an effort to assess its genetic potential. The parentage of each animal through two generations must be recorded. All data is then registered in the electronic database of the BLSDAA.

About 1000 tissue samples each year are collected by an ear punch and sent to a laboratory in New Zealand for DNA analysis. This provides data on paternity, maternity, the level of inbreeding, and a genetic relationship matrix for all tested animals. It also provides data on the genetic background by evaluating the percentages of genetic composition across three bloodlines: English red deer; East European red deer, and wapiti. This work started in 2018 with the costs met by the Ministry of Agriculture. On the basis of the registered data the Association is entitled to issue zootechnical certificates for animals.

From 2008, the Association has organized an annual international deer farmers' conference and an international competition for the cast and cut antlers of breeding stags which is attended by representatives from more than 20 countries. The antlers are evaluated using a modified CIC (Conseil Internationale de Chasse) system (with measurements verified) by the certified experts. Measures are supported by the Ministry of Agriculture.

Antler competition results 2020 (BLSDAA)

№	Farm/Owner	Country	Animal Nr.	Mod. CIC points	Characteristics of antlers			
					Antler type	Length (cm)	weight (kg)	Numb. of tines
1	Deer farming Germany (Hans Honold)	DE	Trompeter	318,63	Engl.	99.25	16.8	40
2	LLC"Game Estate" Dnepr-Kholm"(S. Lvov)	RU	011/08	298,18	Engl.	85.5	24.12	53

Red deer competition 2020 GRAND PRIX antlers 2020.07.23.

(continued)

Red deer competition 2020 GRAND PRIX antlers 2020.07.23.

№	Farm/Owner	Country	Animal Nr.	Mod. CIC points		Characteristics of antlers			Numb. of tines
						Antler type	Length (cm)	weight (kg)	
3	Safariparks MORE (MB)(D.Paeglitis)	LV	144-Roko	285,6		Engl.	90	15.6	43
4	Deerbreeding Schober (F. Schober)	AUT	Zeus	277,18		EEU	120.75	17.1	27
5	LLC"Game Estate" Dnepr-Kholm"(S. Lvov)	RU	1401	273,37		Engl.	79.25	17.26	32
6	LLC"Game Estate" Dnepr-Kholm"(S. Lvov)	RU	1410	271,66		Engl.	81.7	17.46	37
7	Deerbreeding Schober (F. Schober)	AUT	Blau 17	266,03		EEU/ Engl.	110.5	14.64	36
8	LLC"Game Estate" Dnepr-Kholm"(S. Lvov)	RU	1511	265,11		Engl.	74.5	15.3	36
9	Safariparks MORE (AJ)(D.Paeglitis)	LV	2170	264,725	EEU	122.5	13.7	22	
10	Amila (Roman Mykytiuk)	U	2212	260,85	Engl.	70.15	18.4	27	
11	Deerbreeding Schober (F. Schober)	AUT	Blau 6	259,71	EEU	114.75	14.6	27	
12	Deer Farming Germany (Hans Honold)	DE	Max—33/ 15	255,63	Engl.	96.05	12.7	36	
13	LLC"Game Estate" Dnepr-Kholm"(S. Lvov)	RU	011/10	250,93	Engl.	97.9	8.15	24	
	Antler type— expert opinion								
	Engl.—english								
	EEU—east european								

2.3   Income Streams

For farms with extensive breeding areas income is derived from hunting, sales of venison usually as whole carcasses (to restaurants, shops, or individuals), and ecotourism.

For farms with intensive breeding areas the income is from sales of venison and hard antler, tourism, as well as sales of breeding stock. The sale of breeding stock is more specialized and from those intensive breeding farms with paddock systems and handling facilities (breeding farms) which can sell genetics as well as tourism, venison, and hard antler.

The highest revenue item is the sale of breeding animals with average prices for pregnant red deer females of 500–2000 EUR, for breeding stags 1000 EUR for each year of age with trophy quality animals carrying a bonus of 1000 EUR/kg hard antler weight. For stags with antlers weighing more than 15 kg prices are subject to discussion.

The established markets for sales of breeding animals from Latvia are Estonia, Lithuania, Russia, Belarus, Ukraine, and Azerbaijan.

The wholesale prices for deer carcasses in Latvia are 1.5–2 EUR/kg.— Restaurants, shops??

And the price for hard antlers 15–20 EUR/kg.

The above figures mean that the approximate annual income revenue for a high-quality breeding/hunting farm with 400–500 animals and about 300 ha land would be:

- Animal sales—70,000 EUR;
- Tourism—50,000 EUR
- Trophy hunting—25,000 EUR
- Meat and antler trade—18,000 EUR
- Total: 163000 EUR/year

2.4   Scientific Activities

The BLSDAA and four breeding farms cooperate with the Latvian Agriculture University and JSV "BIOLAT" in the implementation of a 3 year collaborative scientific project financed by ERAF. This project is an investigation into the modern sustainable production and processing of deer to achieve high added value food, and products for the cosmetic and pharmaceutical industries.

The objective is to ensure the efficient sustainable processing of carcasses including extending the shelf life of fresh venison and developing the technology to produce and process by-products.

It is planned to develop innovative technology to produce high-quality raw material from the primary processing of by-products which can then be further processed into higher added value products such as tendons, blood, tails, genitals, embryos, placentas, pituitary glands, and internal organs for the traditional medicine trade.

Project title	Modern bio-economic waste-less processing of deer, producing the end product—Raw material with a high added value for food, cosmetics, and pharmaceutical industry.
Project objective	To ensure waste-less processing of red deer carcasses, extending shelf life of fresh venison and developing slaughtering technology to produce and process by-products. It is planned to develop an innovative slaughtering technology to produce high-quality raw material from deer slaughtering (primary treatment) by-products, which potentially can be processed into products with a high added value—Tendons, blood, tails, genitals, embryos, placentas, hypophysis, internal organs.

References

Lange W (1970) Wild und Jagt in Lettland. Verlag hare von hirscheydt in Hannover-Dohren
Skriba G (2011) Staltbriežu izcelsme, izplatība un audzēšana Latvijā. Rīga

The Management of Enclosed Deer in Poland: A Personal Experience

Bartłomiej Dmuchowski

Abstract

This is a discussion of modern deer farming in Poland from the middle of the twentieth century to the present from the personal experiences of one of Europe's most dedicated deer farmers. Early research efforts into antler development and artificial breeding technology were at world level and were followed by the establishment of state-of-the-art deer farms. However, an absence of support from the government and continuing conflicts with the hunting community have created a situation where deer farming in Poland is now in a still difficult and unstable legal and organizational situation and at a low level.

Keywords

Deer farming · Poland · Semen collection · Hybrids · Venison · Hunting

1 The Rise and Fall of Modern Deer Farming in Poland

When discussing the situation in Poland, I will focus on contemporary times, because it is difficult to find important historical or traditional connections. Yes, in the past, there were animal parks in Poland, for example, at the royal courts, and other similar more or less successful attempts to keep wild animals, including deer, in various types of enclosures. In practice, however, nothing survived physically that could give rise to the development of modern deer farming, as was the case in other countries as, for example, in the UK with established parks as at Woburn or Warnham.

B. Dmuchowski (✉)
Company FHU MILU, Mrągowo, Poland
e-mail: milu@milu.com.pl

© The Author(s), under exclusive license to Springer Nature Switzerland AG 2022   235
J. Fletcher (ed.), *The Management of Enclosed and Domesticated Deer*,
https://doi.org/10.1007/978-3-031-05386-3_12

Therefore, one should start with the establishment in 1956 in Popielno of the Research Station of the Institute of Genetics and Animal Breeding of the Polish Academy of Sciences, where, among other things, a small experimental deer farm was created. It aimed to conduct research into the biology of these animals. Of more than 250 works written here, the majority were published in scientific journals of a global scope, cited and mentioned in many publications. The most important achievements include the work of Prof. Zbigniew Jaczewski on the mechanisms of the growth of antlers, and in particular the demonstration of the relationship between seasonal changes in day length and annual antler cycles. On the farm in Popielno for the first time in the world Dr. Andrzej Krzywiński obtained semen from a red deer by the use of an artificial vagina, and here also the first red deer calves were born from artificial insemination. Dr. Krzywiński, however, had broader ambitions and ideas for developing research and applying it in the field of modern deer farming, whose dynamic development in New Zealand and Britain he was aware of. This was more than the Station in Popielno could contain. His passion and persistence led to the creation of the Research Station in Kosewo Górne, which with its 100 ha of fenced pastures gave much greater opportunities and allowed research into early breeding technologies for deer farms. Scientific work on semen collection and storage, insemination, and embryo transfer continued here. Behavioral observations were developed, and methods for taming deer and carrying out hybridizations (between red, wapiti, maral, and milu (Père David deer) x red). Based on the original methods of rearing and taming calves "on a bottle," experimental herds of red deer, fallow deer, and mouflon expanded. The basis for this were the animals from Popielno, zoos and similar centers and foreign farms, but also calves which were deliberately searched for in the wild and taken from various hunting grounds across the country. Some deer were also accidentally found as calves and hand reared. The first offspring were mass-reared "on a bottle" and such tame animals formed the basis for the expansion of the basic herds. As the number of deer increased, various variants and elements of deer and fallow deer farming technology were tested in the relatively harsh climate and natural conditions of north-eastern Poland.

The results of the scientific work in the Research Station in Kosewo Górne, and of the experience of farm breeding, formed a strong basis for thinking about the practical implementations and the development of deer and fallow deer farming similar to what was happening in New Zealand. The theoretical work, which involved New Zealand specialists, was quite advanced and extensive and discussed at high levels. However, it was never realized; its prospects for success—quite realistic from a practical point of view—were based not on weak and fragmented private farms, but on huge possibilities (large areas, substantial infrastructure, state subsidies, etc.) which were typical of the then dominant Polish State Farms. History effectively trashed these plans, as they came at the end of the increasingly inefficient system of communist economy. Shortly after this Poland entered the democratic path, and, after 1991, the economy began to dynamically switch to the market system, along with agriculture, whose state-owned part was privatized. During this period of rapid changes, no one considered nor had the possibility to take up the topic of any deer farms, let alone those on a larger scale.

Luckily, the Polish Academy of Sciences (PAS) survived the transformation and with it the Research Station in Kosewo Górne was moved only within PAS to the Institute of Parasitology (IP). The market economy in Poland stabilized and the ideas for the implementation of deer farm breeding recurred, adopted this time by private farms, already operating on market terms. The dynamic development of the market economy was not always accompanied by the necessary changes in regulations and for many years there was no legal option to run private farms with wild animals, including deer. Despite the change of the system, wild animals including game remained under the legal ownership of the State (as they still do to this day). The only way out (besides setting up a zoo or a scientific center) was to obtain a permit from the Ministry of Agriculture and Forestry, issued rather reluctantly, for "farming of game," but only for narrowly defined purposes and under strict supervision. This did not allow the development of farming as such, although the first farms were set up, often without the mentioned permit. The desire to try something new in agriculture prevailed and the capital capacity of private entities was increasing.

The situation worsened because the new initiatives to legally regulate deer farming as a new area in agriculture were shattered by the interests of the hunting lobby, masked by concern for the protection of the environment and "genetic purity" of wild populations. Hunting in Poland during communism was monopolized by the state. It was doing well, being the apple of the eye of the ruling elite, seemingly allowing large masses of the public to participate in it, but in reality holding everything in its hands, including animal ownership, profits from its acquisition, from organizing hunting events and related side perks. The venison market in the country practically did not exist, as most of the venison was exported through a monopolized network of purchasing and processing plants. The currency thus obtained served the internal needs of the state. This led to the decline in the rich Polish culinary traditions related to venison, which were practically non-existent on the domestic market, supported only by some modest amateurs among the hunters.

I describe this background because we still grapple with the consequences of the past. The organization of the hunting system has survived as one of the few communist relics, protected by private companies often run by the people of the old system. The monopoly of the market for the venison coming from hunting has stayed in the hands of several companies, but is now privatized, most often with the participation of capital from the west where, of course, most of the venison still goes. This situation creates the lowest possible prices for purchased venison, balancing on the edge of profitability and only just enabling the operation of Hunting Clubs and State Forests, the main and only entities dealing with Polish hunting and which supply venison to collection points. Nobody cared about the domestic market, because nobody ever thought in the long-term what might happen if everything went well. There was practically no promotion of venison, so the demand was small too; during communist times, the consumers forgot what it looked like, what it tasted like, and how to prepare it.

It is worth mentioning, to show the essence of this problem, the boom in ostrich farming in those years. Mainly the sellers of breeding material profited then. Lack of educated recipients of the product in the country, an absence of regulations enabling the creation of small slaughterhouses and processing plants, etc., eventually led to the collapse of most ostrich farms, leaving only a few of the largest on the battlefield, with a sufficient scale of production to overcome the logistic and financial barriers of slaughter and export.

Venison from hunting was available at game collection and processing points, but with difficulty and of poor quality, mainly frozen and erratically, restricted by the hunting season. In addition, its price was strangely disproportionate to the purchase prices offered to the hunters for their carcasses. Thus, the black market functioned on the side, mainly among hunters and their associates, as hunters could in part, a little cheaper, buy back the hunted game "for their own use." It was not always "for their own," and the meat often went to friends and acquaintances, including restaurants. This continues to this day. Meanwhile, the hunting lobby, for fear of losing the monopoly, blocked the legalization of deer farms breeding in Poland. This was not only because of concerns for their venison market, but also the fear that private farms would offer the possibility of "trophy" hunting. Arguments about the safety of farms, the possibility for cooperation in the promotion of venison and the reconstruction of the internal market for it, or cooperation in the field of genetic improvement of deer did nothing to help. For many years, things seemed hopeless.

The door only opened with the entry of Poland to the European Union (EU) and the need to adapt national regulations to EU standards. In 1998, motivated mainly by this need, a group of enthusiasts, in cooperation with the IP PAS Research Station in Kosewo Górne, founded the Polish Deer Breeders Association (PDBA) and were admitted to FEDFA the same year. Thanks to its activity, and with the support of FEDFA, it became possible to legalize deer farming in Poland. However, there was no opportunity, nor time, nor will of the involved parties to create a comprehensive legal structure. Instead, to avoid the sure blockade by the hunting lobby of any attempts to interfere with their "business," "deer and fallow deer farmed for the production of meat and skins" were put on the list of farm animals, that is, in agriculture and not hunting, with all the consequences for breeding and meat, which henceforth was classified automatically as red meat and not game. This solution seriously affected the future development of the industry, and the lack of precise regulations to this day creates many misunderstandings and problems at the border between the two areas.

When the new regulations were introduced, there were already about 20–30 farms in the country, mostly small ones, with a total of about 700 fallow deer, 350 red deer, and 60 sika deer. After the legalization, new farms were established quickly, based on breeding animals from the Station in Kosewo Górne and increasing imports from western European farms (thanks to the opening of the EU borders). The development of farms was aided by the possibility of obtaining—in addition to area subsidies (for pastures)—other EU subsidies now available in this type of agricultural activity. The initially consolidated spirit of cooperation among breeders unfortunately began to break due to the internal interests of individual, more significant farmers, including the members of the association and its board. In pursuit of quick profit or the prospects of it, the hard-won vision of promotion, market development, organization of breeding, etc., was giving way to short-sighted interests. This consisted mainly in the selling of livestock to new farms, benefiting from very good prices resulting from the advantage of demand over supply, with the least formal restrictions. The organization and adaptation of national regulations to the real needs of deer breeders were postponed. This related especially to the regulations related to slaughter, processing, and direct sales, which quickly proved too restrictive and not very practical. In addition, the government did not quite know how, nor want, to use the possibility for more accessible interpretation of the EU regulations for small farmers, trying instead to be the EU "star," and obtain further funds and markets. The regulations were therefore interpreted too strictly and mainly for the benefit of large industry. Eventually, only the well-developed agricultural production sectors and large processing plants geared to operate them could meet the challenge. Thus, the government had apparently resolved the problem of safe food production but growing problems and conflicts within the industry shattered and weakened it. This ultimately led to the disintegration of the Association and the creation by some of its members in 2007, of a second organization called the Polish Branch Deer Breeders Association (PBDBA), which took the initiative for several years, including within FEDFA. It also drowned in conflicts over time, and disappeared around 2016, leaving the heavily weakened PDBA on the battlefield, struggling to start over again. During those years, the deer farming situation had moved from the boom phase—for reasons similar to the history of ostrich farming described earlier—to the end phase, i.e. an excess of supply over demand. The demand had concerned mainly breeding animals, i.e. hinds and the relatively few stags needed for reproduction; and the saturation of the market led to the accumulation of surplus stags. High hopes for direct export of animals for breeding beyond the eastern border of the EU (into Russia, Ukraine, Belarus) unfortunately evaporated quickly too. Export barriers were difficult to overcome and, with time, as the political situation deteriorated, became completely impossible for Polish farmers. It was only possible to sell through intermediaries, mainly from Lithuania and Latvia, which had less strained relations with the East, but this was reflected in less and less attractive prices.

Unresolved problems with slaughter, processing and selling meat made themselves felt painfully again. The situation at that time is well illustrated by the history of the creation and decline of the Rudzie and Słoja Deer Farm (2006–2016). A spectacular European-scale project, planned for three farms of several hundred hectares, several thousand red deer, with its own slaughterhouse and processing plant, began to collapse before it had reached halfway to its goal (two farms and 3.5 thousand deer). Restrictive requirements did not permit the launch of the slaughterhouse before the first two farms had reached their full production. This coupled with the aforementioned problems of falling demand for livestock at home and abroad, meant that these two farms lost their outlets and faced the risk of animals accumulating. A lifebuoy was thrown by the Bomafar company, which created, with great difficulty, the first—and until now the only—farmed deer slaughterhouse in Poland capable of receiving live deer for slaughter, with export rights to the EU, but this failed to save the situation. The clash with the realities of the market was painful. It turned out that it was not possible to quickly sell larger quantities of meat on the unprepared domestic market at profitable prices, and entering other western markets was not as easy as it had previously seemed.

The only market ready for farmed venison was the English market. However, this also turned out to be transparent, and buyers quickly learned about the difficult situation in Poland and offered lower and lower prices. The interests of protecting the hard-won market against competition and the protection of existing suppliers from New Zealand and from supplementary purchases from domestic farmers prevailed. With the shrinking market for livestock and no profitability for the meat production, the owner of the Rudzie and Słoja farms decided to gradually reduce the herds, which led to the closure of the Słoja farm and the sale of the Rudzie farm with the remaining deer. Two of the most modern and largest farms in Europe ceased to exist, and their only legacy was the highly regarded breeding material at home and exported abroad (Lithuania, Latvia, France, England, Switzerland, Belarus, Russia, Ukraine, and even Azerbaijan). As a result the Bomafar slaughterhouse lost its main supplier and had to limit its operations to occasional slaughter. It also soon became apparent that the success of this genuine deer slaughterhouse was not met with enthusiasm domestically against a background of growing problems with sales. Attempts to cooperate with PDBA as a partner in organizing regular supplies of livestock failed; the unofficial reason being PDBA's fear of creating a monopoly of sales and with prices dictated. Farmers willing to cooperate tried to deal with the slaughterhouse individually, which however eroded their strength in price negotiations.

Overall, not only the lack of cooperation among farmers and negligence in failing to promote the product got in the way, but also the industry's lack of preparation. Only a few farmers had handling yards and thus most lacked the technical ability to deliver the right quality—animals of equal age, suitably tame and accustomed to being handled and without antlers—as live animals for slaughter. Although it is legally permitted to deliver carcasses of deer killed on the farm, this is restricted to farms which can deliver within 1 h of slaughter. Other problems are distances from the abattoir, and the unprofitable handling and delivery of small batches. The slaughterhouse in turn, due to the lack of regular supplies of slaughter animals was not able to find an outlet at satisfactory prices, especially domestically. Neither could breeders, increasingly on their own, find markets while unsatisfactory prices did not encourage larger production, the creation of new farms nor the improvement of the organization and technical level of farms. And so, a vicious circle occurred, still continuing today, causing only the occasional opening of the Bomafar slaughter-house, mainly for slaughtering their own animals, for the needs of their own restaurant and workers and the local domestic market. For similar reasons, the legal option available from the beginning to slaughter and bleed animals on farms (by free bullet shooting or slaughter by stunning in the crush) and the transport of the ungutted carcasses to the existing livestock slaughterhouses located within 1 h is very rarely used by farms. The main reasons for this are the problems with ante-mortem tests, their high costs, and the reluctance of private slaughterhouses to cooperate, and if at all, at high unit costs, because accepting even (usually) small batches, forces the current cycle of slaughter of mainstream livestock (cattle, pigs) to stop. And the price reference on the market remains still the dramatically low price of deer carcasses in the skin at game collection points.

Against this background, a decline in deer farming has been visible in Poland for several years now. It is hard to give accurate figures; information is incomplete and scattered. The only comprehensive numbers, largely based on estimates, come from PDBA, the latest dated at the end of 2016 (Table 1). Official data are available only from the State Veterinary Inspection, which supervises farms from the veterinary side, including welfare. Unfortunately, the publicly available data concern only the number of supervised farms. Here, in turn, the number of 792 deer and fallow deer farms (in total) was given at the end of 2019. Adding the estimated fairly large number of farms not registered at all, the discrepancies are considerable, which is not a sign of good internal organization of the industry nor care of the state for it. It seems that most farms eluding PDBA estimates are small, hobby farms, remaining outside the structures of the association, usually breeding fallow deer for fun, with a less labor-intensive way of managing their land but obtaining area payments for grassland. At the very end of the list of motivations is selling live animals or meat.

As a rule, farms that do not think about their outlets at the outset inevitably have problems sooner or later. Technically, they are not able to get rid of surplus live animals (no handling yards, low prevalence of anesthetic weapons, and no legal anesthesia in forms suitable for deer). Slaughter for meat by shooting is therefore often the only viable solution, unfortunately considered more and more often, even by the owners, as ethically controversial... This is another vicious circle causing increasing problems. In addition, there are no clear rules related to slaughter for personal use or the legality of selling surplus, farmed deer carcasses to game collection points, which in addition offer dramatically low prices. But some desperate farmers do it anyway. Interestingly, generated surplus animals, despite the lack of clear official possibilities, slowly "disappear" somewhere.

The structure of farms is dominated by small and medium farms, where fallow deer are much more popular. Most farms, due to the highly debatable profitability of farming, try to obtain the highest possible area payments and choose the status of organic farms, despite lower efficiency of such production and no market indications as to the possibility of higher prices for meat. Most often they are run extensively, with the deer kept in one group, set stocked with little or no grazing rotation.

There are a dozen or so larger farms that use more advanced herd management methods, with regular grazing rotation, weaning of calves, cutting off hard antlers for stags, mating with an individual stag or specific stags rotated between seasons, tagging, weighing, and collecting data. Two of the largest farms have attempted artificial insemination. Their feeding system is based on grazing in the pasture season (ca 180–200 days), with hay, haylage, and occasionally root crops for the rest of the year. It is generally necessary, due to the relatively harsh climate, to provide concentrated feed (cereals).

The awareness of the need for prophylactic worming of animals is quite established, but practice is different, because most farms do not have a handling yard that allows the use of injectable or pour-on agents, and the domestic market lacks ready-made granulated deworming feeds or agents for use in food or water,

Table 1 Results of a survey conducted by the Federation of European Deer Farmers' Associations in 2016

17.10.2016 Update to 31.12.2016			FEDFA Survey Synopsis										
			Austria	Belgium	Czech	Germany	Latvia	Lithuania	Poland	Russia	Spain	Swiss	UK
1.1	No of members		934	50	120	1900	35	25	200	20		220	135
1.2	Non-members		700	1500	350	4200	60	320	200	?		300	100
2	No of deer	total	18300	3900	3200	120500	4900	5411	18100	22500	667600	9350	11800
2.1		red	8700	3000	1000	8000	4000	2774	6000	15000	650000	850	10000
2.2		fallow	8600	800	2000	102000	900	2012	12000	3500	17600	8150	1000
2.3		sikha	1000	100	200	10500	0	625	100	4000	0	350	800
3	Average farm size		6	3	5++	2.25	200	50	10	200	1000	3	45
4.1	Only red		400	30	30	1200	15	0	12	0	2000+	57	85
4.2	Av no of animals		20	100	20		180	163	700	n/a	300	18	100
4.3													
4.4	Only fallow		450	20	60	4400	5	1	160	0	0	420	2
4.5	Av no of animals		18	40	30		60	112	80	n/a	n/a	20	150
5.1	Venison consumed		9600	n/i	6000	56000	960	735	1800	n/i		n/i	7000
5.2	Farm venison prod		900	n/i	100	2600	48	4.5	500	n/i	11250[a]	n/i	4500
5.3	Venison imports		4400	n/i	n/i	25400	n/i	0	0	n/i		n/i	2500
6.1	Purpose—meat		85%	70%	10%		15%	16%	60%	100%		80%	80%
6.2	Purpose—live sale		10%	70%	10%		30%	8%	37%	100%	20%	20%	20%

6.3 Purpose—trophy	5%	1%	illegal	45%	16%	illegal		100%	5%
6.4 Agrotur or hobby		10%	80%	30%	60%	3%			
7.1 Park-farms	1%	0%	2%	15%	88%	80%	20%	100%	55%
7.2 Farms	99%	100%	98%	85%	12%	20%	80%	0%	45%
7.3 Legal distinction	No	No	Yes	Yes	No	No	Yes	No	Yes
8 Park-farms									
8.1 – wild?	Yes	No	Yes	Yes	Yes	No		No	Yes
8.2 – tagged?	No	Yes	No	No	No	No		No	No
8.3 – free bullet?	Yes	Yes	Yes	Yes	Yes	Yes	Yes	Yes	Yes
8.4 – ante-mortem	Vet	Vet	Hunter	Vet or CP	Hunter	Vet		None	Hunter
8.5 – post-mortem	Vet	Vet	Hunter	Vet or CP	Vet	Vet		Vet	Hunter
8.6 – closed period?	No	No	Yes	No	No	No		No	Yes
8.7 – evisceration	Abattoir	Abattoir	Park	Park	Park	Park		Park	Park
8.8 – direct sales (carc)	No	No	Yes	Yes	No	No		No	Yes
8.9 – cutting	Abattoir	Abattoir	Abattoir	Reg ent	Abattoir	Abattoir		Park	Butcher
8.10 – direct sales cuts	Yes	No	No	Yes	Yes	Yes		No	Yes
9 Farms									
9.1 – wild?	Yes	No	No	No	Yes	No		No	No
9.2 – tagged?	No	Yes	No	No	No	No		No	Yes
9.3 – free bullet?	Yes	Yes	Yes	Yes	Yes	Yes	Yes	Yes	Yes
9.4 – ante-mortem	Vet	Vet	Vet	Vet	Hunter	Vet		None	Vet
9.5 – post-mortem	Vet	Vet	Vet	Vet	Vet	Vet		Vet	Vet
9.6 – closed period?	No	No	No	No	No	No		No	No
9.7 – evisceration	Abattoir	Abattoir	Farm	Farm	Farm	Farm	Farm	Farm	Abattoir

(continued)

Table 1 (continued)

17.10.2016 Update to 31.12.2016	FEDFA Survey Synopsis										
	Austria	Belgium	Czech	Germany	Latvia	Lithuania	Poland	Russia	Spain	Swiss	UK
9.8 – direct sales (carc)	No	No	No		No	No	No	No		No	No
9i.9 – cutting	Abattoir	Abattoir	Abattoir		Abattoir	Abattoir	Abattoir	Farm		Abattoir	Abattoir
9.10 – direct sales cuts	Yes	No	No		No	Yes	Yes	No	Yes	No	No
10 Hunting on priv land	Yes	Yes	Yes		yes	No	No	Yes	Yes	No	Zes
11 Export regs	No	Destination	Destination		BALAI	Destination	Destination	BALAI +		N/A	BALAI
12 Prices—live											
12.1 – male red live	1000	2000	1000		2000+	2042	1400	2700		1000	1800
12.2 – female red live	600	500	300		1050	869	600	1400	1200	600	550
12.3 – male fallow live	900	250	500		850	670	400	1350		850	500
12.4 – female fallow live	400	150	150		350	305	175	1100		500	450
13 Prices—meat											
13.1 – red carcass in skin	8	10			4	7	2	5+	3	10	4
13.2 – fallow carcass in skin	8	10				7	2	5+		10	4
13.3 – red carcass ex skin								20			
13.4 – fallow carcass ex skin								20			

[a]Corresponds to 222500 deer p/a

based on effective drugs. The second major problem is paratuberculosis, especially among more susceptible red deer. The problem is not much discussed, however, as breeders are generally unaware of it, and if they are, they do not reveal it. This strange practice is aided by the fact that this disease is subject to registration by the state veterinary services, but no longer combated at the expense of the state (as, e.g., tuberculosis). And since registration makes the status of a given farm public and prevents the sale of live animals at home and abroad until the disease is dealt with—which is very difficult—breeders avoid it like fire and the disease spreads. Unfortunately, cattle breeders do the same, which only exacerbates the growing threat. The disease does not cause such havoc in cattle farming, but among deer it can do much harm; the lack of tests suitable for deer, of good practices by farmers, and the lack of good practice and knowledge among veterinarians are not conducive to overcoming it. The situation with brucellosis and tuberculosis is less serious. Farmed deer are not subject to compulsory periodic tests, but those performed compulsorily for export to countries both within and outside the EU are not uncommon and do not indicate a problem so far.

Given the above, the summary cannot be too positive. In Poland, despite quite good adoption of deer farming technology, favorable situations of soil, climate, and natural conditions, it has not been possible to develop deer farming on a larger scale as a new, important branch of agricultural production. The Research Station in Kosewo Górne still exists but is less and less concerned with the problems of deer farming as there is no interest from the association, individual breeders, or the state, which seems to consider deer farming as a costly hobby for a few farmers. The last important achievement for farmers in this field was cooperation with University of Warmia and Mazury in Olsztyn in the release of the publication "Farming and Breeding Deer." However, its release in 2014 as an internal scientific publication of the university for students went unnoticed, and the modest circulation available outside is long exhausted. Scientific research in this area suffers from the poor financial condition of science, recently worsened by an excessive link between its funding and the scoring of scientific articles. Attempts by some agricultural or forestry universities to introduce this subject to the student curriculum are rare and do not go beyond semester long supplementary subjects.

It seems that if deer farming is to survive, its chances lie mainly in small- and medium-sized farms producing for the local market. This is where the efforts of the limping association of breeders continue to go. Fortunately, similar trends emerged in other industries, including cattle farmers who have greater clout. This has finally led to the loosening of regulations and, from spring 2020, the legal authorization of local "agricultural slaughterhouses." Simplified rules, based on the possibility of using a shared room for slaughter and cutting, theoretically significantly reduce construction costs. The implementation of these rules now depends on the quality and completeness of the rest of the accompanying provisions, their interpretation and practical application. Given previous experience, there may be years of trouble ahead. Still, this is not a solution for all farms; for some medium and most small ones, it will be too expensive, and the historical lack of willingness of Poles to cooperate does not bode well for the future possibilities of cooperation among farmers.

Aware of this, the current officers of the Association continue their efforts to reach complete compliance with Directive 853/2004. This Directive permits extensive farms and the use of convenient and cheap solutions similar to those used in hunting: post-mortem testing by a trained person, slaughter by free bullet, evisceration and cutting on farms, local sales to the final recipient/consumer. Overall, the Directive treats the product from this type of extensive farm, or park, not as "farmed venison/red meat" but as "game." Implementation of this would again mean entering the areas of controversy with hunters and the danger of clashing with the many, often unseen and diverse institutions which constitute the hunting lobby. This lobby a few years ago won the possibility of direct sales and probably does not want competition from farms which are potentially able to offer a better-quality product—of known age, more hygienic slaughter, the possibility of using a bio certificate, and no seasonal restrictions on the time of killing. Rather the current game processors prefer to keep buying deer carcasses at low prices from deer farmers who have no other choice. The process of change will not be quick and easy, but I remain hopeful. For now, the addressee of these efforts at the national level, the Ministry of Agriculture, does not seem to understand the situation. The problem is not just Polish but, as it turns out, it concerns many other EU countries from the former communist bloc, having a similar hunting model and belonging to FEDFA like Poland. The effort is therefore twofold, at the national level and at the level of the EU institutions on which FEDFA lobbying currently focuses. These mills grind slowly, and the EU has more serious problems, unfortunately also because of Poland, the former "star" of the EU. . . So, it remains necessary to be hopeful. Remembering that this will only be the beginning of a long way of building a market and product placement.

I look forward to happy times for deer farming in Poland, in which I dipped my fingers from the beginning, devoting most of my professional and family life to it, and with which I am still connected, support and in which I have belief. However, I am aware of the broader processes and growing problems: climate change, social change, criticism of intensive agriculture and the problems plaguing it (epidemics, etc.), the evolution of ethical and consumer attitudes, problems with nature conservation, public education, the wave of populism in politics and other areas. Deer farming in comparison to other agricultural industries seems to still have opportunities in Poland; however, it requires a lot of time and a very skillful development and promotion policy to bypass these and other, yet unknown obstacles on its way.

Further Reading

This contribution represents the personal views of the author accumulated over a lifetime of work in the captive deer industry. Sources which may expand on or substantiate the author's views can be found in a variety of websites: www.popielno.pl; www.kosewopan.pl; www.wetgiw.gov.pl; www.fedfa.com

The Management of Enclosed Deer in Slovakia

Jaroslav Pokorádi

Abstract

The author provides the historical background of Slovakian hunting reserves leading into the development of modern deer farms and the establishment of the Slovakian Association of Deer Farmers (SADF). Conflicts with hunters led to legislation making it illegal to hunt on deer farms but pressures from SADF permitted deer farms to move deer into hunting reserves for hunting. Establishment of game farms with a legal minimum size of 50 ha. created many such hunting areas and deer farms became distinct as small farms numbering 642, of average 3.5 ha carrying an average of only 18 animals. The author's work in artificial breeding is described.

Keywords

Game farms, hunting · Insemination · Venison · Slaughter · Veterinary controls

1 The History of Deer Farming in Slovakia

In its modern history (the Slovak Republic was established after the division of Czechoslovakia on January the 1st, 1993), Slovakia is one of the countries whose agriculture and hunting have deep roots stemming back to the time of the Austro-Hungarian dual monarchy. The hunting grounds were principally managed on individual estates where the predominant quarry consisted of native game—namely red deer, roe deer, wild boar, and small and feathered game.

During the reign of King Saint Stephen (1000–1038), two thirds of the country became the property of the monarch. Royal game hunting grounds were

J. Pokorádi (✉)
FEDFA—Federation of European Deer Farmers Associations, Prague, Czech Republic

established—previously everyone could hunt within the territory. In the thirteenth century, a hunting trail led from the Hron River to Liptov. In 1265 and 1270, the monarch forbade the serfs from hunting deer, wild boar, and big game. In 1729, the nobility enforced a decree from the monarch on hunting and birding. The Charles III Decree ordered the protection of game for the first time (e.g. protection of pheasants, and a ban on hunting pregnant game). In 1786, Maria Theresa and Joseph II. placed hunting rights under strict protection. Poaching was considered theft. Francis II tightened this law in 1802, thereby determining the dates of protection or close seasons for certain species of game and prohibiting the hunting of female game at the time of mating and rearing young. The final adjustment of hunting in Hungary was carried out during the reign of Francis Joseph I. In the nineteenth century, large game stocks began to be decimated by excessive hunting and poaching, which is why expansive landowners started building game houses. Game was imported—including hitherto foreign species such as fallow deer and mouflon. Prince Christián Hohenlohe, Count Karol Forgách, the Pállfies, the Stummers, the Zasy, and the Habsburgs were historically very significant in this area of activity (http://www.hunt.czechian.net/historia.htm).

If we want to establish the essence of modern deer farming, then we must start primarily with hunting in Slovakia and the establishment of game preserves. In the past, game preserves were understood to be fenced-in hunting areas, so as to prevent the game from escaping, and in this fenced area the game was observed, fed, and selected. The preserves had their own structure and breeding goals, which were stricter than for wildlife. The oldest documented game preserve (game park) in Slovakia is Teplý vrch (Meleghegy) near Rimavská Sobota—which was founded in 1666 and primarily used for breeding and rearing quality deer, although some parts of it were also intended for breeding wild game and later imported deer (https://www.lesy.sk/files/lesnik/2004/Lesnik10/najstarsia.htm). The main purpose of establishing the game preserves was to control the quality of the gene pool of the game, and for these purposes, capturing, handling, and systematic long-term work with the game were necessary. The main goal was to restock and refresh the blood in the reserves.

Today we have 46 separate game preserves in Slovakia which are not tied to a free hunting area, and 66 recognized preserves which are connected by infrastructure to the surrounding districts. In Slovakia, the right to hunt can be exercised only on hunting grounds which have been recognized by the relevant forest district office as a hunting ground. The right to hunt belongs to the owner of the land and can be exercised only in accordance with the Hunting Act and Regulations issued for its implementation. The total hunting area in Slovakia is 4,446,662 ha, the agricultural area is 2331 thousand ha, forest areas 1977 thousand ha (forest cover approximately 44.4%), water areas 48.7 thousand ha, and the area of other land is 90.4 thousand ha. The hunting area in individual years varies depending on the specification and changes in the area of hunting grounds in the process of being created, recognized, and revised (http://www.forestportal.sk/lesne-hospodarstvo/polovnictvo/Stranky/polovne-hospodarstvo-na-slovensku.aspx).

In Slovakia, breeding farms were established in the second half of the twentieth century. The importance of establishing farms with species such as red deer (*Cervus elaphus*) and fallow deer (*Dama dama*) was in principle the same as in the menageries, only with significantly higher intensification of game care and its selection criteria—which became stricter. The older name for deer farms was "Animal" farms. The meaning was mainly associated with the use of animals, taking biological samples, and accurate measurement of zootechnical parameters of the respective animal's body. These animals were used predominantly for the enrichment of hunting grounds with top game, which was mostly imported from abroad (for example, fallow deer were transported mainly from Hungary). As with other types of animal breeding, it was important to systematically release game from game farms into hunting grounds, where the gene pool of the game was to be enriched and at the same time to increase its trophy quality. In Slovakia, the development of farms with farmed game species occurred during the time of the Czechoslovak Socialist Republic for a variety of reasons which entailed the intensive care of imported animals: quarantine, controlled breeding, and movement to game parks or for release into the wild. One of the oldest farms in Slovakia is a farm near the villages of Velčice and Neverice near the Neverice Agricultural Cooperative which was founded in 1986 with approximately 700 fallow deer and mouflon. A substantial part of the imported fallow deer came from the vicinity of Gyula in Hungary and these formed the basis of the genetics of breeding fallow deer on this animal farm. Over the course of 20 years, other mouflon farms were established, occupying mainly old orchards. Among the first farms established for red deer was a farm near Stará Ľubovňa.

Gradually, however, the main importance of farms with farmed game changed from the provision of new genetics and the importation of genetics to provide a top quality gene pool for hunting grounds. Instead farms were transformed at the beginning of the twenty-first century to produce young—or game—as quality animal protein. There was very immense pressure from hunters not to include farms in hunting in Slovakia, especially at the end of the twentieth century. This resulted in the amendment of legislation whereby game farms—and subsequently farms with farmed game—were no longer permitted to provide hunting, and thus the farms were not considered as hunting grounds nor could farmed game be used for hunting.. At the same time, however, there was growing pressure to regulate body conformation, liveweight, carcass weight, and animal health. This encouraged a more professional approach and the more intensive farming of game.

At the beginning of the twenty-first century, Slovak game farmers were scattered and uncoordinated; some were members of the Czech Association for Deer Breeders. This situation resulted in the establishment of the Slovak Association of Deer Farming (SADF) in 2008 by 6 founding members. The author was elected president, and later confirmed by the general assembly. Gradually, the association gained in strength and the number of members grew to 69 within a year. After a meeting with the Minister of Agriculture, Environment, and Regional Development of the Slovak Republic—Vladimír Chovan, the association also received state support. The main goal of the association is to connect breeders and supporters of deer farming, to ensure their mutual awareness, to defend the interests of members in proceedings with state and other bodies, and to mediate contacts with similar bodies in the Slovak Republic and abroad, including providing research work with regard to animal protection—namely deer—keeping registers of animals and a register of farms wishing to cooperate with the association. We achieved this almost immediately when we convinced legislators in the National Council of the Slovak Republic to have a progressive and modern market animal segment, such as that of farmed game. A further success was ensuring that farmed game can also (under certain conditions) be released into a game park (which is a fenced-off hunting ground) for a fee, thus bringing the market economy back into the hunting/farming segment. It should be noted that in the surrounding countries this process is legally forbidden, but is done under the cover of darkness and illegally. In Slovakia, we wanted to promote transparency and a market mechanism, as is the case in other developed agricultural countries such as New Zealand. The amendment to the Hunting Act has enabled this status from 2009 until the present. As a new entity representing game farmers in Slovakia, we initiated the FEDFA meeting in Brussels at their regular working meeting, which is attended by the presidents of the national deer breeders' associations, and in 2008—after approval by FEDFA member states—we became full members of FEDFA (Federation of European Deer Farmers Associations) on 1.1.2009.

Based on these circumstances, we—as new members—requested that we be permitted to host the next conference—which takes place every year in a different country according to the agreement. We were accommodated, and under the auspices of the Government of the Slovak Republic and the Ministry of Agriculture of the Slovak Republic, the association decided to organize an autumn international conference of FEDFA farms with our members featuring 155 participants from 14 European countries—as the largest conference ever held under FEDFA. The main section was organized in Smolenice Castle with scientific and professional lectures, which were enriched with excursions throughout Slovakia. Slovakia offered Europe a new perspective on deer farm breeding—using the most modern biotechnics which we can provide to EU countries and to Russia from Slovakia. The number of members in the association who were interested in breeding mostly red deer has gradually increased, and in the 10 years since its establishment, 235 entities became full members (2018).

Alongside the establishment of SADF, a legislative and investment process was formed for the establishment of the first artificial breeding service (insemination station) for red deer, fallow deer, and mouflon—established in the village of Vištuk in 2008. The reference number of approval by the State Veterinary and Food Administration of the Slovak Republic was SK ISJ-001. This centre consists of 96 hectares, divided into 18 sections, joined by a central corridor with an adjoining hall for the control and handling of game, including the internal rearing of young. Individual red deer (average 350), fallow deer (average 100), and mouflons (average 120) are bred on the farm part of the insemination station. The insemination station also includes two laboratory wings, one genetic centre/DNA laboratory and another reproduction laboratory. We began to successfully apply in practice reproductive techniques known predominantly from the bovine environment, mainly a synchronized controlled artificial assisted insemination technique allowing the insertion of a high-quality insemination dose (at least 20 million live sperm with progressive motility in one insemination dose) directly into the uterine body. We also managed to bring to life the attainment and cryopreservation of embryos, and their subsequent use for recipients. The selection of donors involved a complicated selection key in order to achieve the highest genetic quality of individuals from different perspectives, especially in terms of health and genetic makeup, as well as eliminating the predisposition to various defects. The insemination station in Vištuk—in cooperation with the University of Veterinary Medicine in Košice— was the first in Europe to perform an embryo transfer of a mouflon embryo which was deep-frozen in liquid nitrogen and successfully transferred to the domestic sheep recipient. In 10 years of operation in the European field of deer farming, 4800 artificial inseminations, 996 ejaculate samples from top red deer, fallow deer, and mouflon donors have been carried out, with the production of insemination doses and embryos totalling more than 12,500 IDs and 350 embryos (2008–2018). The insemination station has managed to create an excellent centre for the supervision of genetics, and through its hygiene standards, and through the provision of consultancy services for farmers by preparing plans and applying deer farming technology, has achieved an increase in the quality of deer not only in Slovakia, but also throughout Europe and Russia.

2   Contemporary Deer Farming in Slovakia

Today (7.12.2020) there are 642 registered game farms in the Slovak Republic (https://www.svps.sk/zvierata/Zoznamy_schvalene.asp?cmd=resetall& Zoznamy=ostatne&Sekcia=37&Cinnost=0&Podsekcia=0).   Slovakia   has undergone a change—namely from the obligation to report (which was not mandatory for everyone), to mandatory registration with an approval process managed centrally by the State Veterinary and Food Administration of the Slovak Republic in the transferred control function of the Regional Veterinary and Food Administration (RVPS). It was this mandatory measure that resulted in the determination of the exact number of breeders, especially including breeders of ratites. Today we know the exact number of registered breeders, but the exact numbers of animals bred, their sex and age composition comes only from internal information from breeders. Today, we know statistically that 11,345 animals were bred on the 642? farms in Slovakia, representing 46% fallow deer, 28% red deer, 22% mouflon, and 4% other small species. The SADF statistics further show that the average farm size is 3.5 hectares, with the lowest parameters unlimited, and relating to the minimum area per animal per species and age category, as mentioned below. The largest registered farm is currently 96 hectares at the breeding services in Vistuk. The average number of animals per farm is 18.

Over the last 10 years, the philosophy and meaning of breeding—especially of red deer and fallow deer on farms—has also changed. The amendment to the

Hunting Act—which set the minimum size of a game park to 50 hectares—still applies today, therefore the importance of having farms to produce quality trophy game for hunting in Slovakia has more or less disappeared. The fundamental difference between game farms and farms today is not the size of the area, mandatory high fencing to prevent the escape of animals, or intensive care of animals (veterinary care, nutrition and feeding, selection), but only the fact that a game farm is a hunting area and the farm is not. Today, a weapon of a certain calibre can be used on a farm, but only for killing or stunning animals, unless another method of stunning the animal can be applied for slaughter. The second, legislative difference is that game farms are subject to the Act on Hunting and Farms, mainly the Act on Veterinary Care and its implementing regulations (Decree). To put it simply: the hunting fraternity is interested in the game farm as a fenced area for hunting and the veterinary authorities are interested in the farm section. Of course, the state veterinary service also controls the production of game and game products in the game farm. For the protection of farmers (who have higher year-round costs for breeding and care of game), we initiated countless SADF meetings with experts at the Ministry of Agriculture and Rural Development of the Slovak Republic, where common materials were drawn up to serve as the basis for implementing regulations for breeding and the registration of deer farms. This process of the differentiation between game farms and farms—including the production of game in game parks and deer farming was specified by the implementation of Decree no. 178/2012, Collection of Laws of the Ministry of Agriculture and Rural Development of the Slovak Republic of 8 June 2012 on the identification, registration, and conditions of deer farming (https://www.svps.sk/dokumenty/legislativa/v_178_2012.pdf, https://www.sadf.sk/vybor). It was necessary to lay down rules on the basis of which farmed game could be kept and for what purpose.

In English, meat which comes from the shooting of wild game is beautifully distinguished as venison subject to treatment as "game meat", in our country the equivalent is "divina"—as in "meat from a wild animal". And farm meat is also known as venison but subject to treatment as "red meat", which does not exist in Slovak. Personally, I have been using the treatment of game venison for the equivalent of the word "divina" for 15 years. Both divina and wild game meat venison are synonymous according to the linguistic institute. I think that this division is justified, as their differentiation has a major impact on the price of game, which must be higher, because the quality is also higher than divina—from the slaughter and the high degree of hygiene, the age structure of the slaughtered animals, the process of bleeding and evisceration, right up to the process of maturing the meat.

Today in Slovakia, farmed game is understood to mean a farm animal, which prior to 2018 had not yet been precisely defined. Therefore, through my work and the help of our SADF association farmed deer is the intended description for breeders who have the defined goal of breeding for the purposes of the production of breeding animals, or obtaining products—including the production of animals for slaughter, or sports or other uses. The purposes of breeding defined in this way can divide farmers into those who want to breed game for others, the so-called utility, or production farms, and thus who produce breeding animals that meet the conditions of recognized breeding standards. These breeders have the right to request livestock unit headage payments or other forms of subsidies from the state, because they meet the criteria of animal identification and performance control, and are registered in the stud book. And since we have mostly extensive farms in Slovakia—which are intended primarily for game production—they do not have to meet the strict criteria of breeding animals. The aim was to give farmers a choice regarding farming methods and the concentration of technology that reflects their main goal. They have the right to choose for themselves, as opposed to being guided by a state directive. Here, I see one of the main benefits of our association in this process; that we have given farmers a choice, a liberal approach does not build walls, but instead provides ways to achieve the main goal of breeding. Of course, under certain circumstances, farm-reared trophy game may continue to enter the hunting park for a fee, but only in accordance with the Hunting Act. Breeding farms are intended for the production of a young population with declared traits and characteristics that have a known origin and pedigree. These animals are used for export to countries for the establishment of new farms, or as a supplement to genetics with new individuals (the so-called blood flow). In this way, quality red deer and fallow deer were shipped to the Russian Federation, Belarus, Serbia, and EU countries such as France, Spain, Lithuania, Germany, Latvia, Hungary, Austria, Poland, Croatia, and the Czech Republic. Within Slovakia, they serve mainly to enhance emerging farms, and also as a shedding of unrelated blood to prevent inbreeding depression.

The adoption of precise rules in the implementing decree in 2012 set out the details of farm registration and game identification, game registration, farming conditions, special housing, feeding, watering, and the handling of farmed cervids, mouflons, and wild game, as well as the conditions for the killing and slaughtering of farmed game, and the placing on the market of gamed products.

For the simplicity and fulfilment of the desired goals, namely an accurate and clear breeding environment, we have determined the categories of deer and mouflon—namely the category up to 12 months, from 12 to 24 months, and over 24 months. The Slovak Association of Deer Breeders and the Ministry of Agriculture and Rural Development of the Slovak Republic have adopted precise and detailed conditions under which it is possible to breed farmed game and its offspring, and we determined the maximum concentration of animals, which—when including other parameters—will guarantee animal welfare in every respect. See table Special conditions of living, feeding, water supply, handling of red deer and in the farm breeding:

Species of animal	Concentration of animals on outer pastures (pcs/ha)	Lowest drinking water demand (l/day)	Smallest housing dimensions (m²/pc)	Maximum number of animals with internal housing (m/pcs)	Minimum feed trough length (m/pc)	Minimum corridor width (m)
Red deer within 12 months	12	10	4	50	0.7	3
Red deer over 12 months	6	15	5	30	0.9	4
Fallow deer within 12 months	20	7	2	50	0.5	3
Fallow deer over 12 months	15	10	3	30	0.7	4
Roe deer within 12 months	25	1	1.5	50	0.3	2.5
Roe deer over 12 months	20	1.5	3	30	0.5	4
Mouflon deer within 12 months	25	5	2	60	0.5	3
Mouflon deer over 12 months	15	8	3	30	0.7	5

According to the above table (https://www.svps.sk/dokumenty/legislativa/v_1 78_2012.pdf, https://www.sadf.sk/vybor) it can be seen that with red deer, the maximum concentration per hectare is 6 deer over 12 months, while practice shows that the average concentration in Slovak farms is 2.5 per hectare. This is mainly due to the degree of extensive breeding. Water consumption is also very important—something underestimated by many farmers, so it was determined exactly that an adult deer needs at least 15 l of water per day. From the point of view of the technology of breeding red deer and fallow deer, the technological factor is equally important—the width of the corridor or alley that must be at least 4 m for red deer, ideally 6 m with a narrowing to 4 m. A very wide—or conversely a very narrow corridor—can cause serious injuries when turning the animals in the herd or in a narrow alley, weaker individuals may be trampled on. Personally, I think that although this decree is not perfect, we have helped many farmers to build farms for farmed game and at the same time we have prevented many animals from being exposed to unnecessary injury and damage to health thanks to controlled legislation. All farming conditions in Slovakia were and are made in the strict context of animal welfare. A very important role in this process is played by the control activities of the Regional Veterinary and Food Administration—the state veterinary supervision board which controls the conditions of veterinary care and animal welfare—both at the respective establishment and during operations.

In Slovakia, the conditions for deer breeding are applied in such a way that the quality of living conditions of game is permanently ensured in farming so that the protection and well-being of the game are also assured, the nutritional provision of game needs is met, and provision is made to avoid farmed game suffering pain, compromised welfare, or injury. In deer farming, compliance with veterinary requirements is required and veterinary supervision is ensured. Farmed game must be provided with unlimited access to a suitable water source and feed according to the nature of the farm and the type of game. Deer farming is different from animal breeding in that the health status, condition and well-being of game, daily inspection of the technical conditions of equipment, biotechnological equipment, and the quality of used feed are checked daily. Deer farming enclosures must be designed to prevent the escape of game as well as the intrusion of wild game and stray dogs. The lowest height of the enclosure fence for deer is 240 cm, with a maximum span between posts of 5 m. When working with and handling game, it is necessary to minimize stress factors and to use only humane ways of handling the game. Sick and injured animals must be housed separately. Dead slaughtered or casualty game is considered an animal by-product and is disposed of in an abattoir. A place or facility for the collection of animal by-products must be set up on each farm.

In farming, conditions are provided for the equipment for handling game and also the way in which the deer are handled to ensure that the handling of game is treated with care, in the case of bolting occurring, the game is slowly chased, taking into account the physiological possibilities of the game; dogs may only be used if they are trained for this purpose and have been tested as working dogs using lanes/corridors as set out in the table. Driving lanes for individual handling of game lead to a loading ramp or a crush where the immobilization of the animal is ensured by compression and it is restrained by means of a manual movable wall or a hydraulic arm. This type of handling is especially necessary in the husbandry and breeding of animals where close contact with the animals is required. According to the type of animal husbandry technology employed and if it is intensive—animals should learn handling from an early age through the intensive rearing of deer, which are weaned from mothers aged 3–4 months and trained for handling and restraint in a crush. This model has the consequent effect that in adulthood, individuals are calm and stress-free during handling and thus endure the necessary tasks such as zootechnical measurements, weight measurements, the application of drugs and other preparations, as well as artificial insemination in females and ejaculate collection in males.

As I mentioned, the major importance of farmed deer breeding is meat production, but it is necessary to determine the conditions of production and marketing. The State Veterinarian Dr. Rudolf Janto is responsible for the State Veterinary and Food Administration of the Slovak Republic, who I thank for his professional and rational approach. Together, we solved several issues from slaughtering game directly on the farm with a gun with the determination of the projectile and conditions, to maturing meat and hygienic conditions. On the basis of these consultations, we have included in the regulations that where the conditions for killing and slaughtering breeding game prevent movement to a slaughterhouse for transport due to the safety of staff or due to animal welfare being endangered during transport then the game can be slaughtered directly on the farm. The slaughtering of animals must be carried out in such a way as to minimize the suffering to the animals. Information on the slaughter of breeding game is required by the regional veterinary and food administration, who must be notified of the date, time, and manner of slaughter. Slaughtering may only be carried out by a professionally qualified person. Farmed game is killed at a designated location using a crush. The slaughterhouse for killing game with a firearm must have an approved ballistic license. The method of slaughter must be authorized by the State Veterinary and Food Administration of the Slovak Republic; only a qualified person may kill the game with a firearm and in a manner which causes immediate loss of consciousness and death to the animals; when using a firearm, special regulations apply. Other permitted methods of killing individual species of game are penetrating devices causing severe and irreversible brain damage, non-penetrating devices causing serious brain damage, electrical stunners—the lowest current of which must produce an immediate state of unconsciousness and cardiac arrest, the use of carbon monoxide or carbon dioxide chambers, or any other humane method of slaughter approved by the Regional Veterinary and Food Administration. After being slaughtered on the farm, the game must be bled imme-diately and eviscerated within 1 h at the latest.

When placing farmed products on the market, the conditions must be met in terms of an inspection being carried out by an official veterinarian or an authorized veterinarian before the game is slaughtered. This inspection must take place no more than 24 h before slaughter. If the animals arrive at the slaughterhouse alive, the inspection may take place no more than 3 days before slaughter. The pre-mortem inspection of farmed game additionally includes a check of records and documenta-tion, including food chain information. At the same time, these conditions are based on the establishment of minimum requirements for European directives and regulations, but also the specifics of Slovak deer farming have to be preserved. Our main goal—and also my personal one—is to provide all people in Slovakia with high-quality game and high-quality animal protein which is of better quality than poultry, pork, or beef. It is also accepted in the political sphere that farmed game—where no other breeding is possible—is an alternative animal sector that needs to be accepted and supported like other major animal sectors.

3   Targets and the Future of Deer Farming in Slovakia

In its recent history, Slovakia has taken huge steps over the last 15 years towards the organized breeding of deer, thus creating a de facto new animal sector as a specialized concept for mountain and upland areas with the provision of quality animal protein. Farmed game should not replace existing sheep, goat, or beef breeds, but should merely complement and expand the range of animal husbandry for human consumption.

Thanks to the defined goals and the introduction of applied science in deer breeding, we were able to advance and increase production not only on a quantitative, but also on a qualitative basis. We broadcast these goals at the annual professional seminars and we managed to apply scientific knowledge directly to individual farms, by connecting the Slovak University of Agriculture in Nitra, practitioners from the zootechnical environment, and farmers. The SADF Expert Committee deserves a great deal of credit for the concept and fulfilment of the goals. Jannet Prétiová is in charge of legal regulations, their amendments, and implementation in practice. Peter Chudej is in charge of professional advice and research for farmed game nutrition. Katka Gaťášová Žideková is in charge of the association's external marketing and promotion. These are the people who are practically supported by Radovan Kasarda with his knowledge based on scientific-research—covering farm breeding and breeding work. Radoslav Židek is in control of genetics and breeding, fusing the latest genetic analysis into real breeds for higher resistance and genetic condition. Zuzana Krchníková is in charge of farm breeding veterinary care and the precise application of processes to ensure the health and well-being of farmed game. And the last member of the committee of our association is myself—who as president represents the association externally. In addition to the "political" function, I am an example of the connection between all the aspects of the animal and its life on the farm (https://www.sadf.sk/vybor). Today, we routinely use biotechnology methodologies that I have verified. This required almost 13 years of optimization and application experience. The aim was to use biotechnics such as oestrus synchronization, artificial insemination, embryo recovery, and embryo transfer in farm breeding. The goal has been fulfilled, we have optimized the processes and the work continues. It is necessary to realize that the above-mentioned biotechnics should increase the quality of breeding, improve and streamline selection criteria, and achieve a genetic leap in the monitored parameters—mainly health and stability, maintaining the body frame/the standard of the body, increasing weight gain (higher yield in the carcass) in commercial farms and—last but not least: the high level of sexual health.

In Slovakia—and thus also for our partners throughout Europe and in Russia and Belarus—we have established the so-called minimum standards for insemination rates. Today, the sperm used must meet strict criteria regarding health, quantity, and quality. From a health point of view, each donor must be examined for TSIs— sexually transmitted infections, and we have identified the so-called Mollicutes (Chlamydia, Mycoplasmas, Ureaplasms) in ejaculate. Therefore, we were the first in the world to genetically test for the presence of microorganisms in the ejaculate of red deer and fallow deer to be used for the production of insemination doses. These types of bacteria cause serious reproductive problems such as infertility, embryonic mortality, and postpartum sepsis in young. From the point of view of ID (insemination dose) quality, we set a minimum concentration of 160 million sperm per millilitre, which represents 40 million/ID. The qualitative parameters are such that the minimum progressive mobility must be 50%, and an average motility of above 70 µm/s. In direct terms, this means that a red deer or fallow deer ovum is inseminated with 20 million live sperm per insemination dose. These are the quality standards that we have applied to our farms. However, they can only be applied when animals are calm so that to guarantee that the insemination success rate is above 60%, the animals must be kept stress-free. Our insemination station has also optimized the post-mortem processing process, which is of great environmental importance not only for farmed game breeds, but also for protected animal species: we have managed to process and archive IDs of wolves, Tatra chamois, and other protected animal species which can serve as a gene pool for protected species for future generations. The importance for red deer or fallow deer is enormous, because by processing the best examples from the wild (the so-called wild genes) from across Europe, we can create a gene bank to ensure gene variability and to prevent inbreeding depression. Via these means, we optimized the processing process in compliance with similar concentrations and health parameters. In this way, rare red deer from the deep forests of the Carpathians in Romania, Ukraine, and also in Slovakia have been archived for the future.

The transformation of extensive farms (fewer than 2 animals per hectare) into a semi-intensive or intensive enterprise remains a target. The main reason is the fact that it is not possible to apply biotechnological procedures in extensive models and thus effectively improve the quality of breeding. Higher breeding efficiency means—among other things—economic growth and higher profit by improving the ratio of costs to revenues. The advantage of intensive or semi-intensive farms is the contact with the animal—which is on a daily basis—and thus the handling and application of zootechnical operations is all the easier. Animal welfare is directly dependent on the so-called Training of animals, which means stress-free handling during feeding, restraint vaccination, weighing the animal from weaning or from an early age, thus the relationship between the farmer and the deer is positive for both. We want to apply this method of relationship to as many farms as possible, which directly helps the application of artificial insemination, as a means of maintaining a wide range of herd. This is a prerequisite for the diversity of genes and the long-lasting health level of the herd on specific farms.

Today, in 2020—even after the elections to the National Council of the Slovak Republic and the new government of the Slovak Republic—we found the same common goals which serve to help farmers and the development of farmed game farming in Slovakia. The Act on Breeding and Breeding of Animals and on Amendments to Certain Acts will be amended for 2021, when for the first time in the history of Slovakia, farmed deer breeding and its existence are to be enshrined as a breeding activity, as well as for livestock such as cattle and sheep. This law chiefly regulates the rights and obligations of persons carrying out breeding or the breeding of livestock, other livestock or farmed game. It is a milestone that will elevate deer breeding to a level we could only have dreamed about until 2020.

In 2020, the Slovak Association of Deer Breeders—on the basis of a unanimous vote by all participating countries—at the last 7th World Deer Congress in August 2018 in the Altai, Russian Federation—was set to organize the subsequent VIII. World Deer Congress this year. Due to the global COVID-19 pandemic, we were forced to postpone this international world congress until 2022, depending on how the global situation around the pandemic and the possibility of free travel plays out. This would be the first time this congress has taken place in Europe. In addition to excursions around farms, our association also wants to provide professional sections, workshops, and society-wide interest in the promotion of farmed deer and game among the citizens of Europe, including Slovakia.

The Management of Enclosed Deer in Spain

John Fletcher

Abstract

The very large extent of hunting areas in Spain is described and the number of areas which are fenced to contain deer is higher than in any other European country. The objective is almost entirely to provide hunting with very few farms providing venison as a primary crop. Red deer are the predominant species. Most hunting is by driving deer with packs of dogs towards rifles in a system known as the 'monteria'. Some farms harvest growing, velvet, antler.

Keywords

Hunting · 'Monteria' · Cork · Nature reserves · Hunting enclosures, venison

1 Historical Background

In common with almost all areas of Europe, deer numbers in Spain declined during the nineteenth century and especially during the Civil War 1936–1939 and until at least the middle of the twentieth century (Carranza 2010). After the 1950s, deer were reintroduced into areas where they had become less numerous by catching wild deer, often from estates owned by public administration, as well as by the introduction of deer from other parts of Europe. This was especially true of red deer but was also the case for fallow and roe. These repopulations were motivated by the demand for hunting and concern for the status of wild ungulates. Dating from the 1960s, public hunting areas known as Game Reserves were also established to promote the growth of wild ungulate populations, and, by 2011, numbered 49 and covered 3.5% of the

J. Fletcher (✉)
Reediehill Deer Farm, Auchtermuchty, UK
e-mail: johnfletcher@deervet.com

Spanish land area (Pita Fernández et al. 2012). However, these officially subsidised Game Reserves represent only a very small proportion of the hunting area of Spain.

In order to contain the deer many large estates were fenced and this remains the case at the present time. Declines in human rural populations and increasing urbanisation associated with declines in some traditional farming especially extensive livestock production have led to an expansion of hunting areas and a growth in the popularity of hunting as well as an increase in the populations of red and roe deer.

2   The Extent of Hunting Areas and Enclosures

The very large proportion of Spain which is dedicated to hunting remains a remarkable feature of rural Spain. The ownership of the right to hunt is distinct from land ownership so that hunting rights may often belong to someone other than the landowner.

Spanish government statistics show that approximately 29 million hectares of ground are used for hunting (large and small game) representing 87% of the Spanish land area. This comprises 32,800 'cotos' or hunting areas. Of these 82.6% are private ('cotos privados de caza') and a proportion of these are enclosed by a deer fence. This proportion was put at 2.7% amounting to around 1 million hectares in 2013 (personal communication Carlos Otero Muerza, Director APROCA, Spanish Landowners' Association for Hunting and Nature Conservation, Madrid). During the decade up to 2014, hunting areas were reported to have expanded by 12% (https://revistajaraysedal.es/espana-territorio-de-caza/).

There are reckoned to be 14,000 hunting estates carrying out 'big game' hunting of which around 10,000 are hunting deer and around 150,000 red deer and 24,000 fallow are being killed annually. Of those 10,000 hunting areas around 1250 are fenced.

The greatest extent of enclosed hunting is in the Sierra Morena. These mountains extend over parts of Andalucia, Extremadura, and Castilla la Mancha and include many large estates and it is estimated that around 25% of this area is fenced. Indeed statistics show that in this region 65% of the hunting areas are hunting deer and up to 90% of those are fenced (Spanish government statistics www.mapa.gob.es).

Spanish government data state that there are 258 big game 'farms', that is enclosing red deer, fallow deer, roe deer, wild boar, or mouflon. Of these there are 111 'deer farms'. Most of these exist for hunting and make commercial sales of live deer or they may rear deer for releasing into other properties in the same ownership which they are legally permitted to do in order to improve the quality of the deer within the hunting area. There are also 81 deer farms maintained as amenity parks, for teaching and training by institutes or branches of government rather than for hunting, and 52 as nature reserves where hunting is not permitted. These parks are monitored by an authorised veterinarian to maintain control over health and movement. www.irec.es (Institute of research of wild life and hunting species of SCIC and Castilla la Mancha University).

In the north of Spain there are also a very few small deer farms which produce venison and more may be starting but information is scarce. There are also one or two deer farms producing antler velvet.

3 Management of Hunting Enclosures

The legal control of hunting in Spain is carried out by the regional administrations. There are legal restrictions on the minimum size of areas which can be used for hunting large ungulates and this is normally 500 hectares but for fenced enclosures this is generally 1000 hectares or, in Andalucia, 2000 hectares.

For most of the fenced enclosures hunting is by far the chief enterprise and red deer are the most highly valued species although the numbers of wild boar are greater. Other sources of income are sales of cork, and the shooting of birds such as partridge. There is a trend towards holding the valuable red deer within the enclosures at higher densities demanding more intensive practices (Carranza 2010). Within enclosures some form of supplementary feeding is normally required especially in arid areas from mid-July to late September. Hay, alfalfa, and cereal based rations are commonly used. In some situations these rations may attract wild boar with the danger that infection with tuberculosis may spread to other species such as red deer. As the climate changes more estates are feeding their deer throughout the year.

For many years, there has been concern in Spain that the movement of deer creates unwelcome genetic diversity amongst red deer. The native population of the sub-species *Cervus elaphus hispanicus*, is probably around 800,000 so introductions seem unlikely to have made a very significant impact. Nevertheless environmentalists remain concerned and oppose the growth of deer farms as they are deemed to carry *'significant risks of introducing foreign and artificially selected genes into natural populations'* (Carranza 2010).

The hunting of deer within the largest enclosures, especially in the south west of Spain where most of these enclosures exist, is, as for free-living unenclosed deer, by the time honoured practice known as the Monteria. Big game of all species, but concentrating especially on red deer and wild boar, is driven by packs of dogs to armed hunters posted so as to provide opportunities to shoot the animals as they move past. Although the concept of shooting moving big game is not familiar to many cultures the Monteria is beneficial in that each territory, normally around 500 hectares, is only hunted once creating disturbance lasting only around 3 h.

References

Carranza J (2010) European ungulates and their management in the 21st century. Cambridge University Press, Cambridge

Pita Fernández M, Casas Bargueño S, Herrero J, Prada C, García R (2012) Game Reserves in Spain: the public management of hunting. Forest Syst 21(3):398–404. www.inia.es/forestsystems 10. 5424/fs/2012213-02343

Part III

Systems for Managing Enclosed Deer in North America Including Their Diseases

Editor's Introduction

The North American section comprises two principal contributions. The first, shorter piece, provides original material on the breeding of the smaller species of deer, their nutrition and handling.

The second is restricted to the farming of elk (*Cervus elaphus canadensis*). It is comprehensive and very valuable; much of the information is not present in other sections of the book but can be usefully extrapolated to other species in other parts of the world therefore it is included in its entirety.

This second chapter is excerpted from *The Elk Farming Handbook* written by four authors led by Ian Thorleifson. *Elk farming remains the largest deer farming industry in Canada* despite many setbacks including the very serious problems of Chronic Wasting Disease.

The Captive Deer Industry in the USA and Mexico: Details on Breeding, Nutrition, and Handling

Michael Bringans

Abstract

Financial data reflects rapid growth and economic importance of the deer breeding industry in the USA. The objectives are almost entirely hunting, principally of whitetail deer, either within the unit or through provision of deer for hunting areas. Bottle raising of young is common and artificial insemination and embryo transfer are of growing importance. Details of feeding regimes are provided for different species and sexes. The Mexican deer farming industry is described and it also revolves around the supply of whitetail bucks and mule deer bucks for hunting as well as the supply by some farmers of deer for repopulating wild areas. Details of the systems of artificial insemination, semen sexing, and embryo transfer by both multiple ovulation embryo transfer (MOET) and in vitro fertilization are described. Handling systems are discussed.

Keywords

Deer breeding · Nutrition of deer · Hunting in enclosures · Whitetail deer · Semen collection · Embryo transfer · Sexing semen · Handling systems

1   Introduction

Captive deer breeding and hunting is a steadily growing industry in the USA and the focus is almost entirely on the provision of deer with large trophies for hunting. While there are some elk farms producing velvet antler and elk meat for sale, and red deer farms producing velvet antler and venison, these are generally considered by-products to the all-important hunting industry which revolves around the whitetail deer (*Odocoileus virginianus*) although other deer species classed as "exotics"

M. Bringans (✉)
Advanced Deer Genetics, Tiny, ON, Canada

are also hunted and make a significant contribution to the income from hunting on enclosed hunting operations.

In 2017, the North American Deer Farmers Association (NADEFA) commissioned the Texas A & M University to publish data on the USA captive deer industry. Data collected related to 2015 and can be compared to a similar survey conducted in 2007 using data from 2005. It concluded that the total impact of the industry on the USA economy, combining the breeding and hunting components, was 1.1 billion USD annually in 2005 but had grown to 7.9 billion USD annually by 2015. This report concluded that:

> The U.S. deer breeding industry has an established presence across the nation, with the majority of operations located in rural areas. In addition, while traditional farms overwhelmingly dominate the hunting industry, the small niche of hunters this market serves continues to increase in importance to the economy. The increase in demand is fueling the growth in the breeding industry. Over $2.6 billion in direct expenditures are poured into the U.-S. economy each year by the deer breeders and sportsmen of this industry. In turn, this generates $7.9 billion of economic activity. All told, these results highlight the fact that the deer breeding industry continues to be an important and vital contributor to the rural economies of the United States. (Economic Impact of US Deer Breeding and Hunting Operations. Research Report 17-4 Agricultural and Food Policy Center, Texas A & M University, May 2017).

The survey also concluded that the economic activity of the deer breeding industry now supports 56,320 jobs, most of which were in rural areas.

The Texas A and M University survey sheds interesting light on the management of both deer breeding operations and those which breed deer and also provide hunts. The study took data from 97 completed surveys of which 59 ran breeding units, 35 had breeding enterprises but also offered hunts, and three carried out only hunting. The average acreages for breeding were small, occupying between 20 and 30 acres but with an average of a further 1500 acres for hunting where this was carried out. The units were divided into smaller enclosures which normally carried around 20 does and one buck. Twenty of the 97 operations surveyed used other cervid species in addition to their whitetail deer, often buying these in. All types of unit surveyed bought and sold deer with the breeding enterprises buying in only two or three animals per annum but selling about 15 deer of each sex as well as fawns and semen. Those with hunting interests also sold around 20 bucks and 25 does each year as well as fawns and semen; they averaged about 50 bucks hunted of which 11% were sent off for taxidermy. The bottle raising of fawns seems to have been common as was the collection of semen and artificial insemination (see below), as well as DNA testing and also testing for Chronic Wasting Disease.

2   Breeding Management

The deer species being farmed in the USA are mainly seasonal breeders. The onset of the rut is dictated by changing daylight lengths. The peak of elk breeding in the Northern Hemisphere is in September. The peak of red deer breeding is in the beginning of October, and fallow deer peak in late October.

Whitetail deer will generally rut earlier in the northern states than the southern states. The range of onset of the rut in whitetail is from late October to December.

Desert mule deer will normally breed late December to January, while the Rocky Mountain mule deer will breed in November.

An exception to this seasonality is seen in Axis (*Axis axis*) deer, which tend to breed all year round.

Selective breeding is related to demands of the hunting industry.

2.1 Natural Breeding

Superior antlered sires are selected to breed with the female herd. Bucks with superior and desirable antler attributes for hunting are selected prior to the rut. In extensive breeding situations, multiple sires are run with the does (a general ratio of one mature buck to three does). Selective culling is managed by hunting to control breeding buck quality and ratios.

With more intensive management in many whitetail ranches, a single sire is selected, based on desirable antler traits, and introduced into the doe herd. The buck will breed the does (in a general ratio of one mature buck: up to 20 females; or one young buck: up to 12 females). The buck is usually removed from the doe herd after 1–2 cycles by dart gun and replaced with a backup buck for a further 1–2 cycles.

2.2 Reproductive Manipulation

Many whitetail breeders in the USA have embraced reproductive manipulation technologies. A.I has had the most significant impact on the industry, but the use of sexed semen and MOET embryo transfer is gaining in popularity. Several hundred whitetail donors are flushed for viable embryos every year in USA now and it has become an accepted practice for those ranchers who have identified superior females for antler production.

Semen collections and artificial insemination were technologies that have had a huge influence on genetic gain in red deer and elk worldwide. These technologies, in recent times, have been introduced into the smaller deer species being most commonly farmed in the USA.

Demand for superior bucks in the whitetail and mule deer hunting industries has driven semen prices and shooter buck prices to a high level.

Artificial insemination of these smaller species and the access to superior trait semen has had a substantial impact on the genetic gain seen in evidence today.

The selection of superior females, based on the antler desirability of their sons, and culling of those does with less desirable characteristics, has had an even greater influence.

Semen is extracted from sires by electroejaculation under sedation. A normal range of 60–120 doses can be expected from one ejaculate from a mature whitetail buck (1–10 billion sperm per ejaculate).

Mule deer do not produce as much semen from electroejaculation (1–4 billion sperm per ejaculate).

A normal insemination dose is 20 million minimum live sperm for transcervical A.I. and 8 million minimum live sperm for laparoscopic A.I.

Synchronization of females for fixed time A.I. is achieved with the use of a progesterone impregnated device placed into the vagina and hormonal injection with withdrawal of this device after a set number of days.

Transcervical A.I. is performed without sedation while the doe is restrained in a chute. A speculum with light is used to visualize the cervical os, and the semen then placed by the operator into the cervix.

Laparoscopic A.I. is performed on sedated does in specially designed cradles. A rigid laparoscope is used to visualize the horns of the uterus and semen is placed into the lumen.

Fresh and frozen semen are used with both techniques.

These are commonly used procedures and in Texas alone, an excess of 10,000 whitetail does will be A.I.'d annually.

Conception results are generally in the 75–85% range with laparoscopic A.I. and 60–70% range with transcervical A.I.

2.3   Sexing Semen

Sexed semen performed by Sexing Technologies in Navasota, Texas is having an impact on the whitetail industry.

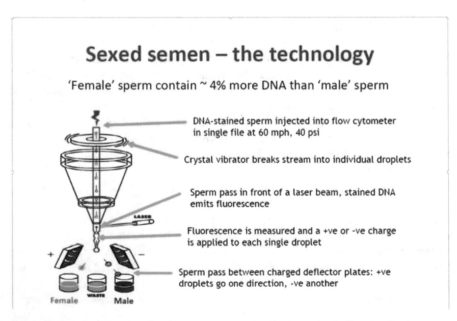

Sexed semen – the technology

'Female' sperm contain ~ 4% more DNA than 'male' sperm

DNA-stained sperm injected into flow cytometer in single file at 60 mph, 40 psi

Crystal vibrator breaks stream into individual droplets

Sperm pass in front of a laser beam, stained DNA emits fluorescence

Fluorescence is measured and a +ve or -ve charge is applied to each single droplet

Sperm pass between charged deflector plates: +ve droplets go one direction, -ve another

Female   WASTE   Male

Female chromosomes in semen are heavier, positively charged and kept to produce heifer calves. Read more *(Dairy Australia)*

Whitetail semen can be successfully sorted with more than 90% of semen giving male progeny.

While there is also the possibility of producing more than 90% female progeny from some superior females (as female replacements in a herd), the biggest financial impact for the USA whitetail industry is using male sexed semen in average does, to produce more bucks for hunting and reduce the economic waste of producing does.

The conceptions are usually 5% less than with conventional unsexed semen.

Whitetail have a twinning tendency in ideal management situations. This may be slightly reduced by the much lower dose sexed semen.

Embryo transfer (E.T.) is a relatively new technology in the whitetail and mule deer industries. It is currently being used in a number of game farm species that are ranched in the USA.

The two main techniques being used are:

1. *MOET (multiple ovulation embryo transfer)*—where donor females are given drugs to induce superovulation and then fertilization is achieved by natural breeding or A.I. (artificial insemination). The fertilized embryos are "flushed" out of the donor uterus surgically or in larger species non-surgically (as in cattle).

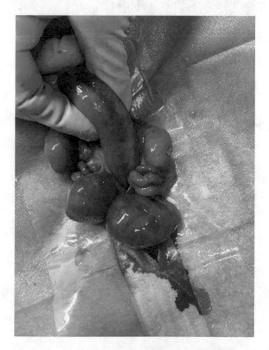

Follicles

Surgical Flushing

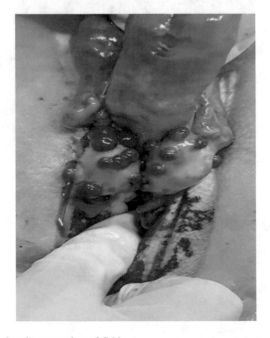

Stimulated ovaries showing a number of C.L's

2. *I.V.F. (in vitro fertilization)*—where oocytes (eggs) are removed from follicles on the ovaries and subsequently fertilized in a laboratory after maturation of the egg.

With both methods, implantation of the embryos is either performed fresh (if recipient females have been synchronized so that their cycle coincides with the donors' cycle) or frozen (if embryos are to be shipped or implanted at a later date). No matter what species are being reproductively manipulated, E.T. (embryo transfer) is a very powerful tool.

First you need to identify the trait you want to improve and then you must identify the females that produce the best progeny for that trait.

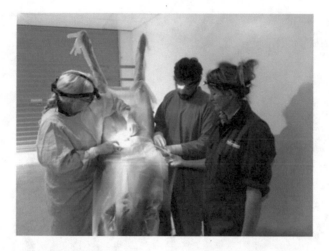

Surgical recovery of embryos

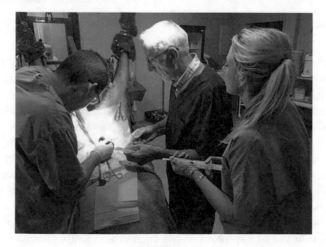

Laparoscopic implantation of red deer in Spain

The four main advantages of E.T. are:

(i) Within Herd—You must know which females are producing consistently superior progeny for the trait you have chosen as economically important. These superior females become donor females so that they can be superovulated or aspirated and the resulting embryos can be implanted into genetically inferior females. These inferior recipient females will then carry a pure offspring of the donor sire and dam, which will be totally unrelated genetically to the recipient herself. The genetic gains seen in herds utilizing E.T. is far superior to A.I. or natural breeding, as long as the selection of the donor females is sound.

Young Red deer showing consistency that only ET and AI can achieve. Ex OLEP Farms Quebec, Canada

(ii) Movement of Genetics—E.T. offers a way of moving species from one country to another, or from one state to another state, where live animal imports may be prohibited.

These embryos will be frozen and transported in a normal liquid nitrogen tank. This will result in the movement of superior genetics to the importing region, and in some cases movement of a totally new subspecies to a country where this species did not exist. This will require a recipient that is genetically close enough to carry an embryo. For example, elk embryos were transported to Australia in the 1980s from New Zealand where red deer were used as the recipient.

(iii) Preservation of a species that is endangered can be achieved with E.T. or re-establishment of a species that has been inadvertently eliminated in an area.

(iv) E.T. can lessen the threat of disease transfer. Embryos can be washed, including the use of a Trypsin wash which digests off the outer layer of the shell of the embryo. This makes embryos a much safer way to introduce a species than by using live animals.

Red deer E.T. has been practiced in New Zealand successfully for many years now. As well as improving the genetics in these herds, it enables the importation and exportation of red deer embryos with superior traits that are sought after internationally.

A number of breeders in North America who are focused on producing high (Safari Club International) SCI scoring red stags for the hunting industry have been importing red deer embryos and implanting them into poorer genetic recipients.

2.4   Smaller Deer Species: Sika, Fallow, Whitetail, and Mule Deer

The smaller deer species are more nervous and flightier. This has always presented more of a challenge as putting these species under stress in preparing them for a superovulation program (which may involve handling them on numerous occasions), can switch the ovulation off. In times of stress nature does not want them to reproduce.

Earlier trials with fallow deer in the 1980s, in New Zealand, used implants to deliver the superovulation drug which resulted in some promise; but only limited numbers were trialed.

In recent years, a trial with blow darting of the superovulation drug was done. Using this technique overcame earlier troubles we were having in producing good quality embryos.

In the same year, we ran a trial with whitetail deer in Wisconsin, USA. This technique surprised us and produced large numbers of good quality embryos.

A recombinant FSH is being trialed in whitetail and mule deer which avoids the extra handling needed for multiple shots of FSH. This technique is showing a lot of promise and accounts for approximately one third of the does superovulated.

3   Conclusion

Many innovative whitetail breeders in the USA have embraced reproductive manipulation technologies. While artificial insemination has had the most significant impact on the industry, the use of sexed semen and MOET embryo transfer is gaining in popularity.

Sika deer ex China

4   Handling

Crucial to the development of successful captive deer breeding and the application of artificial breeding technologies has been the evolution of successful handling systems.

The smaller species of deer raised in the USA, such as whitetail, require specialized handling systems that differ from the standard red deer or elk facilities.

These smaller species have a highly developed "flight response" and so are more prone to "panic" injuries.

Farm layouts and yard design systems have been especially developed to minimize these injuries. It is also common for tranquillizer guns and darts to be used for the delivery of medications to a sick animal, to avoid the stress and danger of running these smaller deer species into a facility to treat them.

The following plans of yard design and farm layout for whitetail deer are provided by Len Jubinville of the "Deerstore" company (http://www.deerstore. com). The yard layouts are designed to maximize easy flow of deer movement and minimize injury.

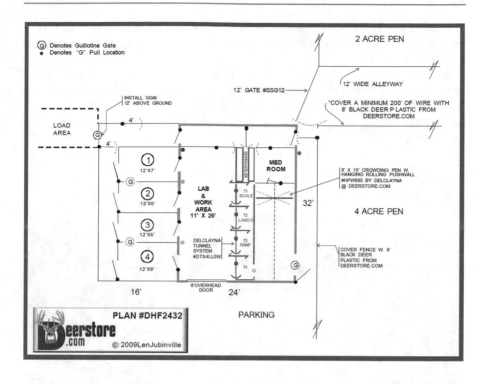

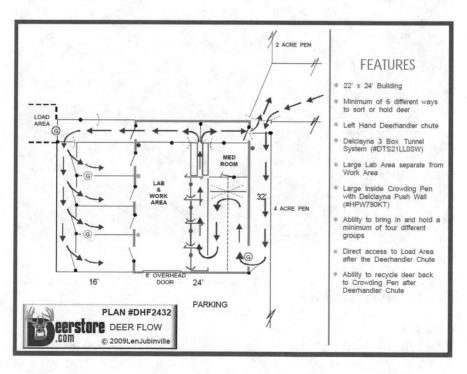

FEATURES

* 22' x 24' Building

* Minimum of 6 different ways to sort or hold deer

* Left Hand Deerhandler chute

* Delclayna 3 Box Tunnel System (#DTS21LL0SW)

* Large Lab Area separate from Work Area

* Large inside Crowding Pen with Delclayna Push Wall (#HPW790KT)

* Ability to bring in and hold a minimum of four different groups

* Direct access to Load Area after the Deerhandler Chute

* Ability to recycle deer back to Crowding Pen after Deerhandler Chute

Purpose built chutes are designed to restrain the deer once inside the facility to enable the operator to perform routine procedures, such as tagging, vaccinations, and de-worming.

Whitetail are also injected intravenously or intramuscularly while in these chutes to be able to perform routine procedures such as laparoscopic whitetail insemination and semen collections. Other smaller deer species, such as mule deer, are also handled in similarly designed facilities. Fallow and sika deer can tolerate the same handling systems.

5   Cervid Nutrition

Chris Stewart
Huntsville, TX, USA

Deer farming as an industry in the USA, generally revolves around the seasonality and subsequent timing of reproduction in their respective geographical areas of the country. Most cervid species on deer farms in the U.S. breed at some time in the fall or winter and fawn or calve in the spring quarter with males growing antler from late April to late August. There is no shortcut to nutrition and its effect on a program's overall success in any measurable category such as antler growth, reproductive success, etc. We use several different rations across 5 different cervid breeding programs to maximize our efficiency and still provide an economic feed cost to the bottom line.

Maintenance Ration

16% CP, 3–4% fat basic vitamin/mineral pack (pelleted) $.

Breeder Rations

16% CP, 4–8% fat w/enhanced vitamin/mineral pack, probiotics and bypass proteins (pellet) $$
17% CP, 8% fat w/ enhanced vitamin/mineral pack, probiotics and bypass proteins (textured) $$$
19% CP, 9% fat w/ enhanced vitamin/mineral pack, probiotics and bypass proteins (textured) $$$

Each individual farm will have its own set of factors that will have to be analyzed to supplement the information below.

Our general farm setup and conditions are broken down by species and sex below:

5.1   Red Deer, Sika, Fallow, Axis

Pens are seeded with some bahia but mostly bermudagrasses (forage). Pasture maintenance includes weed spraying and fertilizer application as needed during the growing season. Soil is sampled yearly and corrected as needed. Our soils are generally deficient in copper so boluses are given to stags in winter January and September and September for hinds. Sika, Fallow, Axis do not receive boluses.

5.2   Stags/Bucks

After weaning, males are placed on the 16–4% cervid breeder ration. They are fed this ration free choice until they are 1.5 years of age (roughly September). Once antlers have finished growing, they are placed on the maintenance ration for 5 months along with free choice forage (alfalfa and fertilized bermudagrass) to get them through the bulk of the winter. Starting in late January/early February, they will be placed back on the 16–4 breeder ration through the growing season on their second set of antlers. If body weights are low, switch to the higher fat ration (such as the pelleted 16–6%) until weights and body conditions are at the program goal. Upon completion of the second set of antlers, males are placed on the maintenance ration with access to quality forage until the final year they are either sold or turned out. The year a male will be harvested for velvet, sold as a stocker, or hunted as a trophy, they are placed on the 16–6% breeder ration in January of that year at free choice rate. Access to a high-quality forage will help males self-regulate their intake of protein source, and after 3–4 months of access to the high fat ration, males are switched to the 16–4 breeder to finish antler growth but not increase Body Condition to a point where it is difficult for the male to be comfortable in the heat and/or humidity associated with late summer in the Southeastern United States.

5.3   Hinds/Cows/Does

Females are placed on the 16–4% breeder ration upon weaning to accelerate musculoskeletal growth through the first year of growth.

Our research has shown that the females will self-regulate their intake of a protein source given adequate forage being made available to them even as a young growing animal. Going into their yearling breeding season, females are placed on the maintenance ration if body weights and BCS are adequate. Any female that is overweight going into breeding season has a greater chance of missing Artificial Insemination procedures.

Ideally a female should be on a rising plane of nutrition (putting body weight on) going into the AI or embryo transfer process for consistent results. Young females will not have offspring pulling their body weight down, so management of their weight and body condition prior to breeding season is important for success at any reproductive technique. As a mature female (2+) years of age, the maintenance ration is fed on a per head rate along with access to high-quality forage for most of the year. Once the female has given birth, we provide the 16–4 or 6% breeder ration to them in order to maximize milk quality to the young and limit the "drawdown" on the female. There is little recovery time after fawning season for the female as offspring are weaned and 2 weeks later the AI or implant process begins, so a healthy weight and body condition at weaning is important to overall breeding success in these species.

5.4   Whitetail

Whitetail pens generally have grass and some natural shade in pens but little access to other sources of browse.

This ensures the rations we feed are not being diluted by any foraging in the pens. Whitetail are given access to high-quality alfalfa hay, or similar substitute.

5.5   Bucks

After weaning, they are placed on a 16%–6% breeder ration through their yearling set of antler growth (~10 months) they are also supplemented with a textured ration as needed to achieve maximum growth of the musculoskeletal system as quickly as possible. Upon completion of the yearling antler set, bucks are placed on 16–4 breeder ration through the winter months.

Bucks consumption will decrease during breeding season even with no access to females. Going into their second set (Feb–March), bucks are placed on a 16–6/8% ration through growing season.

Fat content of pelleted ration is based on body condition score and post rut condition of bucks. Higher fat, breeder rations are fed to bucks through July/August to ensure the "late" growers are finished before switching to a lower fat and more economical ration to go into winter "holding" months.

5.6   Does

Does are treated like bucks for their first year on the farm, with access to a higher fat ration to accelerate early growth and maximize skeletal growth. Mature females are fed a 16–4% breeder ration year round to keep cost of production manageable. After fawning, some years require the does to be put on higher fat ration during lactation to manage weight or body condition concerns. With whitetail, managers should be careful of doe weights toward the end of fawning/weaning of offspring as generally there is a window between weaning and breeding that if not managed, can cause the does to become too fat going into breeding season. We want the does to be putting on body weight still when the breeding season starts.

6   The Deer Industry in Mexico

Eduardo Serrano
Advanced Deer Genetics, Center Point, TX, USA
In North America, there are 38 subspecies of whitetailed deer recognized by wildlife biologists.

Fourteen of them are in Mexico, which means that 37% of recognized subspecies live in Mexican territory. Apart from *whitetail deer*, Mexico has another 3 different species of deer: Mule deer (*Odocoileus hemionus*), Red Brocket deer (*Mazama americana*), and the Gray or Brown Brocket deer *(M. gouazoubira)*. Of the subspecies of whitetailed deer, the Texan whitetailed deer has been the most popular and the subspecies that has been studied the most, due to their bigger body and antler size, followed by the Coues deer, Sinaloa deer, Mexicanus deer, and Yucatanensis deer.

Approximately 40 years ago, the deer population decreased significantly, due to a relaxing of hunting regulations, over-hunting, poaching, and the reduction of natural habitat, especially in the northern states of Mexico. Government, hunters, and conservationists decided to act to bring the deer population back and promote the deer industry. Between 1980 and 1990, deer farms emerged in Mexico, influenced by farmed deer in the USA, especially in Texas. Deer farmers started at a small scale with just a few deer captured from the wild. Today, the top Mexican deer breeders operate at the same scale as US breeders, using the same technology and genetics. Ninety percent of the deer farms are directed to hunting and the preservation of the species, and the other ten percent are for venison production and government programs for reintroduction of deer to zones where the population is low, or the deer have disappeared.

Deer farmers have had to use the newest breeding technologies to meet high demand for hunting. Artificial insemination, using semen collected from the best sires, and embryo transfer have had a huge impact on the industry, not only to increase the numbers of individuals, but also to improve the genetics—looking for the perfect deer, the one with the biggest and the most impressive set of antlers, the one that season after season the hunter will try to catch.

Deer farmers are the biggest contributors to the deer industry, providing high value big bucks for hunting and at the same time working for the preservation of the species. The future of the deer industry in Mexico is based on the promotion of all species and subspecies of deer. A good example is the mule deer industry in the Sonoran Desert—the mule deer has become as popular, and as much in demand by hunters from all around the world, as the Texan whitetail deer.

Elk Farming in North America

Ian Thorleifson, Murray Woodbury, Bruce Friedel, and Tony Pearse

Abstract

A comprehensive account of the farming of elk throughout North America including the history of the industry, the genetics of elk farm stock, the life history and description of the animal, the layout and design of elk farms and handling systems and their use; identification of animals and herd records; health programs and biosecurity. Management for mating, calving, and weaning and the use of artificial insemination and pregnancy diagnosis. Nutrition and pasture management. Velvet production, harvesting, and marketing; sale of co-products, venison, and trophies. Management and economics of hunting. Diseases including the impacts of Chronic Wasting Disease.

This chapter is excerpted from The Elk Farming Handbook (Thorleifson et al. 2020) written by four authors led by Ian Thorleifson.

I. Thorleifson (✉)
HotSun Livestock, Wasagaming, MB, Canada
e-mail: vike@mymts.net

M. Woodbury
University of Saskatchewan, Saskatoon, SK, Canada

B. Friedel
Iron River Wapiti, Iron River, Alberta, Canada

T. Pearse
Deer Industry New Zealand, Wellington, New Zealand

Keywords

Elk · Elk farming · Elk farm management · Velvet harvesting · Trophies ·
Artificial insemination · Nutrition · Disease · Welfare · Chronic wasting disease

Elk farming is based on a remarkable partnership with nature. Historically, wild elk
were distributed throughout the areas of North America where agriculture is now
successful. It is only natural that they would continue to thrive in those areas. Nature
has provided us with animals exceptionally well suited to our farms and, in return,
we the farmers provide the resources to enhance their natural patterns of growth and
reproduction and guide natural selection toward our production goals.

1   Historical Perspective

Before the 1500s, very few domesticated animals were kept in the Americas. Dogs
that came from Asia with the first inhabitants, and native turkeys, cavies, llamas, and
alpacas were all the species commonly husbanded. With the waves of Europeans and
others who migrated to the Americas came horses, cattle, sheep, and many other
familiar domestics. The potential for raising native wildlife species was quickly
realized by the newer migrants. Bison were virtually saved from extinction by
farmers and conservationists who captured some of the few remaining animals,
mostly as calves, from the wild. In 1908, the US Department of Agriculture
published Bulletin 330, "Deer farming in the United States" which detailed the
potential for husbandry of various species of deer, especially elk, and also
recognized the primary difficulty facing such enterprises, namely the contrary and
restrictive ideas and regulations presented by wildlife managers in the various States
and Provinces. A "North American System of Wildlife Management" had emerged
in response to the uncontrolled and unconscionable slaughter of wildlife species that
occurred in the 1700s and 1800s. A strong tenet of that system was "No commer-
cialization." This philosophy allowed exceptions such as commercial outfitting, but
spurred objections to actual live capture and domestication. Over time, the lure of
profits and preferred food and other benefits have eroded that opposition and allowed
development, but opposition still exists.

The past 50 years have seen a dramatic pace of development in the elk and deer
farming industries in North America. Before the 1970s, there were very few commercial elk farms in operation. "Game farms" raised and sold live animals, but these
operations were more like zoos rather than commercial farms. The impetus to begin
commercial development came from contact with Asian buyers of velvet antler and
other products, trophy hunters and from the pioneering efforts of farmers from
around the world, but particularly New Zealand farmers. North American farmers
realized that a large unsatisfied market existed for elk and the products derived from

those animals. The challenge was to establish management systems and political environments that would allow efficient and profitable production.

Management systems for farmed elk in North America were not difficult to develop. Elk were native to much of the southern half of the continent and proved to be highly domesticable and easily adapted to the farm environment. The New Zealanders readily shared husbandry skills and knowledge, with further information provided by Europeans and South Africans. Most of these first elk farmers were already accomplished stockmen who then adapted the techniques learned in other livestock industries to this new strategy.

The first commercial farmers in the 1980s obtained livestock from the wild when permits could be issued by the wildlife agencies or purchased from zoos and game farms. Markets were readily found for velvet antler, primarily in the Asian community, and for aged bulls on the established hunting farms in the United States. The industry strengthened, and the market for breeding stock boomed. This good fortune, unfortunately, precipitated the most difficult times the industry has faced. Science had not kept pace with the development of the industry. Large numbers of live animals were moving across the country with minimal control or testing. Techniques used for the detection of various diseases and parasites in other species were directly transferred to cervids without assessment of their efficacy or accuracy. As a result, tuberculosis was diagnosed in farmed elk in the late 1980s, and CWD—the prion disease of cervids, in the 1990s. These diseases have proven to be the greatest impediment to industry development, particularly CWD.

Disease and parasites are an ongoing challenge for all livestock producers. Tuberculosis has been virtually eliminated in the farmed cervid industry in North America, but new challenges have emerged, Chronic Wasting Disease. First recognized as a "wasting syndrome" in mule deer in a research facility in Colorado in 1967, CWD has to date only been found in captive and wild cervids in North America, the Republic of Korea, Norway, and Finland. In Canada, CWD was first detected on a Saskatchewan elk farm in 1996. The disease has since been routinely detected in wild cervids and farmed deer and elk in parts of Saskatchewan and Alberta and in 2018 on a red deer farm in Quebec. It is now widespread in many areas of the USA as well. The disease has spread through the movement of infected live animals and animal parts and contact with infected materials in the environment.

Although these diseases make for difficult times for some producers, the net result of dealing with them is a strengthening of every aspect of management, and a hardening of the resolve of North American Elk farmers to see the industry succeed. Industry-supported research continues toward the improvement of disease and parasite detection and management techniques. With these ongoing programs, and the development and broad acceptance of the Code of Practice for the Care and Handling of Farmed Cervids has come recognition of the validity, profitability, and support for the elk farming industry throughout most of North America and abroad.

2   The Source of our Farmed Genetics: Wild North American Elk

The North American elk, or wapiti, is the largest form of the species *Cervus elaphus*. "Elk" is the name by which most North Americans know this majestic deer. "Wapiti," meaning "white rump," is the Shawnee Indian name and the common name preferred by scientists, because the animal known as an "elk" (alg) in Europe is not an elk at all but a close relative of the North American moose. Other breeds of the species, all smaller and known as "red deer," "maral," or other local names, are found throughout the northern hemisphere: in Scotland and continental Europe, North Africa, and Asia.

2.1   Appearance

In general appearance, elk are obviously kin to many other species of deer. However, elk are much larger. Among North American deer, they are second in size only to the moose.

Males—Bulls	420–600 kg
Females—Cows	220–320 kg
Young: Calves	
Birthweight: Males	18–28 kg
Females	16–26 kg
Sexual maturity	15–27 months
Gestation length	247 + 10 days
Estrus cycle	21 days
Breeding season	Sept–Nov
Calving—single, (rarely twins)	May–July

The color of an elk coat ranges from reddish-brown in summer to dark brown in winter. Although it looks white from a distance, on closer inspection the rump color is ivory to orange. In contrast to the rump, the head and neck are dark. Elk have long, blackish hair on the neck that is referred to as a mane. Their hooves are rounded, and their tracks may be confused with those of yearling cattle in range country. Their droppings (scats), like those of other deer, are in the form of pellets in winter, but in summer, when the animals are on new green forage, resemble those of cattle. Closer inspection, however, reveals traces of a pellet structure.

Male elk are notable for their impressively large antlers. It is amazing that these large structures are grown new each year by the animals in a period of a few months in spring and summer. Antlers look particularly large in summer when they are encased in velvet—a covering that protects the growing antlers. In later summer, the velvet is rubbed from the fully-grown antlers and the bony structure is revealed. Newly cleaned antlers are light gray in color but become stained by rubbing and thrashing vegetation during the excitement of the rutting season. (Rutting is an annual state of sexual excitement in the male deer.)

The elk is highly vocal for an ungulate. A person close to a group of elk can hear frequent grunts and squeals as they keep in touch with each other. When alarmed the cows give sharp barks to warn the rest of the group. The whistling roar of rutting bulls is a spine-tingling sound on a frosty autumn morning.

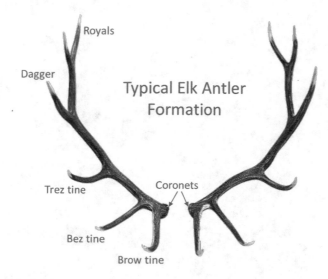

Typical mature antler structure

2.2   Diet

Elk are plant eaters. There are few plants that occur on their range that they do not eat in certain areas under certain conditions. In winter they eat grasses when they can obtain them. However, when the snow gets deep, they readily eat twigs of woody species, even conifers like Douglas fir. In spring, grasses and sedges are favorite foods. As broad-leaved herbaceous plants spring up in early summer, elk include a high proportion of them in their diet. They also consume shrub and tree twigs and leaves. Early summer is the time of year when a wide variety of nutritious food is available for elk; it is also the time when elk cows are providing milk for their newborn calves.

Browsing

As summer passes, the herbaceous plants dry out and elk turn again to dry grasses and browse. When the frosty nights of autumn arrive, leaves begin to fall in trembling aspen forests on the western ranges of the elk. Elk include dry leaves in their diet and continue to eat them until they are buried by snow. When winter comes, elk diets are controlled largely by snow. Elk dig craters in loose snow to expose dry grass and leaves, but when the snow gets too deep or too hard, they must shift their feeding largely to woody twigs. In mountainous terrain, elk must leave areas of deep snow cover and seek locations such as valley bottoms where snow cover is shallow or absent. In areas where deep snow seldom occurs, they may frequent high or low elevation ranges at any time of the year.

2.3 Life History

Elk are sociable animals. They are seldom found without other elk nearby. The herd lifestyle is characteristic of animals that live in open country. However, elk populations today occupy forest or parkland regions, where small groups averaging six or seven animals are common.

Elk are long-lived animals: males survive to an average of 14 years, whereas females may live and breed for 20 years. Although they may travel widely, each elk is strongly attached to certain localities within its home range.

The annual cycle of the elk begins in spring with release from the snows and food shortages of winter. This is the time of calving and increasing of the herds. Calving usually occurs in areas year. Others give birth to their calves in whatever part of their home range they happen to be in when the time comes. Whenever calving occurs, the cows split off from other elk and seek seclusion and cover a few days before giving birth. Elk hide their calves for 10 days or more after they are born. The calves are genetically programmed to remain quiet and concealed as a defense against predators. Later mother and offspring join others in cow/calf bands on the summer range. Beginning in August, the quiet summer life of the elk comes to an end with the start of the rut. The bulls, which have passed a lazy summer in small groups while their antlers grew large and heavy, now move into the cow/calf group and establish harems of cows. In the process, there is considerable fighting among the bulls. Large bulls eventually get control of as many as 20 or 30 cows and drive other males to the fringes of the herds. This does not mean, however, that the young males are totally left out of the breeding. While the large harem masters are running off intruders or rounding up straying females on one side of their group, a young bull may sneak in and mate with a female on the other side.

Following the turmoil of the rut, the bull elk leave the females and move to good foraging areas to recoup their losses in weight and condition before winter. Some go back up the mountains to spend a few more weeks on the nutritious pastures of the alpine zone before snow forces them down. Elk usually, but not always, wait for the coming of snow to move down to the valleys. There is considerable overlap between the winter ranges of bulls and cows. As bulls are larger and more powerful, they can travel and dig through deep snow more readily than the cows, and by doing so they are able to have foraging areas to themselves.

3 Relationship to People

Elk are highly esteemed by hunters and lovers of wildness. They are one of North America's major big game species, with tens of thousands harvested by hunters each year. In wild areas where they are subjected to hunting pressure, they become very wary and difficult to capture, but in parks where they are not hunted by humans, they quickly habituate to the presence of people. Rocky Mountain, Banff and Jasper National Parks are well known for their very tame elk, so tame as to pose danger to unwary or foolish people. These elk are a great attraction for tourists, drawing thousands of elk lovers there.

In mountain areas during winter, elk share valley bottoms with major transportation corridors. This leads to many elk-vehicle collisions, with disastrous results to the elk and to humans and their automobiles. This costly hazard has been controlled in many areas, by the construction of a system of fences, cattle guard gates, and over- and underpasses across highways.

The readiness with which elk can be habituated to people and the value of products derived from them have long aroused considerable interest in domestication and ranching of the animals. In 1908, the US Department of Agriculture published a Bulletin describing the suitability of elk for domestication and the value of their products. One of the most valuable elk products is their antlers. Since ancient times, Oriental people have recognized that medicinal preparations derived from elk antlers are useful for diverse therapeutic purposes. Oriental medicine has consumed large quantities of elk antler, and today more and more people of all backgrounds are recognizing the effectiveness and using those supplements.

In many areas, elk and domestic cattle share the same ranges. Because both eat the same foods and the presence of cattle brings human activity, there is some conflict between the two species. In mountain areas where elk concentrate in valleys that are also important winter range for cattle there is competition for scarce forage and disturbance of elk at a time when they are under stress due to severe weather. Such situations call for close cooperation between ranchers and wildlife managers to keep problems under control. The future welfare of wildlife elk in general depends on cooperation between wildlife authorities and all land managers, including forest industries, oil and mining companies, park managers, and Indian bands, as well as ranchers.

3.1   The Principles of Successful Elk Farming

As the theory and practice of elk farming has developed, several foundation principles have become obvious. To appreciate the success of elk as semi-domesticated farm animals. It is important to review the generic features of the elk we successfully farm. Elk are sophisticated ruminants, very efficient converters of forage and browse. They are also strongly social herding animals, traits which make them well suited to a farming situation. If well-fed and comfortable, they are amazingly free of disease and stress. A key feature of this unique deer species is the close correlation between life cycles and seasonality.

3.2   Seasonality

In temperate environments, elk have a strong seasonal pattern of growth. Maximum gain of body tissue occurs in spring, summer, and fall. Weight may be lost,

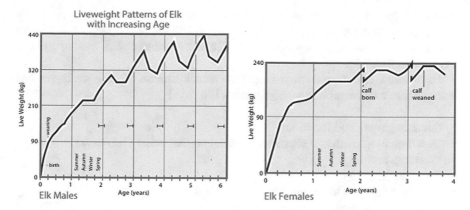

Fig. 1 Live weight patterns of elk with increasing age

particularly by mature bulls during autumn, (the rut or breeding season) and little or no weight gain takes place over winter. This is classically illustrated in Fig 1.

Extensive experimental work now shows that deer growth cycles can be defined as patterns of weight change that occur under proper nutrition and can be influenced by changes in daylight length. The increasing daylight length of spring and summer stimulates appetite and growth and decreasing days of autumn and winter reduce appetite and growth. The cycle of active reproduction is the opposite, or reciprocal of the growth cycle. Further work has since shown that physiological systems the deer might employ during the seasonal extremes, (i.e., a raised metabolic rate during summer compared with winter) are a direct consequence of the food intake cycle.For on-farm management of growth and production we need to be aware that elk have:

- A yearly cycle of growth potential best expressed in the young animal
- A seasonal cycle of food intake—increased in summer, restricted in winter
- An all year round requirement for quality feed and
- An ability to digest feed better in summer rather than winter—so high quality, highly digestible feeds with adequate fiber content are critical in winter.

The annual growth curves show there is a strong seasonal change in live weight. This is shown in adult bulls as the ability to lay down stores of body fat and mobilize it during the autumn breeding season.

This is particularly true in males greater than 2 years of age. Breeding bulls can lose up to 30% body weight during this time of rut. Clearly, this has a major implication for lean venison production. A slaughter target prior to winter at 1 or

2 years of age is ideal for young males. Slaughter for older bulls is appropriate only during post-breeding season, once condition and fitness have recovered if over fatness is to be avoided. Surprisingly, elk females yield excellent quality meat at any time in their life.

Farm management production targets on which to base feeding strategies and management then become:June–September (100 days)

- Summer growth and lactation
- Best time for growth and slaughter September–December (100 days),
- Weaning, breeding
- Potential for growth
- Slaughter options December–March (100 days)
- Winter—supplemental feeding
- Quality feed with good digestibility and adequate fiber
- Spring calving (peak June 6)
- Rapid growth of young stock and antlers

It is no accident that these seasonal features of elk biology have a parallel fit with the seasonal task of feeding elk to seasonal requirements that are predictable and repeatable. The hormone melatonin is the primary hormone produced in response to changes in daylight length that leads to the increase or decrease in the production of all of the hormones involved in reproduction, sperm production in males and seasonal estrus in females. Changes of secondary sex characteristics, (coat color, neck and mane development, vocalizing, etc.) are included.

---

4   Domesticated Elk

Deer species that are successfully farmed share several common features that contribute to that success. Normally, deer that adapt well to traditional or modern farming systems show the following features:

- They are largely a herd animal, social in nature and behavior and are gregarious. An elk herd is controlled by experienced dominant females who alert the herd and lead in the education and experience of calves and herd movement and acceptance of routines.
- Deer are multi-plant species browsers and grazers. The most commonly farmed deer are "intermediate" feeders, opportunistic, adaptable and in part selective over a wide range of plants and pastures. Elk have proven to be highly adaptable to a variety of farm environments.

- The behavior of elk, its polygamous nature, and productivity mean there are powerful tools available for selection and improvement. A dominant or superior male can mate many females in the season allowing a high female: male ratio to be established in a farmed herd. In turn, large numbers of surplus males are available for production end points, (velvet, meat, or hunting) allowing truly superior animals to be retained for breeding.
- Farmed elk have great longevity compared to traditional livestock with a 12–15 year breeding life a realistic target. Herds can quickly expand in numbers or there is the potential to supply surplus breeding females to a growing industry. With a low replacement rate required, selection for superiority—mothering ability, growth rate, antler genetics can have rapid returns in herd improvement and in return on investment.
- Elk under good nutrition thrive, are healthy and carry natural resistance to many diseases. As they have evolved as wildlife their relationship with man on farms shows little dependence on intervention, provided the basics of exercise, food, water, and shelter are present.
- Elk have proven to be highly adaptable, responsive to good routine and management and can be handled easily and systematically with appropriate facilities and equipment. Most elk farmers share a close relationship and interaction with their animals, without compromising the spirit and natural behavior of the animals.

5   Suitability for Farming

Elk are well adapted to the North American climate and integrate readily into farming management systems. Dietary needs are easily satisfied with a requirement for fiber in their diet via browse or quality hay supplementation. Current farming emphasis is on the supply of breeding stock and velvet antler production. Selection for genetic improvement in body size, growth rate, and antler production has been rapid with truly remarkable antlers being produced by the leading breeders.

Elk have proven extremely responsive to on-farm situations if due regard is given to their size and speed that the animals can move under intensive handling conditions. Elk tend to be the most pasture-quiet of all farmed deer species and rarely track fence lines, jump at fences or create pasture damage through wallowing such as red deer are more prone to do.

Traditionally the North American species is recognized as four subspecies or breeds, relating to geographical origin, body size, and antler shape. Terms include Manitoban, Rocky Mountain, Roosevelt and Tule elk. In wildlife relocations, there has been much transplanting of subspecies and interbreeding. The North American Elk Breeders Association (NAEBA) is committed to blood typing and DNA analysis of farmed elk to identify parentage and pedigree and to define productive traits and genetics for an improved breeding registry.

Elk are highly trainable—pail feeding and calling results can be impressive

6   Farm Design and Fencing

6.1   Farm Selection

Farmed elk are truly so adaptable that the first question to ask in the process of selecting an elk farm is "where would I like to live?" Very few areas are totally unsuitable for elk farming, but many areas would be more economically feasible. In particular, the preferred option would be a location in an established agricultural area with access to productive soils, supplies, and services.

Other desirable characteristics include selecting:

- Coarser soil types—tending to silt/sand textures, rather than clay. This allows better drainage and workability.
- Soil pH 6–7, with a balance of soil nutrients. This can be built up but natural soil conditions such as these are ideal.

- Gently sloping or rolling land also improves drainage and gives a variety of options for location of buildings and facilities. Variety also gives the elk a choice of sites for feeding, resting, or other activities. A sheltered south-facing slope provides an ideal wintering location.
- A predictable supply of drinking water may be sourced from wells, streams, or ponds.
- Areas of trees provide shade, shelter, and browse. These are particularly desirable during calving season, in summer heat, and in winter.
- Existing buildings that may be converted to serve as winter shelter and handling facilities, or feed and equipment storage areas.
- Quieter locations, away from school pens, industrial activities or other sources of abrupt noise and movement.

Two things need consideration in each pasture: water and shelter. Water can be in dugouts, but a clean water source of some type is preferable. Dugouts can present dangers in terms of water quality, falling through thin ice, and drowning. Your water needs will be most pressing in the summertime so properly maintained aboveground waterlines are adequate. Adult elk do fine with only clean snow in the wintertime especially if high moisture haylage or silage are fed but young growing stock do much better with clean water all year around.

The last point to remember is "Start Small—Think Big." Plan for expansion in your farming operation, both in pasture area and in handling area.

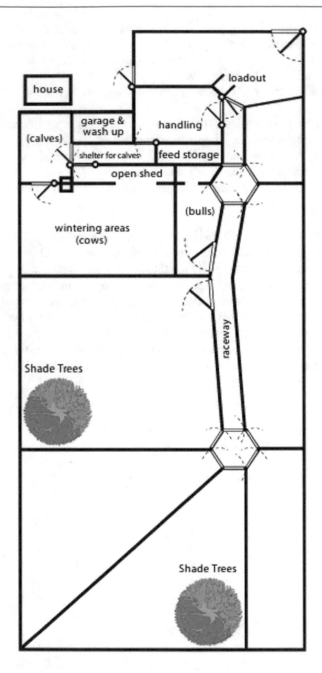

However, where boundary fences must be built to very high standards and expense it is often wise to consider fencing a larger area than the stock numbers may initially need and develop this area over time with less expensive yet effective subdivision fencing. These considerations can often be recognized in advance by taking time in planning layout using farm maps, aerial photographs, and taking time for a farm walk-about in the planning stages.

The handling pens should be on well-drained, gently sloping ground. You need access to the pens with a truck and trailer for loading and transport. From the pens the central raceway extends out to the pastures. You may use one or several alleys, depending on your land. Often, during stock movement, or a variety of on-farm activities, the alleys themselves can be used for sorting or holding areas. A specially designed "keyhole" can be effective in providing rest areas for stock, the ability to sort or draft groups, or simply as vehicle turning areas or for feeding stations. Pastures feed into the alleys at keyholes. Keyholes are essential for one-person operations.

6.2   Farm Layout

The key concept to bear in mind in laying out an elk farm is *flow*. Think of a pool of water. Now try moving that pool around your farm. This is the way an elk herd works. When planning the layout of an elk farm a useful analogy is to compare an elk herd to flowing water. Elk move as a unit flowing across pastures, through gateways, down slopes, around corners. They bottle up in square or dead-end corners and under such pressure can be forced through holes or weaknesses in or under the fence or handling areas.

To properly plan a layout, start with a map and hike or ride through the farm, with "flow" in mind. Plan with the idea that your farm will be a "one-person" operation. The two keys to farm layout are alleys or raceways and handling facilities. Alleys join all the paddocks together, allowing elk to move easily from any one paddock to another or into the handling pens. Handling pens should always be as central as possible. Raceway widths may range from 5 to 10 m (15–36 feet), with larger widths accommodating larger herds of elk and larger farm equipment. A one-person operation would favor narrower alleys. Pasture size should vary, with smaller areas near the handling pens and take account of local geography (trees, water courses, access, and security). Overall, for efficient pasture and stock management, the more pasture subdivision the better.

6.3   Fencing

The investment in elk fence in relation to overall establishment of a farm is relatively high (15—25% of budget) but critical for stock security and farm management. The boundary fence exists as much to contain farm stock and prevent escape as to keep predators, native deer species, and the inquisitive out. In most provinces, there are established regulations for post spacing, fence height, construction and materials, and maintenance requirements. The standard for elk is 8' tall (20-96-6) page wire. Posts are recommended at 6–8 m spacing (20'), but this distance may vary. Even on straight, level ground, spacing should not be more than 9 m (30').

Modern elk farm fencing has been revolutionized by the development of the hi-tensile woven wire netting fence. The strength and integrity of the netting is maintained by the high tensile nature of the wire giving high breaking strain loading and a construction that allows contraction and expansion under the extremes of climate without reducing wire properties or fence life.

Netting is given its function and strength from the development of a special locking tie that joins horizontal wire and vertical stays. This gives rigidity to the construction at a single point but allows the fence elasticity to repel animal pressure, and the whole netting area acts as a unit. Livestock can bounce off it but generally cannot distort netting spacing, cannot crawl through between wires, or in a well-constructed and properly maintained fence, crawl under it.

The dimensions of wire spacing have been carefully planned and tested so that they are closer at the base and mid-fence where the main pressures are applied. This construction repels many predators and will withstand shock from stock pressure either in groups in laneways or single animals charging the fence for any reason. The spacing with a 150 mm vertical stay will stop calves from slipping through fence to find outside cover or shelter, and the product is ideal for game farm fencing in every situation possible. Legislators in some US States have adopted special concession in their typical rules for such fence. These recognize the superior properties of the high tensile netting which allows cost savings in posts used, through increased allowable distance between them, and gauge and fastening of other wires. These recommendations have all been tested over some years and farmers are advised that additional shortcutting or trying to save on construction and costs is a false economy in the long term.

The final feature is that the netting is manufactured to high specifications of production including heavy zinc galvanizing to ensure long life and durability. Hi-tensile netting fence has a record of remaining rust and corrosion free for many years even in quite hostile coastal areas.

Sound, maintenance-free fencing can be relatively easily constructed with attention to a few basic principles, remembering that the integrity of any fence is only as great as the weakest point. A shortcut in cost savings or erection technique can be enormously costly through accident or escape of animals, or a simple deterioration in fence quality. Conditions in much of North America are extreme with heat, humidity, drifting snow, ice storms and frost heave, and conditions of spring thaw all creating their own pressure on post rigidity, alignment, and wire elasticity and strength.

7   Handling Facilities

An essential part of efficiently handling elk is properly designed and built handling facilities. The key word is smooth. Smooth entry, smooth corners, smooth handling. After twenty years of experience, the industry has also included three more essential "S" words, sweeps, slides and squeeze. Sweeps and slides refers to the style of construction that has been proven to be the best by far. Elk will flow through the rounded curves and around the corners of facilities described later in this section. They move into roofed areas made up of a series of steel and wooden cubes, or modules. Each side of these modules is a gate. The gate between modules is a slider gate, the side of each is a sweep gate. To move elk from one to another, you slide open the divider gate, then sweep the elk into the next module. At no time is the human in the pen with the elk, and the elk can see all around making them much calmer and easier to handle. The elk move through one module after another until they reach the squeeze.

"Sweeps and Slides"

A squeeze chute may be a "cow box" with two sweep gates forming a rectangular box about the length and width of an adult cow's body, with sliding access doors on the sides and small swinging doors on both ends. The best so far are the very carefully designed hydraulic squeeze chutes. Once a farm has more than two or three bulls, a hydraulic squeeze is almost mandatory.

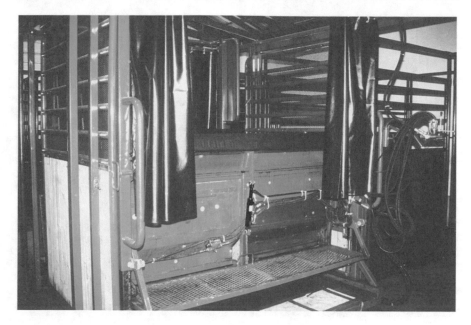

Hydraulic squeeze

Successful management of elk requires that the animals can be collected from pastures or winter feedlots at any time for any reason without difficulty or the creation of stress in the elk or the farmer. Once mustered they need to be sorted into appropriate size groups for handling in a purpose-built facility.

Stockmanship is often more important than a set pattern of pen design, but there are standard basic principles required to meet the combination of the elk's natural response to this somewhat artificial situation, and the necessary management task.

Pens and facilities can be as varied as the farm setting, existing buildings, and myriad of construction materials available. However, there are basic principles that, once understood and incorporated, use the natural instincts of elk to move or flow away from pressure, and thereby add to the farmer's management objectives without stress or risk to safety. Farmed elk are creatures of routine and will, over time, quickly reinforce both positive and negative experiences.

The key points in design and handling are:

- Simplicity and routine
- Control and pressure
- Letting the elk react naturally
- Curves and corners
- Safety and avoidance of overcrowding
- Efficiency and patience

Entry to handling facility

In the entry above, the elk enter on the left from a long curved raceway, then the sweep gate closes, and the elk naturally want to return to where they entered, but they now move into the curved entryway on the right and into the sweeps and slides facility.

8 Animal Identification and Record-Keeping

8.1 Individual ID and Ear Tags

Good management of elk farming requires some form of individual stock identification—a unique number that records the animal against a central registry. Tags often include a lot of information—herd number, import number, etc. and may not be suitable for on-farm management systems. Successful elk farming requires some knowledge of what individuals are—age, gender (at a young age), breed type or breeding origin and a unique management number that can be easily read in pens and in the pasture or winter pen situation.

It is recommended that farmers use a double tagging system for permanent lifetime ID The management tag should be a large plastic type, double sided with management number on both sides. This tag can display a farm code, including a jurisdiction code (e.g., A for Alberta, REF for Roughrider Elk Farm) and a unique individual ID number. The ID number can indicate year of birth, either as a straight numeric 3 or 4 digit sequence with the first one or two digits indicating year of birth, or as an alpha-numeric sequence with a year letter as the last character. Alternatively, the year of birth may be stamped at the base of the tag stem, or the farm code and an ID number may show on the front tag, with year of birth and ID number on the back tag. In addition to this management tag, a smaller metal, dangle or RFID tag should be affixed permanently in the opposite ear. The number on that tag should be totally unique, applied to that animal alone, and this tag must be as loss-proof as possible.

Sex differences in time become obvious with age, but a standard convention of (with the animal facing away from you) left ear for females, right ear for males is common. Ear marks denoting age are ineffective as ears can become torn, and the actual placement of ear marks is both traumatic and variable in position.

A separate color for an ear tag can be used to highlight year of birth, but this somewhat limits the amount of information to be carried. Experience suggests that color has a more applied use in defining breeding types and different strains of stock as the result of on-farm genetic improvement programs. These can include programs recording individual sire bulls or additional improved genetics.

The convention in record-keeping is to record the color of the tag and then the numbers (G123 is Green 123) or the number and the International Year Letter (123G is number 123 born in the year 2019, for which the year letter is G).

Many countries are moving to implement full traceability systems for all food, including cervid livestock and products. The internationally accepted basis for these

Fig. 2 Official Federal tags from behind

systems is unique identification device numbers allocated by the International Standards Organization (ISO). The tags shown in Figs. 1 and 2 are one option approved in Canada for elk. Several Canadian provinces have mandatory tamper-proof tagging systems and an official individual register required to ensure ownership and breed purity. All systems combine to form the basis of animal health monitoring schemes such as TB and brucellosis testing. With accurate and meaningful tagging, the true value of record-keeping can be appreciated for both on-farm management tasks and for maintaining accurate data from slaughter animals and for live sale opportunities.

Alternative identification systems i.e. freeze or hot branding, tattooing, or microchip ID have not been entirely successful in elk for a variety of reasons, as most are impracticable or unreliable. Radiofrequency ear tags (RFID) are now readily available for cervids and other livestock. These are programmed with unique numbers assigned by the International Standards Organization (ISO) and can be read with an electronic tag reader, and the data transferred to a computer.

The most permanent and possibly most accurate ID is a DNA profile, established in an accredited laboratory using hair, blood, semen, antlers or most any body part, and recorded on a registration certificate along with parentage and performance records Hair samples can be the easiest way to harvest samples for DNA work (Fig. 3), but they must include the follicle—a clump of skin and tissue at the base of the hair and they must be very carefully handled to ensure they are not mixed with another sample. A Tissue Sampler is a much better choice. It includes a small container that safely holds one sample punched from the ear and ready for labeling and transport.

DNA is nature's ID, the genetic base which defines what an animal is. Many jurisdictions are now moving to mandatory DNA registration of all, or at least all breeding livestock, to allow permanent ID, and to facilitate the matching of products such as antler and meat, and offspring, to the parents and individual producers. The challenge to industry is to participate fully in this development and to support research that will make this "ultimate ID" technology more useful, accurate, and reasonably priced.

Fig. 3 Above: Hair sample for DNA analysis. Be sure to include hair follicles on hairs

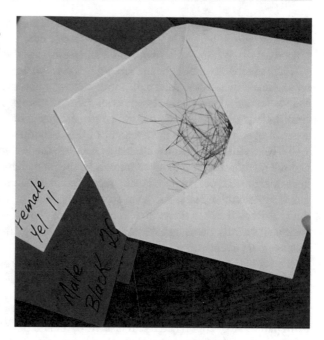

8.2   Tagging Calves

Any system of Identification relies on good management to be effective. DNA matching in elk can be an effective tool in confirming what the good manager already suspects or has recorded. But, because the currently—used techniques depend on a very limited number of markers (measurements of genetically based characteristics that differentiate one elk from another), DNA has limitations in its power to sort through an entire herd and accurately identify the sires and dams of each calf. This difficulty increases as a herd becomes more closely related or linebred and increases with the number of possible sires and dams that could be the parents of each calf. The most effective management techniques to avoid these problems include single sire breeding and tagging calves at an early enough age to match them to their dams.

8.3   Herd Records

Records can be classified as those important for assessing stock genetics and breeding and those involved in a response to management. For example, it is possible and often desirable to record growth rates of young animals from weaning to 15 months of age on a monthly or 6 week basis. For selection criteria only, weaning and 15 month weight and the gain between those two points is genetically significant, but the intervening weights give an accurate check on seasonality, growth, and management.

Many computer programs are available to aid data collection and record-keeping. It pays to consult widely in the industry before selecting the one that best suits your objectives or, develop your own with a spreadsheet and database program.

Field notebooks should be the first line of work in any recording i.e. to record the management events and data, e.g., antler casting and birth dates as they happen, not as recollections. However, field notebooks should always be backed up on computer, record cards or simply recorded in a farm management diary—depending on the scale of operation and the involvement of the individual.

To be useful, records must ultimately relate to the individual animal identification and the relationship of it to other animals in the immediate cohort, i.e., to age group, herd, and breed type. An individual's record chart, related to the important aspects of genetics, should include most or all the information on the Sample Individual Animal Record Sheet (Page 86). Such a record should also include animal health treatment records, notes on temperament, TB status, etc. that also relate to significant management events.

Velvet antler, breeding records, calving performances and individual growth rates should be kept as separate files and have more relevance to age and group than an individual pedigree from a management view.

Record keeping for its own sake ultimately becomes a frustration. Records only form part of an overall objective. It is as important to have that vision focused first as it is to collect the tools that can aid that development. Good property records will certainly allow effective use of the expertise that is available to farmers through Department of Agriculture specialists, consultants, veterinarians, and feed analysis laboratories. Information is relatively easy to collect and store, elk farmers tend to be an enthusiastic group that have a need to know and there is a huge array of information and interpretations available (Fig. 4).

Fig. 4 Cow herd

How Do Your Elk Compare?

Production Parameters of Farmed Elk - Based on Surveys of Alberta Producers 1990 - 91

Average					Averages	Minimum	Maximum
Age of Dam (mother)	Two years (first calves)		Adult Cows		4.7		16
Weight of Dam (lbs)	483	471	531	534	526	352	660
Calving Rate (%)	81	81	96	96	93		
Weaning Rate (%)	73	73	91	91	88		
Calf Gender (sex)	**Male**	**Female**	**Male**	**Female**	**all**		
Calving Date	11-Jun	11-Jun	06-Jun	06-Jun	07-Jun	21-Apr	10-Sep
Birthweight (lbs)	42	40	38.5	36	38.9	25	49
Weight @ 100 days (lbs)	195	180	210	190	197	125	277
Growth Rate (lb/day)	1.52	1.43	1.69	1.52	1.58	0.86	2.35
Weight @ 200 days (adj. lbs)	279	251	286	269	275	197	375
Weight Gain (lbs/day)	0.84	0.75	1.06	0.95	0.95	0.44	1.61
Date Exposed to Bull	11-Aug	19-Aug	28-Jul	03-Aug	05-Aug	01-Jan	15-Sep
End of Breeding	17-Nov	22-Nov	27-Nov	26-Nov	24-Nov	14-Oct	31-Dec
No. days Exposed	97.9	94.4	121.4	115.3	111.4	48	365
Bull/Cow Ratio	18.2	18.2	22.3	22.6	21	4	47
Age of Bull	4.2	3.6	4.3	4.4	4.3	2	10
Fenced Acres per Elk	2.9	3.4	3.1	2.9	2.9	0.7	15.5

Frequency of multiple births: one in 271 calvings

Antler Production

Age	1994 Alberta Average wt (lbs)		Target wt (lbs) for improvement
Spiker	2		
Two year olds	8	2 year	10
Three year olds	11	3 year	14
Four year olds	14	4 year	18
Five years & older	18	5 year	22
		6 year +	26

Herd health

---

9   Herd Health

9.1   Basic Principles

Farmed elk are remarkably free of disease given good management. The industry has been fortunate to develop in parallel with innovative herd-based approaches to disease management and increased awareness of welfare in domesticated animals.

The term "herd health" focuses on maintaining health status and the determinants of health rather than those of disease. Focusing on what keeps elk healthy instead of chasing disease after it occurs can be described as preventative medicine. This concept identifies the potential risks for disease occurrence and establishes procedures that reduce the risks and effects of disease on production in the herd. We have learned to manage away from disease and use procedures that contribute to disease prevention such as vaccination.

An important factor in a herd health program is the prevention and/or reduction of stress. Acute (short duration, sudden onset) stress due to animal handling events is one type of stress and is unavoidable in livestock production. The chronic stress imposed by poor nutrition, inappropriate management of social groups, subclinical disease, harsh environmental conditions, as well as frequent and repeated disturbances from predators (and humans) can lead to reduced resistance to disease and subsequent poor health. Chronic poor health and immune system depression not only leads to financial loss and waste but is also an animal welfare concern, and it is largely avoidable!

9.2   Animal Welfare

The *Recommended Code of Practice for the Care and Handling of Farmed Deer (Cervidae)* now forms the basis of cervid industry production standards in Canada. First developed in 1996 in collaboration with the National Farm Animal Council of Canada, it was among the first of such guidelines in the world. The Code is available online through the National Farm Animal Care Council (NFACC) at http://www.nfacc.ca/pdfs/codes/deer_code_of_practice.pdf

Adoption of the principles of animal care outlined in the Code gives new emphasis to the industry direction and credibility as well as an acknowledgment of its responsibility in a modern farming and marketing environment. Consequently, preventative herd health programs, rather than treatment of disease, are emphasized. This enlightened position aids in the enjoyment and reward of farming and, at its heart, is the reality that farmed deer themselves are highly adaptable, healthy animals in the first place.

9.3   Herd Health Programs

To assess health status, accurate records of management practices and measures of production, (e.g., breeding programs, feeding regimens, supplementary feeding, weight gains), should be kept. Herd health programs are built around optimizing production whether it is calf production or kilograms of antlers and meat. Management factors that might be used as indicators of health status are:

- Growth rates and weight for age data
- Clinical health records

- Mortality records and slaughter data
- Antler production
- Presence of specific diseases tied to poor management
- Chemical, biochemical, immunological, and physical disorders resulting from poor management

9.4   Seasonality and Elk Biology

The most important general animal health principle is that elk are fed according to their seasonal requirements. Winter nutrition differs from summer nutrition. Each phase of the annual production cycle has specific nutritional requirements. Ignoring this biological fact will inevitably lead to problems.

The art of preventative management is to practice routine programs according to season or time of year (Fig. 5). Variations on the routine are dependent on the stock class and age and any regional influences i.e. local climate, predation risk, soil mineral deficiencies, and local health threats. Herd health programs dealing with vaccination or routine parasite treatment are customized to the situation on individual farms.

A stress-free farm environment with good feeding management is further enhanced by knowledge of what can go wrong and the prevention of parasitism, infectious diseases, e.g., bacterial and viral diseases, mineral imbalances, animal predation, mechanical damage, or accidents.

9.5   Disease Resistance

Healthy animals are not simply healthy because they are never challenged by the agents of disease. All animals have innate defence mechanisms that protect them from disease threats and we as producers must learn how to support and take advantage of their natural disease resistance. Stress reduction and supplying good nutrition are examples of how we support the immune systems of elk. Vaccination is another means of conditioning an animal's immune system to protect it, and its herd mates against specific diseases. These are preventative activities aimed at lowering the risk of disease.

Attention to genetic selection of desirable genotypes can also be used to lower disease risks. A good example of both genetic resistance and susceptibility to disease can be seen in New Zealand red deer and their relationship to Malignant Catarrhal Fever (MCF), a disease of cervids carried by sheep. Thirty years ago, death losses from MCF on deer farms were quite common. The incidence of MCF on NZ farms is much less these days and there are certainly no fewer sheep around. This occurred even though blood testing surveys for antibodies against MCF show that most, if not all, yearling deer are positive for MCF antibodies, indicating exposure and infection have taken place. Why aren't they all sick or dead?

It's because resistance (and conversely susceptibility) to MCF virus infection is, in general, a genetically heritable trait. Whether it is because of selective breeding programs by red deer producers, or because of the natural deletion of susceptible individuals from the breeding pool, it seems that NZ red deer are more resistant to disease caused by MCF than they once were.

This concept can be used in elk breeding programs as a disease prevention strategy. We are currently using this approach to deal with CWD on North American elk farms. It has been found that certain genetic types of elk are more resistant to CWD, and breeding programs favoring these genotypes are currently being developed in a preventative health approach.

9.6   Biosecurity and Herd Certification

"Biosecurity standards" are a set of practices used to minimize the presence of pests and the transmission of pathogens in animal populations, including pathogen introduction, spread within the populations and release. Biosecurity and testing for disease are both fundamental to trade in live elk and elk products whether it be locally or globally. Disease surveillance and testing for specific diseases is a means of verifying the absence of disease in your industry and on your farm. These concepts are what builds quality assurance and trust between trading partners.They are based on Four Key Principles:

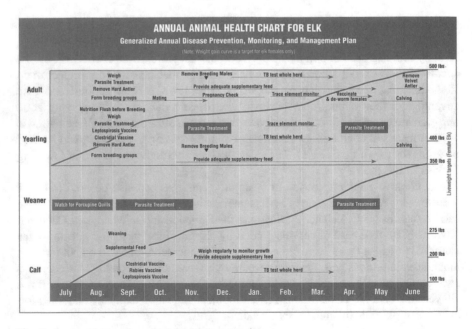

Fig. 5 A general procedure chart for routine programs

1. Management of the farm
2. Management of animal health
3. Management of people
4. Protocols and record-keeping

Biosecurity is not a new concept. Many of our current practices are biosecurity based. Creating and following a written set of standards has been the industry response to animal health managers and consumers who now demand that food producers maintain high food safety and quality standards and that those producers are able to prove those claims. Biosecurity is simply good management with the added practice of documenting that good management.

All sectors of the food supply chain are developing, publishing, and promoting biosecurity and other sets of standards, including animal welfare assurances. The cervid industry worked with the Canadian Food Inspection Agency (CFIA) and other partners to develop these standards. They are subject to review and revision every few years, so we have not included those in this handbook. The most current versions are presented on the CFIA website, at this URL: https://www.inspection.gc.ca/animals/terrestrial-animals/biosecurity/standards-and-principles/cervid/eng/1490298809224/1490359268963?chap=2

Biosecurity standards can be required as one component of certification programs. Like the standards, certification programs are an industry response to demands from health program managers and from consumers for written and auditable proof of risk management and mitigation. The most relevant such program for the cervid industry is the Chronic Wasting Disease Certification Program. CWD is described later in this chapter. Risk reduction methods and practices are available for most animal health issues, but CWD has so far evaded our attempts to understand it effectively enough to develop methods that ensure our farmed cervids are not infected. Although no evidence has ever emerged to suggest that humans are susceptible to the disease, caution is advisable. The CWD HCP, through documentation, whole herd testing at post-mortem, and other practices including biosecurity, ensures that producers, consumers, and managers are made sufficiently confident to allow trade and safe food production.

The most current version of the CWD Program Standards is available on the CFIA website: https://inspection.gc.ca/animals/terrestrial-animals/diseases/reportable/cwd/herd-certification/2019-national-standards/eng/1565804565857/1565804566138

For producers in the USA, go to the USDA APHIS website at: https://www.aphis.usda.gov/ aphis/ourfocus/animalhealth/animal-disease-information/cervid/cervids-cwd/cervid-cwd

9.7   Parasitism

Parasites on an elk farm are inevitable; they come with the animals and as part of their life cycle they also exist in the environment. They have evolved to parasitize their hosts at levels that sustain the parasite's existence but not necessarily kill the host. Deaths from parasitism usually occur because of other stresses, such as poor nutrition or overcrowding, placed on the host that allow the parasite to gain the upper hand. Overwhelming parasite burdens are the exception rather than the rule and parasites usually co-exist with their hosts as subclinical infections. However, parasites can and will adversely affect the health and productivity of farmed elk, so it is usually cost-effective to prevent or treat heavy infections. The goal of parasite control is to limit parasite infections, so animals remain healthy and clinical illness does not develop. The goal is NOT to eradicate every parasite from every animal in the herd. Not only is total eradication impossible to achieve for any length of time, but the inevitable result would also likely be the development of drug-resistant parasites and the creation of elk with no natural immunity to parasites.

The species of parasite that infects your elk will depend on where in North America you live. Parasites like warm, moist environments so it makes sense that there will be more parasites and problems in warm, southern climates where it rains frequently. In hot, dry areas parasite eggs and larvae do not survive long outside the host. Cold, wintery climates also interfere with parasite life cycles making it difficult for larvae to overwinter in the environment. For this reason, there is no one-size-fits-all control program, and you should consult with your veterinarian to find out what your local needs are. Every farm is different, and programs should be customized made to suit the situation.

9.8   Parasite Control Programs

Most veterinarians recommend traditional internal parasite control programs involving the treatment of animals with anti-parasite drugs at regular intervals. Treatment in the spring before turn-out is aimed at killing internal parasites in mature animals and preventing pasture contamination with parasite eggs. Often a second treatment a month or so after the first is used to kill worms that were picked up during early pasture grazing. Fall treatments are aimed at killing parasites that may have accumulated throughout the summer grazing period.The goals of any parasite control program are:

- To minimize the risk of clinical parasitic disease
- To control parasite egg shedding
- To maintain the effectiveness of drugs and to avoid the development of anthelmintic resistance (Table 1)

Not an exhaustive list, but some of the most common parasites that infect North American farmed elk include lungworm, meningeal and tissue worms, liver flukes, and winter ticks.

Table 1 Considerations for an effective parasite treatment program

• Different parasite treatments (and vaccines etc.) should never be mixed.

• Read the label, or on veterinary advice, treat at the correct dosage.

• Store parasite treatment as the manufacturer suggests. Some have a shelf life and are light-sensitive.

• Check the accuracy of the treatment gun. Under-dosing is a common fault and may encourage parasite drug resistance.

• All parasite treatment/veterinary drugs have a withholding period—check and observe strictly, especially for animals destined for slaughter.

• Parasite treatment is a management technique, not a test of speed. Accurate placement of the pour-on applicator gun against the skin, along the midline is important for successful treatment.

• Application should be done with care—the whole midline back region used to spread the liquid applied from tail to shoulder against the lie of the hair on the skin.

• Injectable avermectins can have a painful or stinging reaction about 30 s after dosing. Bulls and adult hinds can react with aggression to each other, or the handler as a response.

• Correct placement of the needle between the skin and muscle is important. Tent a fold in the neck skin area (lower neck) for delivery. Injecting is not appropriate in the rump, loin, or shoulder areas.

• Pour-ons—it seems important that the animals remain dry 5–6 h following treatment.

• If possible after treatment, move treated young animals onto new pasture after 24 h to minimize risk of infestation and if possible, rest the last pasture for as long as possible before reintroduction of stock. Use adults as follow-up groups.

• Record the event—date, group, and dosage. Develop a system when treating groups to prevent over-dosing or missing animals.

10   Diseases Caused by Clostridium Bacteria

Bacteria in the genus Clostridium cause a variety of diseases. The associated diseases in elk are the same as found in other domesticated ruminants and may be referred to as "blood poisoning," or other local names. Clostridium bacteria release volatile and damaging toxins, often resulting in rapid death of the host animal. Frequently, animals are simply found dead. Under stressful conditions or injury these bacteria, which are normally present in the environment and in the animal's intestinal tract, gain entry to the bloodstream or site of injury, multiply rapidly, and produce potent toxins that overwhelm the animal's defenses.

Some of the most common Clostridial diseases are:

• *Blackleg (Cl. chauvoei)* occurs after trauma to musculature resulting in bruising, providing an anaerobic (low oxygen) environment for bacterial growth and reproduction. It is sometimes called "gas gangrene" (see further description this section).

• *Tetanus (Cl. tetani)* occurs after puncture wounds or exposure of wound site to infective organisms providing an anaerobic (low oxygen) environment for bacterial growth and reproduction. Can also occur from uterine infections.

• *Malignant edema (Cl. septicum)* occurs after puncture wounds or exposure of wound site to infective organisms, resulting in a fatal toxemia.

- *Botulism* (*Cl. botulinum*) occurs after ingestion of decomposing animal or plant material (i.e. dead mouse in hay) contaminated with the bacteria's toxins.
- *Enterotoxemia* (*Cl. perfringens* type C and D) caused by disruption in digestion, often when introduced to new high carbohydrate diets too rapidly—allowing normally present Clostridium organisms in the gastrointestinal tract to rapidly overgrow and proliferate.
- *Redwater disease*, also called bacillary hemoglobinuria, results from a proliferation of *Clostridium hemolyticum* stimulated by damage to the liver (where spores can be present and inactive), often precipitated by liver flukes. Redwater has caused significant problems in some poorly drained pastures and is associated with liver damage from fluke infections. This disease results in intravascular rupture of red blood cells and anemia.

Most elk producers vaccinate annually for these bacteria, with very good success. Several other bacterial diseases may appear in cervids, particularly on ground that has had other livestock on it before becoming an elk farm. These can include Leptospirosis, Yersiniosis, Necrobacillosis, Anthrax, Johne's, and Bovine Tuberculosis. The last one mentioned, Bovine TB, affected the industry in its early days. Prevention and surveillance testing, followed by enforced destruction of positive herds, has largely eradicated the disease from farmed cervids, but reservoirs still exist in the wild and in neighboring countries.The most significant animal health issue remaining for the industry is Chronic Wasting Disease. CWD is a prion disease related to but different from BSE. CWD was first identified in captive deer in a Colorado research facility in the late 1960s, and in wild deer in 1981. By the 1990s, it had been reported in surrounding areas in northern Colorado and southern Wyoming. As of August 2020, CWD in free-ranging deer, elk and/or moose has been reported in at least 24 states in the continental United States, as well as two provinces in Canada. In addition, CWD has been reported in reindeer and/or moose in Norway, Finland, and Sweden, and a small number of imported cases have been reported in South Korea. The disease has also been found in farmed deer and elk.

It is possible that CWD may also occur in other areas without strong animal surveillance systems, but that cases have not been detected yet. Once CWD is established in an area, the risk can remain for a long time in the environment. The affected areas are likely to continue to expand.

Across its range, the overall occurrence of CWD in free-ranging deer and elk is relatively low. However, in several locations where the disease is established, infection rates may exceed 10% (1 in 10), and localized infection rates of more than 25% (1 in 4) have been reported. The infection rates among some captive deer can be much higher, with a rate of 79% (nearly 4 in 5) reported from at least one captive herd. Mule deer appear to be the most susceptible species, followed by White-tailed deer. In most elk herds that have been found positive, infection rates are low, with a few puzzling exceptions.

That word "puzzling" is a key to understanding the greatest impacts of CWD. Uncertainty surrounding the disease has led to severe restrictions on commerce and huge negative publicity. As this is really not a Herd Health issue, a separate section is included later in this publication.

11 Breeding

11.1 Seasonality and Management of Reproduction

Temperate cervid species exhibit highly seasonal patterns of reproduction that are profoundly influenced by the changing conditions of photoperiod, or regular seasonal changes in the ratio of night and day length. These patterns have evolved to maximize survival of offspring within highly seasonal environments, and overall, fit the requirements of elk farming extremely well.

All these changes are ultimately initiated by the release of the hormone melatonin in the body. Its release rate and circulating levels change in response to increasing and decreasing nighttime length. Red deer, fallow and elk are short-day breeders; i.e., the influences that induce estrus cycles in females, and the physical changes in males are prompted by an increase in melatonin secretion as night lengths increase toward autumn. The effects of seasonality differ outwardly in cows and bulls and for both are a critical part of the annual cycle.Bulls begin to be eager and capable of breeding in late summer. Bulls that are normally reasonable and pleasant to handle at velvetting become quite unpredictable and can easily cause injuries to themselves, or in an intensive farm situation to fences, buildings, trees, and vehicles. Only the best are selected, and all the rest are best segregated away from breeding areas until after the rut (Fig. 6).

11.2 Female Seasonality

Females too have their reproductive status controlled by the same day length change and melatonin secretion endocrine pathway with the ovaries being the ultimate target organ of hormonal action. The onset of the breeding season occurs late in summer as photoperiod decreases. Most farmed elk are naturally bred, but AI and ET can deliver the best genetics.If cows are not bred, then the reproductive cycle of ovulation, CL formation, and breakdown can continue for 4–6 months, all under typical hormonal control until the springtime daylight changes cease the cycling activity. The estrus cycle takes 19–21 days for farmed elk.

There are few outward physical signs of estrus in female elk. On-farm, estrus cows, when handled, may stand and, if pushed on the rump, squat slightly and can show a vigorous tail wagging, but this behavior is too imprecise to establish timing for an artificial breeding program.

Fig. 6 Rutting bulls sparring

Female Reproduction Characteristics	
Cycle length (days)	19–21
Onset of first estrus	Early September
Estrus duration	8–12 h
Gestation length (days)	245 ± 10
Breeding ratio	1 male:30 females
AI technique	Cervical
AI success rate	40–80%
Timed insemination	56–64 post CIDR

11.3   Artificial Breeding Techniques

There are several reasons why the use of artificial insemination and embryo transfer are becoming more common in elk. Artificial insemination provides a method of exploiting the genetic superiority of the male. Superior males are identified through records on growth and antler production in themselves and their offspring. Through artificial insemination and semen preservation techniques, a sire can continue to produce offspring after his death or in his advanced years after his health has declined. Artificial insemination also allows more females to be serviced and increases the number of offspring produced. Embryo transfer offers the same

Fig. 7 Artificial insemination

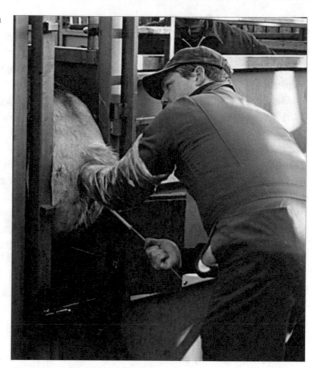

advantages for superior female genetics. These techniques also provide an easier method of movement or transport of genetic material.

Techniques now exist to extend the capability of an elite bull by synchronizing cows either side of the 20 days first cycle period and offering five or six cows in estrus on alternative days. This technique also forms a reliable basis for fixed-time artificial insemination (AI) procedures. For natural breeding, removing CIDRs from five or six cows treated every second day gives the bull a small number of cycling cows to mate and thus spreads the potential to mate many females in a relatively defined period (Fig. 7).

11.4   Semen Collection

The techniques and equipment used for semen collection are similar to those used in cattle and sheep. Collecting semen from elk requires professionally designed handling facilities. Animals can be collected while under general anesthesia, but the quality and concentration of the semen will be decreased. Semen collection for cervids is done on the farm rather than in centralized dedicated facilities.

11.5   End of Breeding

Even in times of an expanding industry and reasonable returns for live sales, gaining a few extra late-born calves by extending the breeding season is a management headache at weaning and for the average growth of the young stock.

Generally, primary sires are given a four to five week period from late August to early October covering two full cycles in most cases, and then sires exchanged for younger bulls or clean-up bulls for a further 20–30 days, and then all breeding bulls removed to allow a definite end to the breeding season. After primary sire breeding, individual groups can be pooled and multi-sire breeding with young stock used. There may be a reduction in conception in the first year of adapting from extended breeding, but from then on calving patterns will continue to concentrate in the first cycle and thus with better lactation and feeding, weaning weights and cow performance improve. This management best suits an early weaning program, but also has an application on smaller properties where calves may be left unweaned by choice.

11.6   Females and Breeding, Pregnancy, Calving

Once minimum threshold weights have been achieved in the first 15 months of growth, elk females should always be either pregnant or lactating. Reproductive success, measured by calves weaned to total number of cows mated is generally very high in North America; (adults 92–96%, yearlings 84–92%), but problems have occurred if yearling heifers are not well grown.

Once born, most calves survive unless victims of opportune predation by coyotes, bear, etc. There can be a 2–4% loss of young calves through difficult birth problems, mismothering or misadventure.Conception rates, and the spread of calving, can be negatively affected by mismanagement both at calving, and at breeding.

Table 2 represents a calving/breeding planning calendar and anticipated calving dates for elk cows. Breeding success can be improved through the following management techniques:

(a) Early weaning 10–14 days minimum prior to joining with a bull. Late weaning, i.e. removing young stock in late October when the bull goes out, will not significantly reduce the number of cows that are ultimately bred and become pregnant, but there is a tendency under farmed conditions for levels of 60–65% only to conceive to the first cycle instead of the 80% plus possible.

Weaning stress at the time of first estrus seems to delay or upset this first cycle and as a result the mean calving date can fall by 8–10 days. Early pre-rut weaning gives immediate relief from the high feed demands of lactation and after a 48 h settling period cows are quickly able to regain lost body condition and be on a slowly rising plane of nutrition over breeding until early winter. Any sudden change or restriction in diet during the three cycles may increase embryonic mortality and further reduce calving percentage. There is anecdotal evidence that dry cows, or cows that have lost

Table 2 Calving planner

Elk calving planner
Breeding and Calving Dates

Breeding date	Event	Expected calving date
		245 ± 10 days
Aug 20	Begin weaning	April 23
Sept 1	Synchronize females	May 5
Sept 4	Primary sires in	May 8
Sept 10	Silent estrus	May 14
Sept 20	Peak of first cycle	May 24
Sept 30	AI	June 2
Oct 10	Peak of second cycle AI	June 12
Primary sires out, clean-up bull(s) in		
Oct 30	3rd cycle	July 3
Nov 10		July 14
All bulls out—End of breeding		

calves may not conceive if they are too fat at breeding. Far better to lose weight over the summer and regain during the fall.

(b) Keep cow groups as stable as possible in terms of mixing animals that have not been together before, and if purchasing new animals, have them introduced well before the breeding season begins (particularly for yearling heifers).

(c) Keep breeding groups away from newly weaned calves and as free as possible from disturbance.

(d) Yearling heifers: It is often more practical to run yearling heifers together. Once male yearling bulls are separated from the young animal group (by June at the latest) yearling females may be slow to come into estrus in sympathy with the adults. The introduction of one or two adult cows per 10 or 15 yearlings seems to bounce these yearlings into earlier estrus, provided they are well grown. If yearlings cycle late and a defined end of breeding date used, the chances of missing breeding altogether increase. This induced synchrony and advanced ovulation appear to work well.

The common techniques of estrus synchronization, using vasectomized teaser bulls, CIDRs or running entire bulls in adjacent pastures immediately prior to breeding all seem to work very well in adult and yearling females. The common objective is to reduce the calving interval, start to finish and have as many cows as possible calving in the first 20 days of the season.

11.7 Pregnancy Determination

The simplest and most effective method of pregnancy determination in elk is rectal palpation at 6 weeks or more post breeding. The uterus of a non-pregnant elk lies entirely within the pelvis. The pregnant uterus feels like a smaller bovine equivalent, with less than ten cotyledons. The technique of rectal probe ultrasound scanning to determine pregnancy at an early age, or to fetal age to distinguish cycles of pregnancy, e.g., pregnant to an AI treatment, or pregnant to the backup bull is becoming quite common and forms a backup to many breeding programs.

The primary purpose is to distinguish between pregnant and non-pregnant cows. Scanning is done by a professional operator using a linear rectal 3 or 5 mHz probe and a portable scanner. These machines, or particularly the probe head itself, are expensive. For safe and effective use, the elk must be fully restrained inside a mechanical or hydraulic squeeze so there is no chance of damage to it or the probe. Ideally, this would be done in early December, at least 10 days after the finish of breeding and the backup bull has been withdrawn.

In natural breeding, there will be considerable cross over as the cycles of individuals overlap, but some fetal aging might be possible. The technique that gives a simple pregnant/non-pregnant response is extremely quick and reliable and will in time become a common marketing tool, or a prerequisite for female live sales in the early part of winter.

The more subtle use is for pregnancy confirmation in an AI program to distinguish between pregnancy from AI, and pregnancy from later backup bulls. Two scannings can be used, the first at 29–35 days after the initial fixed time insemination, which will only confirm treatment pregnancies and the second at 54 days where fetus can be seen and the backup bull pregnancies and dry cows identified. Alternatively, a scan at 40–45 days will allow distinction between treatment and chaser backup pregnancies.

An alternative, and very accurate pregnancy detection technique, is a blood test for PSPB, Pregnancy Specific Protein B, available at specialized laboratories.

12   Nutrition and Pasture Management

One can generalize regarding the Nutrition Requirements of elk—they generally thrive with similar nutrition to cattle, sheep, and other ruminants. But there are noteworthy differences.

12.1   Calcium, Phosphorus, and Other Minerals

The metabolism of calcium and phosphorus is interrelated. These two minerals are considered simultaneously when designing or evaluating a ration. Both minerals are important constituents of bone. Calcium is also essential for nerve impulse transmission, muscle contraction, secretion of some hormones and enzyme activation. Phosphorus is an essential constituent of enzymes involved in fat, carbohydrate, and energy metabolism.

The actual amount of calcium and phosphorus required by elk increases greatly during growth, lactation, and antler growth. A minimum of 0.7% calcium and 0.4%

phosphorus is suggested for elk rations. Of equal importance is to ensure a calcium:
phosphorus ratio of at least 1.5:1. How wide a ratio is tolerated by elk has not been
researched although other ruminant species can tolerate calcium to phosphorus ratios
as high as 5:1.

Calcium levels are relatively high in hay, especially legumes, but phosphorus
content is quite low. Grains are the opposite, being relatively high in phosphorus and
low in calcium. Rations comprised of hay/haylage and natural grains and oilseeds or
meals will not provide enough minerals and vitamins. Hay analysis is required to
decide which type of mineral/vitamin supplement is most suitable. Access to browse
(twigs and leaves) improves natural mineral intake on pastures.

Commercially prepared mineral-vitamin supplements are extremely variable in
both nutrient content and price. Follow these general guidelines:

- Choose either a 2:1 (2× as much calcium as phosphorus), or 1:1 (equal calcium
  and phosphorus levels) depending upon forage analysis and grain feeding levels.
  Avoid low calcium, high phosphorus minerals as they tend to be both unpalatable
  and expensive
- Trace mineral levels for copper, zinc, and manganese are particularly important in
  feeding elk. Elk, unlike sheep, require supplemental copper. A feed or mineral
  designed for sheep will not contain any added copper. Conversely, the copper
  levels required by elk and other cervids would kill sheep.
- Next check the levels of vitamins A, D, and E and of selenium. All these nutrients
  are expensive and will add to the cost. Selenium and vitamin E levels should be
  increased in late winter/early spring; feed vitamin levels decrease during storage
  and will be quite low by this time. Supplemental vitamin E feeding can decrease
  during pasturing; fresh grass is a good source of natural vitamins.
- Check the feed tag for salt levels (if present, will be listed as sodium or Na). Some
  minerals include salt, others do not. Additional salt will be necessary if no other
  source is provided, particularly in spring and summer. Elk will not voluntarily
  consume much salt in winter.
- Consider the advice of some nutritionists who suggest that you are safer feeding
  the total mineral requirements rather than assuming your elk are getting a portion
  of their needs from their feed.

12.2   Evolving Micronutrient Targets for Elk Nutrition

Continual improvement is a mantra for most well-run businesses. In the elk farming
business, we have an advantage over many others—we are still relatively new,
developing new markets, discovering new genetics and improving our management
practises at a very rapid rate. Our innovations can be very rewarding. In the past
60 years, we have learned a lot about how to feed our elk. We started off feeding
them like sheep or cattle—what else would we do? But elk are not fast cattle nor
long-legged sheep, or directly comparable to any other than cervid species. They are
quite unique in many ways, particularly in their miraculous ability to grow antlers

the way they do every year. Or course, as with any exceptional ability, there are prices to pay—in this case in the nutritional consumption and utilization that cervids undergo while preparing to grow and actually growing antlers. What other creature robs its skeleton and muscles for nutrients in order to grow another body part? Such miraculous growth requires exceptional nutrition, and micronutrients are critical components.

Not just antler growth requires adequate levels of micronutrients. Every aspect of growth, gestation during pregnancy, lactation, and body maintenance requires inputs of adequate and appropriate levels of micronutrients. Each micronutrient contributes significantly to various body functions. Copper seems to stand out for cervids as being particularly important. In plants, copper is essential for enzyme activation, hormone regulation and energy reactions within the plant. It can also play a role in disease resistance, is recognized to have antimicrobial properties and it stimulates enzyme activity of various types. At the cellular level, copper is essential for transfer of electrical energy associated with cell communications.

Beyond the documented research on the activity of copper in all animals, our most experienced producers are learning by observation that adequate copper levels in the diet of farmed elk are essential for optimum reproductive performance and growth, as well as production of the ever larger and more spectacular antlers we are seeing on our farms. Advancements in both scientific and "local" knowledge have led to considerable changes in the recommendations for nutrient levels for all types of livestock, but particularly for farmed cervids—for the reasons we mentioned above. It is very useful to stay current with evolving knowledge and recommendations and to incorporate the best and most profitable practises into our everyday consistent farm management. Follow the recommendations in Table 3.

Developing a Feeding Strategy
To develop a feeding strategy for your elk, start by looking at your basic assets:

- Livestock
- Land, both fenced pasture and crop production areas• buildings and facilities
- Available crops
- Available human resources

Your livestock, your elk, are classic "mixed feeders." This means that they both graze and browse, selecting the most nutritious grasses, legumes, forbs, leaves, buds, fruits, and twigs of an endless variety of plants. The shape, structure, and function of their teeth and digestive systems allow them to successfully use this great variety. Elk also have strong seasonal peaks and lows in appetite and nutrient needs, as shown in Table 4. Fortunately, their appetites, growth rates and lactation all peak at the same time as pasture production peaks in temperate climates.

In healthy wild elk populations, herds will move considerable distances to find the feed their bodies need and want. Our aim as producers of farmed elk must be to

Table 3 2020 target nutrient requirements for elk whole ration (dry matter basis)'

Total/whole ration (DM)	Adult	Velvet	3–6 months	6–9 months	9–12 months	Mid gestation	Late gestation	Lactation <4 months	Lactation 4–8 months
CP %	7–10	16	18–20	16–18	12–14	14–16	14–16	14–16	12–14
DE (Mcal/kg)	2.20	3.40	3.10	2.90	2.70	2.40	2.70	2.90	2.80
TDN (%)	50–52	55	68	64	59	57	59	64	61
Fat % min.	3	3	3	3	3	3	3	3	3
ADF % min-max	25–45	25–45	16–35	20–40	20–45	20–45	20–45	20–40	20–40
Calcium %	0.35	1.40	0.60	0.55	0.50	0.50	0.50	0.70	0.60
Potassium %	0.65	1.0	0.65	0.65	0.65	0.65	0.65	1.0	1.0
Magnesium %	0.20	0.40	0.25	0.25	0.25	0.25	0.25	0.25	0.25
Sodium %	0.06	0.10	0.10	0.10	0.08	0.06	0.08	0.10	0.10
Sulfur %	0.08	0.15	0.10	0.10	0.10	0.10	0.10	0.10	0.10
Copper (ppm)	25	35	30	30	30	30	30	35	30
Molybdenum (ppm)	<5	<5	<5	<5	<5	<5	<5	<5	<5
Manganese (ppm)	40	40	40	40	40	40	40	40	40
Zinc (ppm)	50	150	100	100	100	100	100	100	100
Iron (ppm)	50	200	200	200	200	200	200	200	200
Iodine (ppm)	0.30	1.0	0.50	0.50	0.50	0.50	0.50	0.50	0.50
Cobalt (ppm)	0.20	0.30	0.20	0.20	0.20	0.20	0.20	0.20	0.20
Selenium (ppm)	0.40	0.40	0.40	0.40	0.40	0.40	0.40	0.40	0.40
Vitamin A (iu/kg)	2900	4400	4000	4000	4000	4400	4400	4400	4400
Vitamin D (iu/kg)	550	1100	1000	1000	1000	1100	1100	1100	1100
Vitamin E (iu/kg)	22	44	33	33	33	44	44	44	44

Table adapted with permission from Bruce Friedel, Iron River Elk, Alberta. Note that selenium availability is highly variable geographically—test your soils and feeds. CP = crude protein, DE = digestible energy, TDN = total digestible nitrogen and ADF = available digestible fiber

Table 4 Elk Velvet Antler (EVA) prices paid in Canadian dollars per pound at farm gate for Western Canadian frozen unprocessed product

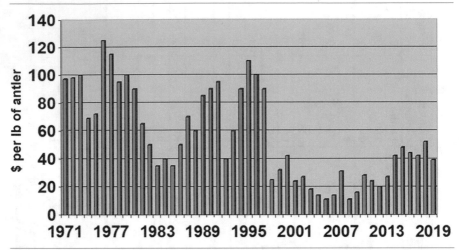

provide all the nutrition our elk require in the feeds, we make available. These requirements must also be provided in a palatable and reasonably priced form. This is quite a balancing act, one that requires our regular and continued attention. Feed quality and requirements change substantially with the seasons, local conditions (including soil, water, terrain type and climate) and type and source of the feed. Feed analysis is an essential tool in balancing rations. If we buy all our feed, we can often get away with requiring the suppliers to do all the analysis and balancing according to our instructions. If we produce our own feed, we may become quite skilful at assessing feed quality and adding the nutrients we are confident will be lacking. Accumulating this kind of knowledge requires a lot of feed analysis and consultation with neighbors in the early years of our management of a piece of land and a herd of elk.

Elk, with careful management, will do quite well in a feedlot situation. However, the cost of growing pasture will generally be substantially less than the cost of stored feed. Try it yourself—even calculating initial land purchase and fencing costs, spread over 20 or 25 years, plus annual maintenance and fertilizer, and assuming a conservative figure of five tons of dry matter (DM) consumed per acre, pasture should cost less than 3¢ /lb., or $60/ton. Compared to the cost of other feed, to maximize pasture area and production is the best start to our feeding strategy. Pasture management is covered in greater detail in later in this chapter.

The next step in our feeding strategy is to establish a supplementary feeding program. When we say, "supplemental feeding," the winter season always comes to mind. We tend to think that our livestock can always get by on pasture from May to October, without giving much thought to the quality of that pasture. Early in the season, rapidly growing vegetation may be "washy"-up to 80% moisture. An elk fills its rumen but does not get enough nutrients to meet its daily needs. Later in the season, elk may be grazing "belly-deep" grass which is so high in lignin and other indigestible fibers that it becomes unpalatable and nutritionally inadequate.

Pastures made up of a variety of grasses, forbs, legumes, shrubs, and trees are always the least expensive source of elk feed. Good management can provide adequate nutrition from grazing and browsing for much of the year. However, when even our best efforts fall short of nutritional requirements, we need to provide supplemental feeds.

Harvesting Corn Silage

The most important supplemental feed we provide is forage. The species best suited to feeding your elk will depend primarily on where you operate your farm. Climatic conditions vary with latitude and altitude. If the climate suggests a particular crop or mix of species, there are other characteristics of forages to consider. Plants contain two kinds of carbohydrates. *Non-structural* carbohydrates are starches and sugars that help the plant store energy and are easily digested by livestock. *Structural* carbohydrates include the cellulose and hemicellulose fibers found in cell walls. Cellulose and hemicellulose work with lignin to hold the leaves and stems together and help the plant stand up. Rumen microbes digest hemicellulose more easily than cellulose but lignin is virtually indigestible. When interpreting the results of a feed test, neutral detergent fiber (NDF) measures the amount of cellulose, hemicellulose and lignin. High NDF levels limit animal nutrition, as their rumen is filled, but with low-digestibility materials. Acid detergent fiber (ADF) is the amount of less digestible cellulose and lignin but excludes hemicellulose. Digestibility declines as NDF and ADF increase.

In perennial grasses, cellulose, hemicellulose and lignin levels increase steadily as the plant grows and matures. As the plant gets taller, it requires more structural integrity to keep standing. ADF and NDF increase and digestibility decreases as grasses mature. Non-structural carbohydrates and protein levels rise initially, peak, and decline after grass has headed out. These changes result in a gradual decline of the nutritional value of pasture as grass matures. Rotational grazing or mowing practices that keep grass vegetative by "clipping" and preventing it from heading out help maintain the nutritional quality of the pasture later into the growing season.

In contrast, annual crops such as cereals have been selected for vastly increased grain yield. The amounts of structural carbohydrates and lignin still increase as the plant grows and matures. But thanks to millennia of selection for increased seed yield, the percentage of non-structural starch in the whole plant continues to increase after seed set, through kernel filling, before slowing during ripening. Whole-plant starch content is low at the late milk stage in barley (3%) and oat (2.7%), but much higher in the whole plant at hard dough (25% starch in whole- plant barley, 14.5% in whole-plant oat) and ripe barley (24.6%) and oat (12.8%).These effects are even more pronounced in corn. In addition to higher starch content, corn also produces 50% greater silage yields than barley—under the right growing conditions.

Corn can produce higher silage yields and starch content than barley because of a few fundamental differences between the two plants. Because corn is open-pollinated, it is easier to hybridize than self-pollinated barley. Barley and corn also use slightly different photosynthetic processes to convert carbon dioxide and water into carbohydrates. Photosynthesis has light and dark stages. During the daytime (light stage), plants absorb energy from sunlight and temporarily accumulate it in energy storage molecules. At night (dark stage), the stored energy is used to convert carbon dioxide and water into structural and non-structural carbohydrates. Most of the earth's plant species-including most tame grasses, barley, other small grains and oilseeds—are "cool-season" plants that all use the same dark stage process.

A minority are "warm-season" plants such as corn, sorghum, maize, green foxtail, kochia and blue grama, and use a more energetically efficient dark stage process. That means that warm-season plants such as corn can put more photosynthetic energy into starch production, plant growth and crop yield than cool-season plants such as barley. Warm-season plants also tolerate heat and moderate drought better than cool-season plants. But cool-season plants also have advantages. They do better during cool, wet growing seasons and produce more protein than warm- season plants.

Much of Canada is suited to cool-season plants, but plant breeding companies have developed corn hybrids that require fewer corn heat units (CHU) during the growing season. The minimum CHU rating for corn hybrids has fallen from 2300 to 2000 over the past 40 years. But corn silage may still be a risky crop. In a cooler-than-average growing season or high elevation, even a low CHU hybrid may not reach the growth stage where starch content adequately offsets structural carbohydrate (NDF and ADF) levels. This would have a negative impact on nutritional values.

Regardless of crop or species mix chosen, if the decision is to harvest and store for supplemental feeding rather than to graze in place or in swaths, all forages are stored and fed as hay, haylage, baleage, or silage. These names mostly differentiate stored forage based on moisture content, ranging from 15% to 65% water content vs. dry matter. Where weather conditions allow adequate drying, hay is generally the least cost/least work alternative, but silage is the choice for low cost, high nutritional efficiency but much more work in storing and feeding.

If forage is stored at a higher moisture content (18% +), it must be protected from spoilage. This is accomplished by restricting airflow with plastic wrap, concrete, or hard packing of piled forage. The limited amount of air (oxygen) left in the wrapped or sealed forage permits a limited fermentation process that increases palatability and quality for feeding elk. If the forage is chopped and piled or stored in large plastic tubes, it is typically stored at around 45–60% moisture and referred to as silage. If the forage is cut and conditioned, left to wilt for about 24 h, then baled without chopping and wrapped with plastic, it is referred to as "haylage" or "baleage."

Haylage is typically stored at 20–45% moisture. Its palatability is excellent, but nutrient availability is less than with chopped silage. Both silage and haylage offer advantages and disadvantages compared to hay. The higher acceptable moisture contents allow forage to be safely stored much more quickly than with hay. This is an obvious advantage when high humidity or frequent rainfall is a problem. Palatability and nutrient availability are superior in the softer and fermented higher moisture products. However, costs for storage and harvesting equipment are usually higher for silage or haylage than for dry hay. In addition, silage may spoil or freeze quickly when removed from the pile, necessitating more frequent feeding than dry hay.

Silage for elk is typically chopped grass and legumes or cereal crops. The most cost-effective and successful approach may be to hire custom harvesters to cut, chop and store all one season's production in a few days and store it in tightly packed and covered piles, pits, or large plastic tubes. A loader is used to fill a forage wagon or TMR (total mixed ration) wagon, and the silage is augured into feed bunks or onto the ground. Feeding on the ground has associated hazards. Parasites and disease organisms are consumed with the feed. It is essential to feed in a different area each day and avoid any heavily soiled areas. Try to put the feed on fresh grass or snow each day.

No matter what storage method is chosen, elk require good quality "YTRG" (young, tender, rapidly growing) forage. Chopping does not make coarse, poor-quality forage into excellent feed. However, chopping does improve nutrient availability and voluntary feed intake.

A bale shredder will do an excellent job of preparing long hay or haylage. The chopped feed, like silage, can be blown into feed troughs or onto the ground. This is particularly useful when feeding alfalfa, as elk are exceptionally good at separating the leaves and leaving the stems behind. Chopping will encourage them to clean up most of the forage. Moldy feeds of any kind must be avoided. Not only does mold reduce palatability, but the mold spores also ingested and inhaled will tax the animal's immune system, leaving it more prone to infections.

When feed testing identifies shortfalls in nutrition targets, supplements are fed in the form of grain or pellets. Grains provide energy and phosphorus, plus a variety of other nutrients that may not be present in sufficient quantities in forage.

Prepared TMRs, fed as cubes or pellets, are an easy but expensive way for new or small-scale producers to deliver a high-quality balanced ration to their elk. If fed in open troughs, these cubes should be fed daily if any rain is expected. They may also be fed ad lib in creep feeders, but the producer loses the value of interaction with the animals. Adding a grain tank to a shredder also allows you to deliver a TMR.

Bale shredder with grain tank

This interaction between the farmer and his or her elk is an important part of good stockmanship and livestock marketing. In fact, other than the provision of nutrients identified as being deficient in the forage, this "taming" ritual is the most valuable part of grain feeding. The grain is a treat for the elk, and they look forward to being fed. The person who feeds becomes a trusted individual, and further training allows the elk to accept the presence of new people as well. Nothing sells elk like the experience of going with the herdsman to the pasture, calling the elk and having them run up close and comfortable with the prospective buyer.

Purchasing either commercial rations and protein supplements or cereal grains and vegetable protein sources is usually a matter of individual preference. Commercially prepared feeds may be convenient and offer an "all-in-one" package. Commercial feeds can either be custom formulated (according to your specifications or recipe) or purchased as standard stock. Usually, standard feeds will be less expensive although you must decide whether to purchase a beef, dairy, sheep or, in some cases, a specially prepared elk ration. Sheep feeds are unsuitable because they lack added copper. Dairy feeds often are more highly fortified with vitamins and trace minerals than beef feed. Dairy feeds also are formulated for a larger protein to energy ratio than beef feeds. These considerations suggest choosing dairy feeds if specially formulated feeds for elk are not available.

Commercial feeds can be formulated as pellets or as mixed feed. A completely pelleted ration prevents any sorting. Some elk have proven adept at sorting through mixed feed, leaving the finer-grained minerals and other supplements at the bottom of the feeder. Pelleted rations vary in pellet size and density; it is important to find one that your elk will readily consume. Once you have found a feed company and product that your elk readily consume and perform well on, think carefully before changing!

A disadvantage of a commercially prepared feed is that it is impossible to know which feeds are included in the ration unless the company guarantees a fixed formula, and even better, areas where the feed is sourced. Avoiding diseases like CWD requires that feed for elk should not come from an area where CWD is present in wild cervids, especially in years following difficult harvest conditions. Generally, feed formulas vary depending upon ingredient prices. Palatability and nutrient content might change from batch to batch. On the plus side, commercially prepared feeds will be well formulated with precise control over ingredient inclusion rates and uniform mixing of ingredients which may be more difficult to achieve on-farm.

Elk farmers who choose to feed straight grains or protein supplements gain flexibility in their feeding program at the expense of convenience. With two or three sources of feeds on-farm it is easy to custom mix batches to suit the needs of each management group. On the negative side, feeds mixed on-farm may not be as uniform as a commercially mixed feed. Natural grains and oilseeds or protein meals need to be fortified with minerals and vitamins. Unfortunately, finer-grained mineral and protein feeds often settle out of a ration and are left on the bottom of the feeder. Options include adding molasses to bind the mineral; pelleting the mineral and protein feeds together; or purchasing the mineral in a pelleted rather than loose form. Offering minerals only on a self-fed free choice basis is not an adequate way to ensure optimal intake.

Oats are the best grain to use for supplemental or treat feeding. Their unique digestive characteristics make them safer than other grains. If the farmer is always careful to slowly introduce the elk to any new feed, and to increase the grain portion of their ration very slowly, that feeding should never cause problems such as grain overload. Oats come with a caution. Many oat crops are swathed then left on the ground and then swathed to complete drying. Especially if inclement weather forces the farmer to wait a while, this allows wild cervids to help themselves and leave their urine and feces mixed with the crop. In a CWD-affected area, this could result in an animal and financial health disaster! Be sure you know exactly where your oats are coming from and how they were harvested. This is less of a concern with oats or any other grains that are harvested standing, either by chopping for silage or straight-cut and combined without lying on the ground.

A carefully formulated pellet designed using analysis of the feed and water available for your elk is the best choice for supplementation, if you know the history of the ingredients used. These pellets are to be fed in addition to fresh, green leafy legume/grass hay or pasture. Ideally, these pellets would contain 60% mixed grains (barley no more than half that), 30% peas, 2% flax and the remainder being added mineral mix. Add molasses at about 10 liters/ton to enhance palatability.

Feed these pellets mixed with oats (50% oats / 50% pellets).

Feed bulls 4–6 lbs. of the oats/pellets mix per head per day all year around, if they eat up all you give them.

Feed cows 2–5 lbs. from early May to November, depending on pasture quality and availability, then drop slowly to 1–2 lbs./head/day in December to April.

Feed calves 4–8 lbs. from August of their calf year through to December of the next year when they reach about 18 months of age, then treat them as adults. Remember that these feed levels constitute luxury feeding. If the regular forage portion of their ration is as good as it should be, you need not feed these levels.

Always make any changes in ration content or amount quite slowly and carefully, over a period of 1–2 weeks. In addition, keep a trace mineralized salt block available to your livestock all year around. Typically, elk will only use this salt lick extensively in the summer.

Each different area or farm and each different feeding program may need slight adjustments made to the formula described above to balance the levels of nutrients in your pasture or stored feed. Begin this process by sampling your soil, looking both at macronutrients like NPK, and micronutrients as well. Consult your local agronomist, and farmer neighbors as well for advice at this stage, remembering that elk are not the same as cattle or sheep in their requirements. The next step is to test your feed—including your growing pastures and all your stored or supplemental feeds. If you use a great variety, test different batches regularly and balance your supplement accordingly. All nutrients must be balanced for each area and diet. Copper is an especially important nutrient for elk. Several other minerals including molybdenum and iron may interfere with uptake of copper. Supplementation should be increased if no browse is available, as woody browse delivers more useable minerals than forage.

Always make sure that adequate feeding space is provided for each of your elk. Social pressures will exclude lower-ranking elk from feeding areas if space is inadequate. They all tend to sleep at the same time and feed at the same time, so allow at least 2' (60 cm) of feed bunk, trough, or other feeding space per elk in that feeding area.

Bulls spaced at feeder

Elk calves begin to utilize forages at about 20–40 days of age. At this stage, they will often eat soil, apparently to begin the growth of rumen flora (digestive bacteria). Give them fresh topsoil (preferably forest topsoil) in a pile near where their mothers are fed or watered. This will help avoid the situation where they pick up disease organisms from contaminated soil. At about 80–90 days of age, calves will begin to eat grains if they have access to these with their mothers. Spread oats plus supplement pellets out in a long line in feed troughs or on clean grass, in a different spot each day. By weaning time at 100 days, the calves will adapt more easily to a diet of forage and grain without milk.

Supplemental feeding is a regular part of farmed elk management. Done right, it makes all the difference in producing quiet, excellent quality elk livestock and elk products. Be sure to work with all your local sources of information and knowledge, and do not be afraid to ask questions or make suggestions.

Formulating the most appropriate rations involves many choices and decisions. No single approach will be appropriate for every herd. As an owner, you need to find a feed company and nutritionist you trust and feel comfortable with. It is important that you as an owner communicate your desired goals to your nutritionist and continue to seek their advice in fine-tuning rations to meet your herd's needs. Nutrition is a vital cornerstone to healthy, productive elk. As a team you, your veterinarian and nutritionist can help to optimize the health, productive and reproductive status of your herd.

12.3 Some "Best Picks" for Elk Pasture

YTRG Alfalfa and orchard grass mix

1. Alfalfa, the "Queen of Forages," will be the preferred legume in most areas where elk are raised. Alfalfa is the most productive legume in most situations, yielding well over five tons per acre if well managed. Good management includes rotational grazing. Alfalfa should be allowed to grow to a height of 8–10 inches before grazing down to 3 inches, then allowed to recover for 30 to 40 days before grazing again. One recovery period must cover the 30 to 40-day period before the first killing frost of the autumn, to allow root reserves to build up for spring growth. Residue after frost may be grazed in autumn or winter.

   Rapidly growing alfalfa in a pure stand may exhibit an excessive protein level. The by-products of this high protein diet are excreted in urine, possibly resulting in "burns" to the penis sheath or vulva. To avoid this, alfalfa should be grown in a mixed stand with grasses such as timothy or meadow brome, or with forbs such as chicory, and allowed to grow to 8–10 inches before grazing. Aim for an alfalfa content of 30–40% in the stand. Best varieties vary with location and soils, but consider Alfagraze, Rangelander, Stampede, or Algonquin.

2. Orchard grass both are highly palatable when kept grazed to a height of 3–10 inches, and both begin growth early in the spring. This grass will both tolerate earlier and more frequent grazing than alfalfa, and therefore might do better when planted with clovers or in a pure stand with applications of nitrogen fertilizer.
3. Meadow Brome Grass is a proven standard for prairie and parkland pastures. Its longevity, palatability, and rapid regrowth after harvesting makes it a highly preferred variety, particularly in a mix with alfalfa. Talk to your local specialist when choosing the right variety for your area.
4. For pastures in more southern areas, good pasture species include mixed native prairie forbs and grasses, red river crab grass and Bermuda grass.
5. Other good choices include small clovers, particularly alsike, red or small white and highly productive soft grasses suited to your area.

13 Elk Products

13.1 Velvet Antler

Antlers and antler harvest form the basis of traditional deer farming industries in China, Korea, and Russia. There is considerable market potential in these and western countries with the recognition that the pharmacological and bioactive properties of prepared soft, or velvet antler, are real and diverse in many therapeutic and tonic medicinal applications.

Strictly speaking, the term "velvet" refers to the skin covering a growing antler. It describes the fuzzy texture provided by many fine hairs growing on the surface. "Velvet antler" refers to the entire antler when it is in the growth phase. At this stage it is soft and lacks the mineralized characteristics of "hard antler" which is nothing less than bone. Frequently "velvet" and "velvet antler" are used interchangeably to refer to the antler in its growth phase. It is at this growth stage that the entire antler is harvested for drying and use in pharmaceutical preparations. The finished antler product from elk is therefore often referred to as "elk velvet antler" (EVA).

A bull elk "in velvet"

Velvet antler is an actively growing, cartilaginous tissue, which is supplied with a complex vascular and nervous system. The skin surface covering the antler is variable in color and often very dark; black or dark brown in red deer and a distinctive light orange-brown, gray, or black in elk. The skin is covered by short hairs growing at right angles to the surface and the whole antler often has a greasy feel from sebaceous gland secretions.

13.2   Antler Branches or Tines

Tine growth follows a distinct pattern. Antler "tines" start as buds developed from the main beam and grow at differential rates compared to the faster growth of the main beam. The "brow tine" is the first branch off the main beam. The "bez tine" (pronounced "bay") can be variable in size and position on the antler and can be absent either on one or both sides. This may not occur every year and it is important to have knowledge of the time of casting and the stage of development to properly determine whether an unusual bez tine is growing, or whether the "trez tine" is emerging. The "bulb" or top of the growing beam is used as the major assessment of whether velvet antler is ready to harvest or not. Typically it thickens above the trez tine (pronounced "tray") and will bulb out depending on age and the extent that "royal tines" will branch off. An average growth period to this stage is 63–65 days after last year's antlers are shed but is extremely variable and an individual feature.

13.3 Producing and Harvesting Elk Velvet Antler

The following broad features of antler growth are important for production and are significant for age, breed type, and individuals.

- Older bulls cast significantly earlier than younger bulls with up to 6 weeks difference to 2 year olds, but individual variation is remarkably high.
- The pattern and timing of tine initiation and growth is broadly similar over all ages.
- Velvet antler weight increases with age until at peak production (elk 8–10 years old) and then stabilizes or gradually declines.
- A general relationship exists between liveweight and velvet antler weight within an age group and a blood line.
- Across breed types, elk yield more velvet per unit liveweight and take longer to grow antler while still producing premium grades.
- Above the trez tine the main beam will grow rapidly relative to the development and differentiation of the royal bulb (up to 2 cm per day) and appreciable gains in weight can be made by closely watching the growth patterns (up to 0.5 kg in 7 days).
- Once the royal bulb flattens, differentiation into royal tine buds is rapid as is calcification of the basal tines. Severe economic penalties exist for overgrown product despite the apparent greater size and weight.
- Older bulls, with greater beam and earlier growth and casting, can be grown out to a greater extent. Young animals should be cut earlier rather than later to preserve quality.
- The whole antler must be assessed in determining cutting stage as brow and bez tines can calcify rapidly as indicated by pointing and upward curvature of the tips.

It can be clearly seen that once the main bulb has creased there is rapid calcification of the lower tines, with little addition in weight in relation to the large drop in quality. The growth in antler length from stage 1–2 is significant in weight terms with little loss in quality.

It pays then to observe bulls daily once the beam has developed from above the trez tine. Older animals, with thick beamed antlers can be left to grow out more than young animals or thin antlers.

13.4 Management of Velvet Antler Harvest

That elk velvet antler is in demand and significant in pharmacology in Oriental culture is both an incentive and a bonus of this farm management operation. Velvet antler is a valuable food product and maintaining unequaled hygiene and welfare standards is important to international marketing and acceptance by the public. The annual growth, and harvest of velvet antler offers elk and deer farmers an attractive product source and financial return but carries with it several management and welfare responsibilities that cannot be ignored.

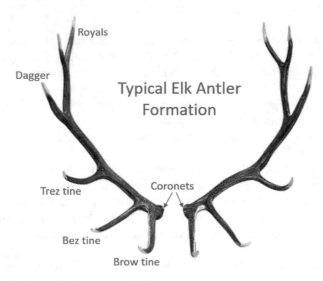

Typical elk antler formation

13.5   Selecting the Right Stage for Cutting

Once the antler has reached about 55 days of growth, the form and growth of the antler should be checked every two days at least, and preferably every 24 h. The best time is in early morning with sunlight behind the animals as the hairs on the velvet skin form a halo effect and the flattening or division of the royal bulb can be seen clearly. If animals are grouped in smaller groups according to casting dates, then there is no need to pen and handle groups at different stages of growth every time.

In deciding when to cut, the most important indicators are your records from years past. Each bull is slightly different regarding: density, growth style, and pattern of calcification. Some must be harvested at a certain stage, others at a different point in development. The number of days since casting is a very useful guide, plus the shape of the royal bulb with its 4th and 5th tines, and the shape and character of the brow tine. The royal bulb should be just starting to "dimple" and divide in young bulls. In older bulls, the division might be as much as several inches, if you know that bull does not calcify early. The brow tine tips should be straight, round, soft, and velvet covered.Ideal harvest time is when the tip of the royal bulb just begins to flatten, or in larger antlers a small crease (0.25 cm) will appear in the bulb area (Fig. 8).

The bulb should be at about 80% of the height from the brow to the trez tine, above the trez. If, however, the brow or bez tines show signs of turning up and

Fig. 8 Ready to harvest—brow and bez tines turning up

pointing, calcification has begun, and the antler should be harvested irrespective of bulb development.

As a rule, younger animals must be harvested earlier, and only thick beamed antlers should be left to grow to the limits of royal bulb development. The farmer should not sacrifice quality for weight and as the task is to determine when the antler has reached maximum growth prior to the onset of rapid calcification at the base, he/she must observe the whole antler.

14   Animal Handling

Bulls, particularly 2- and 3-year olds should be thoroughly familiar with the pens and squeezes or restraints long before harvesting. Training in the handling system and pen familiarization can be achieved during early spring as they cast. With casting dates being recorded, it is appropriate to group animals (20–25 maximum) according to casting date to allow easy assessment of the stages of velvet antler development before removal. Therefore, it is necessary to assemble only those animals ready for harvest over a 10- to 14-day period. Older animals will naturally thereby be grouped separately from younger and later developing bulls.

Elk respond well to a simple established routine. Conduct velvet antler harvest at a similar time each day with one or two handlers with whom the bulls are familiar.

The task becomes non-stressful and efficient and large numbers can be handled under such a routine.

Overcrowding in pens is the primary source of antler damage, particularly in young animals due to rearing and boxing with the front feet, or with bulls being under crowd pressure, lowering their heads in large milling groups and suffering breakage. A spacious pre-pen entry pen allows ample room for quick drafting into smaller groups of three to five animals. Bulls should not be left on their own for extended periods as they easily become stressed.

Alley and entrances to squeezes should have enough width at head height to avoid contact of antlers with walls or entrances, and squeezes should be well-positioned in the handling system so that little effort is needed to encourage the bull to enter. Modular handling systems use the bulls' natural desire to move around corners. Subdued lighting in these pressure areas seems to help. Pens should be thoroughly checked prior to the season for broken gates, latches, projections, and broken fence wires. Aggressive, nervous, or flighty animals should be culled as they can initiate further disruption in an established group.

Velvetting should be avoided in hot, unsettled weather or in the heat of the day or if rapid weather changes such as thunderstorms are forecast. Bulls are also likely to become upset by high winds. Visitors or outside disturbances during velveting should be discouraged and, above all, the procedure must be conducted with respect and regard for animal welfare. There are important legal and ethical requirements to be observed in antler removal techniques and producers always have a responsibility to observe these.

15   Welfare Considerations

Elk farmers assume a deep responsibility in the harvesting of velvet antlers, and it is significant that Canada has produced its own Code of Practice for farming of deer as an industry gold standard and is among the first deer farming countries to do so.

When done correctly with analgesia (pain control), antler removal is thought to be less stressful than common practices like ear tagging. While the actual amputation of antler is rather simple, the peripheral activities of animal handling, restraint, and analgesia are frequently not. There is potential for harm to the animal and to the producer if these aspects of the process are poorly performed. The animal welfare concerns include inappropriate and harmful physical and chemical restraint, and inadequate local anesthesia resulting in a painful amputation of the antler. There are safety concerns over accidental human exposure to anesthetic drugs, and drug residues in animal products derived from treated animals. There are also legal issues over the acquisition, distribution, and use of veterinary drugs in the livestock industry.

Mostly among special interest groups, there is considerable public opposition to the concept of velvet antler harvesting, supported by speculation and an uninformed view of velvet antlers' properties and uses. As farmers, we take the responsibility to not only ensure that velvet removal is the most sophisticated and correct operation on the farm, but to understand and promote the product carefully and respect it and the welfare of the animals equally.

Elk producers wanting to harvest antler from their animals should seek training in the antler removal process through a recognized training program. The instruction should be comprehensive, dealing with the peripheral issues of animal handling and antler product storage as well as the actual removal procedure. The training will allow you to harvest antler under the indirect supervision of your veterinarian and obtain prescription drugs used in antler removal.

16   Physical Restraint for Antler Harvest

The ideal physical restraint device for antler removal is a hydraulic or pneumatic squeeze chute which allows the gentle but firm confinement of the bull with his head and antlers exposed for antler removal. Squeeze chute designs for elk are very sophisticated, with various padded and adjustable features allowing containment of the animal in a safe and comfortable manner. All but the most unmanageable animals of any species can be squeezed and held still without using sedatives. Older experienced animals tend not to panic as much as young stags while being held. All animals benefit from the application of a blindfold. The blindfold reduces random visual stimulation and has a definite calming effect on animals.

Most squeezes allow some degree of head movement. If desired, complete immobilization of the head can be achieved using a halter and lateral tying. A regular horse halter is adapted by attaching ropes to the rings found on each side of the nose band. When the halter is fitted to the animal and the ropes are firmly tied down to cleats on the squeeze, the head is held motionless allowing easier cutting of the antler.

In the recent past, the need for sedatives, halters, and sometimes suitable squeeze chutes was obviated by using electro-immobilization devices to render the animal motionless. Most early elk industry proponents of electro-immobilizing devices recommended first providing physical restraint with a hydraulic squeeze, and then applying the electrical device to the stag to prevent any struggling during antler amputation and to theoretically provide some analgesia. There is no question that electro-immobilization is effective for the purpose of restraint. However, this method of restraint is brutal, inhumane, and should not be used under any circumstances. It provides no pain-killing properties and is obviously painful in and of itself.

17   Analgesia for Antler Removal

Prior to antler removal, some form of anesthesia must be administered to the animal so that amputation is painless and humane. Chemical immobilization with sedative drugs is not really the same as general anesthesia and even heavily sedated animals sometimes need additional pain control in the form of local anesthesia.

Presently, the only universally accepted form of pain control for de-antlering bulls that are not under general anesthesia is the application of nerve blocks to the antler pedicle using lidocaine or a similar local anesthetic solution. The Model Code of Practice for the Welfare of Animals: The Farming of Deer (Australian), The Code of Recommendations and Minimum Standards for the Welfare of Deer During the Removal of Antlers (New Zealand), and The Recommended Code of Practice for the Care and Handling of Farmed Deer (Cervidae) (Canada) have guidelines for the removal of velvet antler that have been created and agreed upon by the respective industry representatives, animal welfare organizations, and veterinary associations. These guidelines call for the application of local anesthetic infiltration techniques to block pain sensation to the antler and pedicle.

A regional nerve block to the antler pedicle requires injections to the major nerves which are the main sources of sensory input to the antler. Often there is minor input or innervation of secondary importance from local nerves. Complete blockage of nervous impulses requires the injection of local anesthetic to this nerve as well. With that in mind, a "ring block" placed around the base of the pedicle obviates the need for precise placement of local anesthetic solution. A series of small overlapping injections using a dose rate of 1.2 ml of lidocaine per centimeter of pedicle circumference (total volume of approximately 15 ml per antler) are placed under the skin around the base of pedicle. In places where the skin is closely adherent to the skull infiltration can be somewhat difficult. Persistence and practice combined with physical injection pressure will overcome these difficulties. The use of multi-dose vaccination-type syringes to place the local anesthetic speeds up the process and provides superior injection pressure to that of an ordinary syringe.

Many elk producers who harvest antlers complain that the use of local anesthetic requires that the animal spends too much time in physical restraint. They maintain that the process of squeezing up the animal for an injection of local anesthetic is at least as stressful to a cervid as having the antler removed rapidly but without any anesthesia or analgesia. Scientific research has demonstrated that approximately 60 seconds after application of a ring block, the small nerve branches to the antler will be sufficiently blocked to allow amputation. By the time the injection of the second pedicle is done, the first antler will usually be ready for removal.

The search for ideal methods of antler removal continues. The ideal method would provide easily applied, rapid, fully effective, long-lasting, reliable analgesia yielding a residue-free antler product at a reasonable cost with maximum safety and minimal stress to the animal. Above all, the humaneness of the method should not be compromised by economic considerations.

18   Antler Removal

Hygiene is paramount, both for equipment and the harvested antler. Saws, tourniquets, recording materials, tags and storage facilities should all be on hand before starting.

The head must be firmly restrained, and the antler fully supported during cutting to prevent any sudden movement of the animal from tearing the antler skin (banana peel effect). After the bull is securely restrained or suitably sedated, tourniquets are applied in a loop at the pedicle level below the coronet. They should be of material that has pliability and some width to comfortably compress blood vessels without crushing or cutting into underlying tissue. They must be clean before application and washed between uses.

A medium toothed saw is advised, and it should also be washed and disinfected between uses. Cutting should be parallel to the coronet, 5–10 mm above it and started from the outside of the antler toward the center. If the animal is held in a squeeze chute, then local anesthetic administered via ring block at the base of the pedicle must be given time for effective pain block to develop. A small surface cut is made first, the level of analgesia checked by lack of reaction from the bull, and the antler cut through smoothly and cleanly from this position. The saw must be sharp and disinfected and the antler supported at all times. Cutting from the center out can leave the potential for the antler to fall at the last minute, or the animal's head to move as the weight is relieved, and risks damage to the pedicle and tearing of the velvet skin. If the animal's head moves at any stage, cease cutting until control is regained.

18.1   The Antler Removal Process (Fig. 9)

Handling and Storing Harvested Antler   After removal, velvet antler should be held for cooling in a clean, shaded insect-proof area, reclined cut end uppermost on a slope of 15°–20° to allow contained blood to be evenly distributed through the antler. Velvet is also commonly hung upside down on racks until cool and then frozen in an inclined (15°–30°) or upright position. If this occurs for too long blood can pool in the upper region and create processing problems (see Fig. 10). Turn antler as it freezes—at least every half hour for 3 h. Great care must be taken to avoid deforming the bulb or tine shape during freezing to ensure the antler is presented in true form for grading. Frozen velvet antler should be stored in airtight plastic bags and handled subsequently with the due care that a valuable edible medicinal product demands.

Antler must be stored frozen and allowed to freeze without deformity of the bulb or tines. After harvest, and once cooled while inclining cut end uppermost, the blood should be evenly spread through the antler tissue. Antler should be cleaned, tagged, date recorded, weighed, and measured for grading, then frozen as quickly as possible and sealed in plastic bags to avoid moisture loss and freezer burn. Antler must be turned as it freezes—at least once every half hour for at least 3 h. Frozen antler must be kept clean and handled carefully. Never bump or drop it—cracks will form which make it useless for slicing after it has dried.

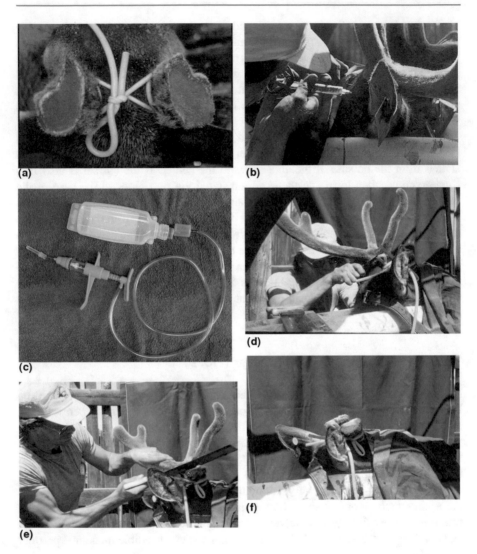

Fig. 9 (a) Here is a correctly placed and effective tourniquet made from surgical rubber tubing. (This picture is, of course, after antler removal). (**b**) This picture shows the placement of local anesthesia in a "ring block" around the base of the pedicle. (**c**) Vaccinator gun with a lidocaine Tetra pak for placing ring blocks. (**d**) Scoring of the antler with a knife along the proposed cut line before using the saw can prevent "banana peeling". (**e**) Sawing through the antler with a coarse saw from medial to lateral, or from inside to outside. Cutting from this direction can result in "banana peeling" if the antler is unsupported during the final saw movements. (**f**) Successful antler removal from a well-secured animal. This removal involved the use of wound powder, but this is an individual choice

Fig. 10 Antler rack

19 Selling Harvested Antler

Several marketing venues exist for frozen antler. Buyers may travel directly to producers' farms, usually accompanied by a local facilitator, and buy antler directly. Several local processing plants are prepared to buy from producers that deliver directly to their processing location. Irrespective of the product end point, quality is reflected in price as the international market is increasingly competitive. New initiatives include capsulated velvet antler pills for local use and as these products gain more acceptance this market should increase rapidly.

Most of the industry associations are active in developing co-operative sales with pooling of product with defined standards. Processed products are generally sold in dried whole stick form, but the drying procedure is a learned and closely protected operation best left to the experts who have invested in the industry.

A good relationship with a processor is particularly useful to have. Good communication on the quality and presentation of the farmer's product, particularly as it relates to retained blood content, calcification, and growth stage at cutting is essential as it is only by this that combined objectives of producer and processor can be achieved.

20   Elk Co-Products

20.1   Tail

Applicable to sika, red deer and elk only because of the presence of a large sebaceous gland on the undersurface of the tail. In Oriental countries, dried deer tail is the most consumed part of the deer next to velvet antler. They are valued based on the size, and relative ease of processing. The finished product is jet black in color, hairless, and has a very distinctive odor. Its value per kilogram makes it one of the most expensive forms of medicine. The dried tail is soaked in a special processed alcoholic liquor for 30 days, the liquid drunk and the tail cooked with or without other herbal or animal medicines and eaten, served with chicken or duck soup. Uses are varied but mainly for sexual dysfunction, uterine disorders, lumbar and leg pain and as a tonic to replenish the Yang system, enriching bone marrow and kidney function.At slaughter, the tail should be removed at the base of the spine and held with animal I.D. until the animal has passed all inspection procedures, then frozen hygienically for later processing.

20.2   Pizzle

The pizzle—penis, prepuce, and testis is processed to a hard dry form with a distinctive translucent red color. Appearance and size are critical to product marketing. The pizzle is usually air dried with a portion of pelvic bone attached to authenticate its origin as this is particularly important for export. It is consumed exclusively by men, soaked and sliced in wine liquor or ground and used in tonic form. It is used for impotence, sterility, and to improve sexuality.

At slaughter, it is necessary to remove the entire penis, testes, and scrotum plus the erector muscle and leave attached to a small slice of the pelvic "aich" bone. The pizzle should be thoroughly cleaned and frozen for processing.

20.3   Sinews (Tendons)

The sinews are dried to a bright natural color with or without the inclusion of the rear (dew) claws. Muscle and fat are removed, and the sinew straightened. Sinews are used as a tonic soup for vital energy restoration and for muscle and eyesight ailments.

20.4   Hard Antler

Hard antler is washed, dried, and ground and reprocessed in the form of antler glue and used to promote blood flow, lactation deficiency in women and body weakness.

20.5   Skins and Leather

The fashion attributes of elk skins have been mentioned earlier and the strength, unusual grain character and ability to shave or split to provide a soft supple leather, combine to increase use in the garment industry both as nappa leather and a buck skin suede. The product does, however, suffer from a variety of faults, mainly mechanical damage due to antler scoring or hoof damage plus the myriad of abrasions and cuts that can arise from poor handling and pen and fence construction. Bacterial damage from poorly preserved skins and tick and fly damage can also occur.

At slaughter, skins can be frozen, salted or treated in the pickled pelt form. In salting, defatting and an adequate salting is essential with the skin left spread to cool prior to treatment.

Quality farm handling is, however, the prerequisite to improving returns and the acceptance of farmed elk skin. The size of red deer and elk skins allows large panel cuts to be made from a single skin and add flexibility to form and design.

These additional products from elk should not be considered as cast-offs or by-products but are a significant and valuable part of elk farming returns.

20.6   Processed Elk Products (Fig. 11)

20.7   Elk Meat and Trophies

Elk meat is a highly desired, unique flavored and tender meat if produced and harvested correctly. Most elk on North American farms are still selected primarily for antler production, rather than for rapid early growth and desirable meat qualities. Markets are changing and developing, and the characteristics of elk meat encourage increased markets, especially given consumer trends.

Nutrient comparisons emphasize the obvious features of elk, i.e., low in fat, "bad" cholesterol, and calories. Protein levels are comparable to other meats. Elk meat is uniquely rich in minerals, particularly iron and phosphorus which, in turn account for its rich dark coloring. This nutritional bonus can be a disadvantage in retail marketing where consumers are used to bright red colored meat, contrasted by (often trimmed) white fat. Farm-raised elk are reared with absolutely no added growth hormones or steroids, since elk are naturally effective producers of meat, and artificial manipulations negatively affect velvet growth and reproductive performance.Based on the results of numerous nutritional analyses of various elk animals

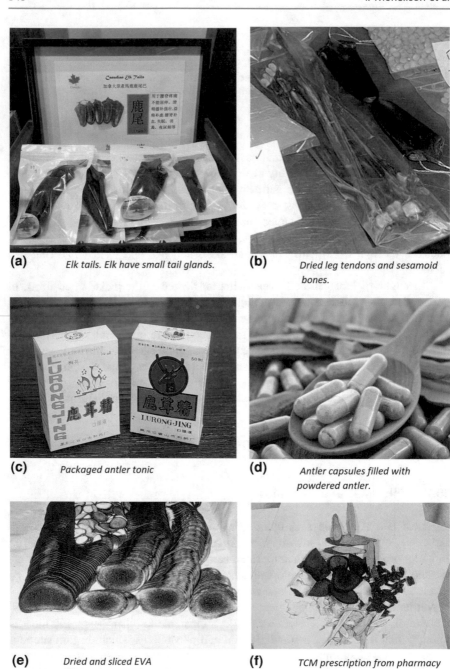

Fig. 11 (**a**) Elk tails. Elk have small tail glands. (**b**) Dried leg tendons and sesamoid bones. (**c**) Packaged antler tonic. (**d**) Antler capsules filled with powdered antler. (**e**) Dried and sliced EVA. (**f**) TCM prescription from pharmacy

and cuts, in comparison to other foods, we can make these precisely worded "Nutrient Content Claims":

"Elk Meat is:

- A powerful source of energy
- Very high in protein
- An excellent source of digestible protein
- Low in fat
- Low in saturated fats
- Free of trans fatty acids
- A source of omega 3 and omega 6 fatty acids
- High in potassium
- Very high in iron"

21 Marketing Position

While its nutritional advantages make it potentially valuable in a health-conscious, "heart smart" dietary role, most farm-raised wapiti suppliers are currently marketing at top of the range prices to the culinary hospitality trade. This strategy involves professional chefs and food distributors promoting the product and educating consumers until the volume of supply can satisfy the demand of retail outlets where use and preparation methods are more variable.

Because of its market positioning and relative scarcity, buyers of elk expect a consistently high standard of supply and product that meets rigid specifications in terms of:

- Muscle cut and size (repeatability),
- Tenderness (and by implication, moisture content)
- Quality of preparation and presentation (chilled/fresh and hygienic—no waste or spoilage). These standards aim to set the farm-raised product as a superior product, separate from the traditional consumer perception of wild-shot venison.

The product must meet the highest standards of:

- On-farm production (on-farm food safety and quality assurance)
- Transportation of both elk and processed product (properly trained transport operators)
- Slaughter and processing (HACCP accreditation of slaughter and processing plants)

The marketing and quality control programs currently in place for New Zealand-produced venison provide the benchmark for farm-raised wapiti in North America. Premiums are available for local product because of its identification, freshness, and relatively uncomplicated supply route. Competitive pricing against New Zealand products is a controversial issue for North American producers, many of whom have established direct links from farm, through local processing plant, to end user. They find that their returns have been eroded by a wholesaler-distributor operation based on a more consistent supply of imported venison. The current small size of the North American industry creates difficulties in formulating an integrated marketing plan, but its smallness is also an advantage in developing tight and profitable relationships individually with a host of outlets not available to competing importers.

Suppliers and processors together can concentrate on value rather than volume by reducing the variability in seasonal supply, adding value through knowledge of local and ethnic eating trends, and producing variations to suit. They can capitalize on education and communication by retaining the full association with the product and slowly developing new markets for increased volumes.

As demand builds, competition based on quality and consistency should increase and enhance producer returns rather than competition based on price, which has a negative effect on producer returns. Therefore, the market position from the outset must be at the top end and elk farming systems and products must continue to be able to support that position.

22   World Marketing

Traditional demands for venison and game meats are centered on Western European and Scandinavian markets. With the advent of large-scale elk and deer farming, feral (wild-shot) venison is being replaced by higher quality, farmed origin product. However, the market is rapidly growing with greater availability of products, and the taste and nutritional attributes of farmed elk and venison are being promoted.

New markets are rapidly emerging. North America has now displaced Europe as the largest consumer of elk and venison in the world, and S.E. Asia is considered to offer the most potential for expansion. Appropriate pre and postmortem health inspection at dedicated slaughter plants allow health requirements and meat inspection standards to be met universally while allowing for kosher and halal acceptance as well.

It is vital that the distinction between farmed elk and wild-shot venison is preserved, to add to the associated perceptions of enhanced quality and consistency in standards.

In emerging markets quality production and top-end market positioning are the targets. Age at slaughter, quality hygiene standards, and the ability to consistently supply product are important. With this approach elk farmers worldwide, irrespective of their stage of industry development, enjoy a growing and appreciative demand for their high quality and healthful meat products in the enviable position of undersupply. In the world marketing of agricultural production, this aspect alone is a unique feature of elk farming and a significant aspect of sustained profitability.

23   On-Farm Meat Harvest

Elk are harvested for two primary products: meat and antlers. These products have been extensively described in previous chapters. Harvest for meat mainly takes place in government inspected slaughter plants that have had their receiving and handling facilities purpose-built or renovated to handle elk in a safe, humane and efficient manner. This requires entry and raceway sides and gates that are tall enough (at least 2 m) and sometimes with covered tops—like a tunnel—to prevent elk from leaping up and over. These facilities do an excellent job of handling elk and other species and meat quality is predictably very good. Consumers and animal welfare experts have approved and accept their methods and products. Harvest method is usually with a captive bolt rifle or a powerful stunning device that renders the animal unconscious, followed rapidly by exsanguination i.e. bleeding out. For many other producers, processors, and consumers, on-farm harvest of animals is preferred.

Farmed elk become very comfortable in farm or ranch situations. Whether they are born and raised on a particular property or moved and allowed to settle in they are relaxed and understand how, given adequate and appropriate care, to function successfully in that environment. However, loading and hauling to a processing plant is stressful, as demonstrated in a recent bison project completed by researchers at South Dakota State University, where they looked at serum cortisol levels as an indicator of stress in slaughter bison. Results are shown in Fig. 12:

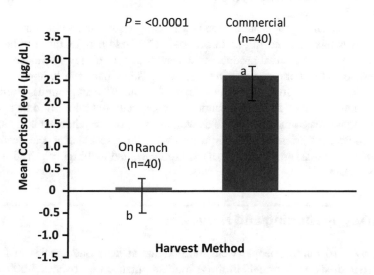

a,b - Means lacking a common superscript are different, $p < 0.05$

Fig. 12 Effect of slaughter method on bison cortisol levels. Courtesy of J Janssen, SDSU, 2020. (**a, b**) Means lacking a common superscript are different, $p < 0.05$

Treatments; On-Ranch = grass-finished bison heifers ($n = 40$) harvested on-ranch by a mobile slaughter unit. Commercial = grass-finished bison heifers ($n = 80$) transported ~720 km and harvested in a commercial facility. All heifers were harvested at ~28 m of age. The results show a strong advantage to on-farm harvesting of animals and is convincing evidence that on-farm harvest is much less stressful to animals destined for the food industry.

The preferred method of harvest is in a large "pen"—at least several acres in area, and a high-powered rifle shot by a skilled and trained professional. The initial shock of a hit in the upper neck/lower skull and spine area is followed quickly by exsanguination. After either termination in the plant or in the field, carcass processing proceeds in a hygienic manner.

These results and observations of behavior before and during both slaughter methods have led animal welfare experts to conclude that producers and consumers should be given the opportunity to choose how they proceed with harvest and marketing to consumers. In most jurisdictions, meat products destined for wide distribution through further processing, and then wholesale distribution and retail sale, are required to come from a larger plant that is inspected in every aspect of its operation by government or third-party service providers who ensure that food safety, animal welfare, and labor standards are observed and enforced. This approach reassures human health managers and regulators as well as consumers who prefer to eat food that moves through these well-documented and trusted channels. The large scale of these operations also provides economies of scale that allow sale at reasonable price levels.

For other consumers, their preference is knowing where their food comes from, and how it was raised, cared for, and harvested. For these people, on-farm harvest allows them to meet the actual producer and view the farm. If they so choose it also allows them to conduct or participate in the harvest. This is particularly important for many people of many diverse religious faiths or cultural backgrounds. Products harvested and sold directly to the consumers are generally not allowed to be resold.

For people interested more in antlers as well as meat, on-farm harvest is again the preferred choice. Elk trophies, valued primarily for their spectacular and beautiful displays of hard antler, are one of the most highly valued and highly prized in the hunting world.

24   Trophy Hunting and Harvesting

In 2018, over 16 million people went hunting for at least one species in North America. Hundreds of thousands of those hunters pursued elk; over 223,000 individual wild land elk hunters in Colorado alone. Those hunters reported harvesting 72,000 elk and spending an average of $2600 USD each on their elk hunting adventures. In addition to these, several thousand elk were hunted and harvested in high-wire enclosures. Varying in size from 50 to 50,000 acres, this form of on-farm harvest is increasing dramatically in availability and popularity around the world.

In North America, harvest of privately owned animals on private land has faced challenges since the early 1900s. The late 1800s witnessed a whole-scale assault on the world's wildlife by commercial and simply greedy hunters. This phenomenon was particularly noticed in North America, where market hunters and others with various motivations drove the bison almost to extinction and most other species, including elk, to much-reduced population numbers. In response, a conservation movement developed that included many forward-thinking and idealistic people. Among other concepts such as National Parks, these thinkers created what is known as the "North American Model of Wildlife Conservation." Included in that Model was a ban on harvest of wildlife for commercial purposes, and acceptance of wildlife being a Publicly Owned resource.

This model still dominates the thinking of many wildlife professionals, but the application of its tenets has evolved as we have moved far beyond the days of over-exploitation, vast areas of "public" land, and colonial authoritarian concepts of government. Consider all these variations:

- Bison were saved from extinction largely by farmers. They have been propagated on farms and re-introduced to wild lands to the extent that there are now more than 30,000 roaming all parts of their former range and another 400,000 on North American farms.
- As part of the Treaty process many indigenous people were granted the privilege of harvesting as much wildlife as they and their families and communities need from "unoccupied" public land or private land to which they have been granted access without licenses or bag limits, and few restrictions on methods.
- Those same indigenous people have the privilege of managing wildlife on their own lands as they see fit.
- Outfitters and guides have been granted permission and licenses to charge fees to hunters who wish to harvest wildlife for meat or trophies.
- Furbearers are not only harvested from the wild for their pelts but also raised on farms for the same purpose.
- Birds originally considered exotic to North America such as pheasants are raised on farms specifically for release and harvest with shotguns.
- A great many other species, both native and exotic, are raised in ranch situations behind high wire for conservation and harvested for trophies, meat, and other by-products.
- Farming and ranching elk and their cousins in high-wire or walled enclosures has been well established for at least 3000 years and is an accepted and profitable component of livestock agriculture in most countries in the world. In North America, laws and regulations governing elk farming reveal highly variable approaches and policies.

Everyone involved has realized two essential truths:

1. On-farm harvest is a preferable alternative to abattoir slaughter for the animal.
2. Consumers are prepared to pay prices for on-farm harvested elk that are profitable for the producer.

Furthermore, many of those consumers wish to harvest the animal themselves, sometimes for religious reasons, sometimes to ensure it is done properly and according to their wishes, sometimes for the experience of harvesting that animal in an efficient and humane way. To satisfy these consumers elk may be harvested on small farms, in pens or restraining devices, or on large-scale hunting ranches or wildlife estates.

25   The Business and Experience of Operating a Wildlife Estate/Hunt Ranch

The concept of a wildlife estate goes back to Europe where for hundreds of years members of the wealthy classes maintained country estates with a castle or manor house surrounded by wildland sometimes enclosed by walls or fences and populated with wildlife species. Usually, the owner would organize hunting and hospitality events for his family and associates. These events were bonding experiences for all involved and led to new family and business ties. Today's wildlife estates operate in similar ways, with the exception that the guests now pay fees for the experience. Hunting ranches are similar in terms of the experience.

Properties used for these operations vary greatly in size and experiences offered. Some are as large as 50,000 acres, rivaling the largest harvest properties in Africa. On these huge ranches hunting is conducted much like in wildland situations. Population densities are maintained at relatively high levels to ensure success, comparable to wildland situations where target species congregate in large numbers. Success includes not just the harvest and ownership of the trophy and meat but also the overall experience which might include the adventure and joy of traveling there, meeting new friendly people with similar interests, sharing meals, toasts and stories, spending glorious days in the bush and fields, looking for just the right elk for you, then humanely and efficiently harvesting that fabulous animal.

A hunt ranch needs to offer all these components to be successful. Your clients become your friends, who come back year after year for the joy, adventure, therapy and the opportunity to harvest a larger or different trophy and refill the freezer with high-quality elk meat.

The experience may not cost a lot more than the expenses one would incur for a wildland hunt. Refer to the cost reported above for the Colorado wildland elk hunters. It was $2600 USD each, with a harvest rate of 25%. On a hunt ranch complete packages start at $4900 USD, with a harvest rate of near 95%. The highest-end opportunities are for SCI Record Book elk with antler racks scoring 550+. The fees for these could exceed $50,000 (Fig. 13).

Fig. 13 A 645-inch total score, non-typical, 88 inches wide

26   Supplying a Hunt Ranch with Elk

Most hunting ranches or wildlife estates have a production farm attached that grows bulls and surplus cows for harvest but very few produce all they need. That need creates opportunities for elk producers to raise and sell bulls. Creating a long-term stable relationship with one or more hunt ranches is the best strategy to develop. You learn what the operator and his clients are looking for which, of course will vary somewhat, but usually there will be some predictability and stability in what they prefer. Looking at the images in this chapter, you can see the wide variation in antler confirmation and type. That gives the producer a wide range of options, but the one option that cannot be ignored is *good management*! Follow this guide:

27   Optimizing Elk Antler Production

Efficient antler production is not something that just happens in spring each year. It is a potential that is developed and maintained in elk bulls, starting with their genetic makeup and established during their years of growing into productive bulls. Concentrate on "developing" your bull calves—*every day of their early lives!*

1. *Get cows bred early* so that calves are born early. Calves born in May make much better use of summer pastures than July calves. Early calves wean more easily than their later-born siblings. Feed your cows free choice elk mineral (or use horse formulations, but not sheep or cattle) all winter, along with mixed grass/legume hay testing 10 to 14% crude protein on a dry matter basis.
2. *Sort bred cows* into calving groups based on their age (year of birth) in April. Fifteen to twenty cows per pasture if possible. Apply fly tags and pour-on Cydectin or Dectomax, and booster vaccinations. Calve in clean pastures with some shelter available—small patches of aspen trees are ideal.
3. *Reduce stress* in calving groups by stocking pregnant cows @ 2 acres/cow. Summer graze at a stocking density minimum of 1 acre per cow/calf pair, depending on pasture production. Rotate pastures when grass is 2 inches high, graze new pastures at 10–12 inches high—about every 14 to 30 days.
4. *Produce high-performance pastures*—in the Parkland area of western Canada, seed 5 lbs. Fleet Meadow Brome +3 lbs. Kay or Okay Orchard grass +3 lbs. deep-rooted alfalfa (Algonquin is a good choice) + 3 lbs. creeping rooted alfalfa (such as Heinrichs or Spredor) + 2 lbs. alsike clover per acre. Fertilize 40 lbs. N + 15 lbs. P + 10 lbs. K + 5 lbs. of copper sulfate per acre. Keep pasture growing by clipping, grazing or mowing—No seed heads!
5. *Flush elk cows prior to breeding*—feed 3 lbs. per head per day of 25% rolled peas and 75% barley with elk mineral in a minimum of 4 feet of bunk space per cow. Calves learn to eat supplement alongside their mothers.
6. *Wean calves before the rut*—between September 4th and 14th. August is too early. Separate bull calves from heifer calves at weaning, and pasture separately.

Try to leave the social groups the same. Put a babysitter or two (healthy, old, and quiet or bottle-fed cows) in with each calf group to supply some brainpower.

7. *Feed calves*—4 lbs. per calf per day of 17% CP DM calf ration (oats & peas) along with 15 to 17% CP DM grass/alfalfa hay or lush pasture. Grow your calves, do not fatten them! Feed grain in bunks off the ground to avoid parasite infections. Deworm with pour-on ivermectin and vaccinate with 8-way clostridia vaccine at weaning and vaccinate again in mid—October.

8. *Weigh your calves*—every time you handle them. Weigh and treat for parasites again in late December—use Eprinex or Dectomax this time. Calves should weigh from 300 to 400 pounds at this weigh-in, having averaged 1 lb./day average daily gain since weaning. At this time reduce supplement to 1–2 lbs./head/day, and reduce CP levels to 10–14% CP DM.

9. *Place a grumpy old herd bull*—(antlers cut off) in with your bull calves. He will round them up and boss them around and virtually eliminate pecking order arguments. This seems to suppress the endocrinological (hormonal) cycle and delays pedicle and antler growth. The best antler is always grown on green grass.

10. *Pasture spiker bulls separately*. If pastures lack quality or quantity, feed adequate supplement (as you would in winter) to ensure continued growth and healthy pedicle and antler development. Fly tag in early May. Remove hard spikes in August before the rut. Separate from all breeding activity unless you want to risk using an exceptionally well-developed spiker on a few females.

Give your rising two-year-old bulls the absolute best care and attention!

- Boost feed supplement beginning in March and continue right through summer and the rut. Feed spikers 5 lbs./head/day of 14–16% CP DM calf ration in bunks with 6 ft. of space/ elk.
- Ensure spikers recover body condition post-rut by feeding 7 lbs./head/day of 14% CP DM ration (oats/peas/mineral ration) with 6 feet of bunk space each and free choice grass/ alfalfa (green and leafy) hay testing 14 to 16% CP DM.
- Give all your bulls a Christmas vacation! Managing as described above should put your rising twos in fine condition by Christmas. Then work with their seasonal cycle of reduced appetite and intake during the winter. Continue to provide green and leafy hay, mostly grass, and reduce supplement to a couple of pounds per head per day until about March 21.
- Boost the feed again in mid-March, slowly increasing the oats/peas/corn/barley/mineral ration (at 12 to 14% CP DM) to a maximum of 10 lbs./head/day by early April. Feed more than once per day if you can afford the time.
- Monitor high-energy ration feeding by observing manure consistency and hoof growth.
  Manure should remain lumpy, not loose. If bulls become lame, hooves appear to be growing too long, or manure becomes loose, back off on the energy. High energy rations are not dangerous if they are introduced or increased slowly.

- Keep protein levels lower. Balanced antler growth rations should not exceed 16% CP DM, including forage. Excess protein is excreted in urine resulting in burns to the penis sheath or vulva.
- Record, record, record! Note button drop dates, calculate approximate cutting dates, carry those with you as you observe antler growth and record notes as the optimum date is decided. After harvest, tag antlers, record weights, "drydown" amount, and any other comments you might have regarding conformation or any comments you hear from the antler buyers. Use those records to cull, cull, cull. We need to eat and sell elk meat, and that is where your poorer producing animals must go. Production at an early age is a particularly good indication of production in later years.
- Select the hard antler candidates by tine length or grow them out and compare. If a bull is well below average for both velvet and trophy production the choice to eat him should be easy.
- Maximizing antler production means maximizing voluntary feed intake. As bulls continue aging the secret to success stays the same—Excellent management! As an excellent manager, you must balance the goal of optimum antler production with costs, profits and the health and longevity of your elk.
- Continuously improve your genetics. With excellent management in place your primary limiting factor will be the genetics in both your bull and cow herds. Given the rate at which productivity is increasing in elk farming you should have very few older animals in your herd. Yes, keep a few "supercows" who consistently give you the best offspring, but consider that those offspring should be even better than the super momma. Read current publications, attend conventions, competitions and other learning opportunities, travel widely and keep up with the best new genetic combinations and production techniques. Never stop learning and improving!

Your goal should be to produce a harvestable animal in as little time as possible. Many producers now have their bulls scoring 400 inches by 4 years of age, consistently, with the top end at 400 + by 2 years of age. This offers three major advantages:

- Your costs to produce revenue are significantly reduced.
- Your herd stays genetically and reproductively young. You are always improving production and performance and consequently working with your best.
- You maximize avoidance of CWD, which has a long incubation period before problems arise.

28 Current Challenges and Future Trends

Management systems have been developed, genetic selection has dramatically improved the quality and productivity of out livestock, strong and increasing markets are demanding our products—the stage should be set for a dramatic run-up in populations and profits on North American elk farms. Instead, we have witnessed a dramatic decline in those metrics in the past two decades. Why?

28.1   The Impacts of CWD

Chronic Wasting Disease was first noted in mule deer in sister Wildlife Research Laboratories in Colorado and Wyoming in the 1960s. At the time, the cause was unknown and was referred to as "a wasting syndrome." By the 1970s, the cause was identified as being associated with abnormal nervous tissues called "prions" and the syndrome was given a name, shortened to CWD. It was also realized that the disease was somehow transmissible, but that realization may have come too late. Many cervids of various types had been captured from the wild, used for various research projects then released—to the wild, to game farms and to zoos. One of these releases became the first diagnosis in Canada—a mule deer in the Toronto Zoo diagnosed CWD positive in 1978. By the 1990s, cases were found in the wild and on cervid farms in the USA and in Canada—farms that had purchased and imported elk from a game farm in South Dakota that had in turn, received some of the releases from the Laboratories. By the year 2000, it was obvious to both the farmed cervid industry and to the Canadian Food Inspection Agency that CWD was established and was being transmitted between farmed elk and wild deer in western Canada. An agreement was reached, and action began. The disease was listed under the Canada Health of Animals Act as a Reportable Disease. An Eradication Program was designed and put into play. The Program made it mandatory to report immediately any evidence of CWD in a farmed animal. That evidence was followed by a complete quarantine on the farm, then quickly by depopulation and testing of the entire herd. Fair market value compensation was awarded to the owners to encourage their cooperation.

Over the next 10 years, 2000 to 2010, over 120,000 farmed cervids were tested for CWD. All of these were tested postmortem, as no live test had yet been developed. From these tests, 470 cervids were found to be positive—0.004%. Sixty-two farmed herds were ordered destroyed, with over 11,000 deer and elk depopulated.

This level of activity attracted attention from the world's media. The Bovine Spongiform Encephalopathy outbreak was huge news in the 1980s and 90s, so the news that another Prion disease, related to BSE, was established in western Canada, and was serious enough to warrant the forced destruction of tens of thousands of farmed animals quickly caught attention. South Korea banned imports of antler from North America in 2001. Until then, that country had purchased and imported most of the antler harvested and sold by US and Canadian producers.

Prices for antler and livestock plummeted. Producers were getting rid of their animals as quickly as they could. Some went for human food, some for dog food, and some were destroyed and buried.

In 2001, there were 111,000 elk and white-tailed deer on 2000 farms in Canada.

In 2011, there were 43,000 elk and WTD on 1000 farms.

In 2020, there are 24,000 elk and WTD on 600 farms.

In 2000, those elk and deer were estimated to be worth $600 million CDN on the open market. In 2020 estimated market value of livestock is $10 million.

How could an industry suffer so badly from a disease that affected less than 1% of the total population? Even in herds with a diagnosed positive and ordered destroyed, the infectivity rate has been less than 4%. The answer is threefold: Uncertainty, Publicity and Political Pressure.

After 60 years of study and accumulate knowledge, we still do not understand prion disease well. They are complex and peculiar compared to most other diseases. But 60 years of increasing knowledge of CWD has provided two key facts:

1. CWD is somehow transmissible to other cervids
2. There is no evidence that CWD is transmissible to humans.

However, in the process of learning those facts we went through many years of uncertainty regarding the risk the disease posed, and how to best deal with it. This caused exaggerated fears in the media and in the minds of their readers and listeners, and pressure placed on politicians to "Do something about it" even though what to do about it was not obvious to anyone. Add to that the dogged persistence of those in the wildlife industry who adhere to the North American Model and who are idealistically opposed to farming cervids and even the livestock industry in general. The political pressure generated by those lobby groups resulted in extreme restrictions of the functioning of the industry and the acceptance of its products for many years.

We are now reaching a point in industry maturity and knowledge of CWD management where more sensible and workable are being made about industry regulation. Now we face another serious CWD challenge that has slowly worsened oner 60 years—CWD in wild cervids surrounding our farms.

CWD was first diagnosed in a wild deer at about the same time as the first in farmed cervids. Since then, it has spread to wild populations in 25 States and 2 Canadian Provinces. In Alberta in 2019, 19,400 wild cervids were postmortem tested for the disease. Of these, 1160 were positive—an infectivity rate of 11.2%. That is three times higher than the rate found on positive and depopulated farms, and 28 times the rate found in the total of farmed cervids tested at slaughter or found dead! The consequences of having that vast reservoir include well-managed farms finding positives despite their efforts. New strategies must be developed and applied.

28.2 Future Trends in the Elk Industry

Like all agricultural industries, the elk industry will continue to evolve after the 2021 publication of this book. Here are some important developments to follow in coming years:

- Increased cooperation and collaboration will develop with other livestock industries, university, government and private research entities toward continual improvement in genetic selection, productivity, best management practices, disease resistance and management, and all other aspects of the farmed cervid industry.

- Traceability systems based on the establishment of an effective identification and inventory program for all farmed, ranched, zoo, or research cervid herds. As well as being essential for animal health management programs and quality assurance schemes, industry can use traceability systems for:
  - Keeping wildlife concerns reassured
  - Ensuring consumers of the health and safety of our products
  - Supplying information to registry systems
  - Linking with partners in the value chain to gather and exchange data on product suitability, quality and value, facilitating best practices for breeding, management, processing and marketing for all partners.
  - Generating equitable program funding for industry development and promotion
- Vertical integration. Health management issues have demonstrated the risks of translocating live creatures, whether they be farmed, wild, or humans. Successful value chains will develop where producers are contracted to supply animals to harvesters, processors, distributors, and marketers located close to each other. Hunt ranches will maintain a breeding operation near the harvest area that supplies all the trophy and meat animals their customer base requests.
- On-farm harvest will be expanded to meet the demands for improved animal welfare and consumer reassurance regarding the management, quality, and safety of the products they buy directly from producers.
- Online stores selling elk meat and other products. The success of retailers like Amazon and many others, plus the COVID-19 shutdowns, have demonstrated the effectiveness and efficiency (not to mention profitability) of this direct-to-consumer marketing approach.
- Processing and extraction of concentrates from EVA that target specific needs and customer bases. Many consumers are not satisfied with eating capsules of antler powder. Many specific targeted components may be extracted, concentrated and offered in more acceptable and effective forms.
- Minimizing the risk of CWD infection.
  - Research and observation of wildlife populations has demonstrated that some cervids are resistant to the disease. Analysis has identified genetic characteristics associated with this resistance, and cervids can be tested to determine if they carry these. The resistance is not total, but it appears to forestall the development of the disease for several years. Breeding strategies and management that emphasize productivity at an early age allow cervids to be harvested before exposure or disease development occurs.
  - Live tests are in the process of validation now. Early results suggest a high level of effectiveness, with room for further enhancement. Live animal testing will allow assurance of customers and regulators that trade in livestock and velvet antler is safe. Postmortem testing after slaughter or natural death will continue to confirm the effectiveness of this program.

- Biosecurity programs will become more accepted and accurate, limiting the risk of importing CWD to a producer's herd.

In a quick half a century elk and deer farming have proven they can become an integral and highly profitable part of North America's agricultural economy. Imagine where we will be by the end of that century, and beyond!

Reference

Thorleifson I, Woodbury M, Pearse T, Friedel B (2020) Elk farming handbook. Woodbury Training and Consulting, Saskatoon

Part IV

Deer Management Systems and Diseases in the Russian Federation

Deer Breeding and the Velvet Antler Industry in Russia

Sergey Aliskerov

Abstract

An account of the historical development and present status of the deer breeding industry within the Russian Federation including reindeer herding in the north, velvet antler production, and provision of deer for hunting. The different species used and population fluctuations within the different geographical regions are described. The commercial preparation of velvet antler for China and Korea as well as the domestic markets are discussed.

Keywords

Russia · Deer breeding · Reindeer husbandry · Velvet antler production, pantocrine

1   Introduction

Deer breeding is a branch of animal husbandry which has grown out of hunting, and which now exists on the territory of Russia in the form of several variants, differing in form and purpose. First of all, there is the most ancient *northern pasture deer breeding*, where the reindeer (*Rangifer tarandus* Linnaeus, 1758) is the object of breeding and which was originally the main source of food, fur, etc., as well as providing a means of transport.

The velvet antler industry is a younger industry that emerged in family farms in Western Siberia and the Primorsky Territory about 150 years ago. Here, the animals used are *red deer* (*Cervus elaphus* Linnaeus, 1758); and the red deer subspecies:—*siberian elk* (*C.e. sibiricus* Severtzov, 1873); *manchurian wapiti* (*C.e. xanthopygus*

S. Aliskerov (✉)
FEDFA - Federation of European Deer Farmers Associations, Prague, Czech Republic

Milne-Edwards, 1867), and *sika deer* (*Cervus nippon* Temminck, 1838) of which there is one subspecies used in Russia (*C..n. hortulorum* Swinhoe, 1864). The product of this sub-industry is the valuable raw materials for medicines—velvet, and also meat, skins, etc.

In the twenty-first century, more intensive deer breeding on farms has been developed with agricultural, hunting, tourist, and esthetic purposes, where mainly European *red deer*, *sika deer*, and *fallow deer* (Cervus dama Linnaeus, 1758) are the objects of breeding.

In hunting enclosures (parks) are kept and bred mainly the same *red deer* and *sika deer*, *fallow deer*, less often—*roe deer: European* (Capreolus capreolus Linnaeus, 1758) *and Siberian* (Capreolus pygargus Pallas, 1771) and *Eurasian elk* (Alces alces Linnaeus, 1758). Besides these native species of the deer family (Cervidae Goldfuss 1821), today the imported *white-tailed* or *Virginia deer* (Odocoileus virginianus) is bred in Russia.

The purpose of this work is to write a small historical overview of the deer breeding industry in Russia and its current state.

Management of reindeer is a separate large topic, and we have devoted our review to examining in more detail the branches of red deer breeding—velvet and farmed deer breeding.

2   Deer Breeding Centers in Russia

2.1   Management of Reindeer

It is believed that the reindeer, as a species, appeared about a million years ago in the territory of present-day North America and from there spread widely across northern Asia and Europe. A particularly large number of deer lived on the earth during the Glacial and Interglacial periods, when, as a result of a strong cold spell, plants and animals gradually retreated to the south. One of these periods was called the "Period of the mammoth and the reindeer" (Borozdin et al. 1990).

The domestication of the reindeer and its transformation from a hunting object into an object of organized animal breeding was one of the forms of economic development of the tundra, forest tundra, and taiga landscapes.

The most probable location for the domestication of reindeer was evidently the Baikal and Altai regions, from where management of Reindeer spread to the north and northwest. Archeological excavations in the Yenisei basin, rock engraving and "deer" stones of Transbaikalia and Northern Mongolia, as well as rock engraving of Karelia indicate that before the common era, management of Reindeer was widespread in Central Asia and northern Europe (Borozdin et al. 1990). Vasilevich G.M. and other authors make proposals about the initial emergence of management of Reindeer in two centers—the Sayan (Sayan-Altai Upland) and the Tunguska (mountainous regions of Transbaikalia and the Amur region), from where it spread by different peoples to the north of Asia and Europe. Later, from the Okhotsk coast and Chukotka, management of Reindeer gradually spread eastward (Borozdin et al. 1990).

These materials show that many years ago people learned to use the reindeer in many ways: they obtained meat, skins, etc. from it and used it as a transport animal. The emergence and development of the management of reindeer covers a huge historical period, during which it became the property of most of the indigenous peoples of the North of Eurasia—from the Kola Peninsula to Chukotka, Kamchatka, Sakhalin, and the Lower Amur (Muhachyov 1990; Syrovatskij and Neustroev 2007). A number of researchers are of the opinion that there are several centers for the development of management of Reindeer: Tunguska, Samo-Nenets, and Chukchi-Koryak (Druri and Mityushev 1963; Pomishin 1990; Syrovatskij and Neustroev 2007).

While domesticating reindeer, people tried to preserve and exploit in them their natural adaptive characteristics, first of all, their ability to independently obtain food for themselves in the pasture all year round, as well as other economically useful skills.

Domestic reindeer are subdivided into *four breed groups: Evenk; Chukchi; Even; Nenets*. They differ in some morphological characteristics (size, weight, color), fertility, behavioral characteristics, and the degree of domestication (Pomishin 1990; Baskin 1970; Syrovatskij and Neustroev 2007).

Providing for the needs of the community, management of Reindeer was origi-nally of a family or clan nature and, herds of reindeer were both large and small. Since there was no veterinary and zootechnical service until the twentieth century, deer often died by tens of thousands from epizootics of anthrax and other diseases. The use of pastures was not optimized and this led to the death of animals from starvation. In general, the management of reindeer was unproductive and the num-bers of livestock fluctuated greatly for the above reasons (Druri and Mityushev 1963).

The sustainable development of management of reindeer began in the 1930s. In the USSR, practically all branches of the national economy were developed at the State level, conditions were created for the successful development of management of reindeer, its transformation into the leading branch of animal breeding in the Far North. According to Drury I.V. (1955), for almost 20 years from 1933 to 1952 the number of reindeer increased by one and a half times, reindeer accounted for more than 70% of the total number of animals in the Arctic Circle in most tundra, and partly in taiga regions. In 1990, the number of reindeer in the USSR was about 2,300,000. With the collapse of the USSR and the aggravated economic problems, the number of reindeer decreased (Syrovatskij and Neustroev 2007).

In the twenty-first century, the management of reindeer remains the main occu-pation of the indigenous peoples of the North, Siberia, and the Far East. Nineteen regions of Russia are classified as deer breeding. Two-thirds of the world's domesticated reindeer come from Russia.

Khanty, Dolgans, Chukchi, Nenets, Mansi, Komi, Koryaks, Sami, Tofalars, Evenki, Eveny and Yukagirs and other nations continue to roam the tundra with their herds (Gurvich 1966; Muhachyov et al. 2004). According to the outcomes of

Table 1 Dynamics of the reindeer population in the federal districts of the Russian Federation for 10 years (as of July 1st) (thousand heads)

	Year 2006	Year 2016
Russian Federation	1664.7	1906.0
Northwestern Federal District	382.2	375.5
Komi Republic	103.0	105.3
Arkhangelsk Region Including Nenets Autonomous Area	212.2 209.9	209.2 207.6
Murmansk Region	67.0	61.1
Ural Federal District	763.5	940.5
Tyumen Region Including: the Khanty-Mansijsk Autonomous District—Yugra	763.5 31.3	940.5 53.7
Yamalo-Nenets Autonomous District	732.2	886.8
Siberian Federal District	72.9	154.4
Republic of Buryatia	0.8	0.3
Republic of Tyva	1.7	3.5
Krasnoyarsk Krai Including: Taimyr (Dolgano-Nenets) Autonomous Okrug Таймырский	69.0 63.6	146.7 -
Evenk Autonomous District	4.8	-
Irkutsk Region	0.8	0.8
Chita Region	0.8	-
Far Eastern Federal District	446.0	435.6
Republic of the Sakha (Yakutia)	186.1	172.8
Khabarovsk Krai	6.4	5.1
Amur Region	6.3	6.2
Kamchatka Region Including Koryak Autonomous District	42.0 35.1	52.3 -
Magadan Region	18.0	17.0
Sakhalin region	0.2	0.1
Chukotka Autonomous District	187.1	182.0

Note: sign «-» means lack of data on the subject

the All-Russian Agricultural Census (Itogi 2008, 2018), the population of reindeer increased by 231,300 heads over a 10-year period, mainly due to increases in the Ural and Siberian federal districts (Table 1).

The increase of livestock is most rapid in small agricultural holdings and peasant (farm) households (Table 2) (Itogi 2008, 2018).

Geography of Russian Management of Reindeer Management of Reindeer in Russia covers a vast territory from Scandinavia to the Bering Strait, from the coasts of the White Sea to the Sea of Okhotsk and Chukotsk. Currently, there are several centers for the development of regional forms of deer breeding, which have developed according to the national way of life of peoples and methods of deer breeding: Sami, Nenets (Komi-Izhemsky), Tungus-Yakut, and Chukot-Koryak.

Table 2 Dynamics of the reindeer population of the Russian Federation in households of different categories (thousand heads)

№	Households category	As of July 2006	As of July 2016
	Households of all categories including:	1664.7	1906.0
	Agricultural holdings. including:	921.1	1133.3
1	Large and medium-sized business	844.6	861.8
2	Small business	73.1	239.4
3	Subsidiary farming of non-agricultural organizations	3.4	31.9
4	Peasant (farm) households	16.0	61.0
5	Individual entrepreneurs	6.2	7.5
6	Private households	721.4	704.4

The Sami (or Lapland) Deer breeding is different in fact that the deer are released in the summer areas unattended.

Nenets Deer breeding unites several nationalities: Nenets, Enets, Nganasans, Kets, Dolgans, Selkups, Komi-Izhemtsy, Khanty, Mansi and covers the territory from the White Sea to the eastern borders of the Taimyr Peninsula. Here, deer are kept in large herds all year round under daily protection and leadership from the movement of the herd. The number of deer here reaches 40% of the total Russian population.

Tunguska-Yakut Deer breeding covers deer breeding areas of the Krasnoyarsk Territory, Irkutsk, Amur, Chita, Magadan regions, the republics of Buryatia, and Sakha (Yakutia). Here, the Evenks (Tungus), Evens (Lamuts), Yukagirs, Kets, Dolgans, Yakuts, etc. are engaged in deer breeding. Due to the wide variety of natural conditions and national characteristics, the greatest diversity of deer breeding is observed here, from round-the-clock protection of large herds to free camp keeping. These deer are the most domesticated and are not afraid of humans. The population is about 20%.

Chukotka-Koryak Deer breeding covers the northeastern part of the country from the Kolyma River. The Chukchi, Koryaks, Itelmens, Orochi, Nivkhs are engaged in deer breeding, practicing round-the-clock protection and grazing of deer "from the hands," that is almost enclosure keeping. Animals are harnessed here only in winter, and these reindeer are the wildest, poorly tamed to transport. In the most favorable years, up to 47% of the Russian livestock was concentrated in this region (Syrovatskij and Neustroev 2007) (Map 1).

Table 3 presents the five largest northern deer breeding households of Russia.

Some of these enterprises have difficulties due to conflicts of interest with oil and gas workers, as deer breeders occupy promising sites of oil and gas development sites (Sever i severyane 2012).

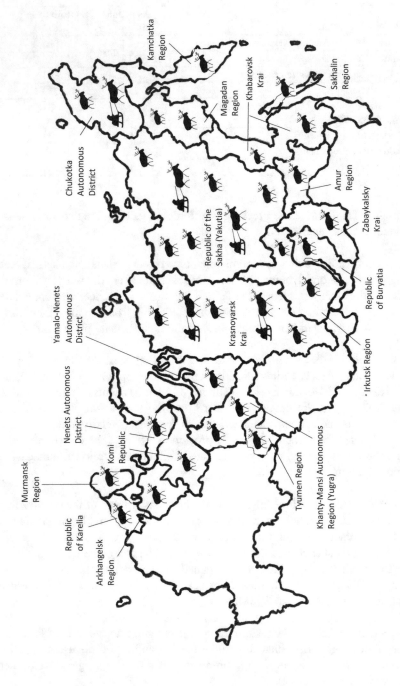

Map 1   The main development regions of management of Reindeer in Russia

Table 3 The five largest reindeer breeding households of Russia

Place	Household	Region	Population of reindeer
1	MUP «Yarsalinskoe»	*Yamalo-Nenets AO,* Yar-Sale village	15,722
2	SPC (collective farm) "Izhma Olenevod and Co"	*Nenets AO,* Iskateley village	14,861
3	SPC «Tundra»	Murmansk region, Lovozero village	12,665
4	SPC «Put Ilicha»	*Nenets AO,* Khorey-Ver village	10,478
5	MUP (municipal unitary enterprise) «Panaevskoe »	*Yamalo-Nenets AO,* Panaevsk village	7600

Prospects for the Development of Reindeer Management in the Russian Federation Over the past 10–15 years, the number of deer in Russia has significantly decreased—by 1.9 times. The reasons for this decrease are the impact of socio-economic reforms. It must be noted that a significant reduction in the number of domesticated deer in recent years has been observed in almost the entire circumpolar region, and this can be considered as a reflection of the global process of civilization's influence on the traditional economy. However, the crisis state of the industry is noted only in Russia, which is mainly due to a sharp deterioration in the social and economic conditions of the northerners.

Reindeer management is the only branch of agriculture in the Arctic region, in which only the indigenous peoples of the North of Russia are practically employed. The uniqueness of management of Reindeer is that it remains not only a branch of agriculture but also a way of life for deer breeder's families. In Russia, it is called an "ethno-preserving industry," the role of which in preserving the traditional cultures of the indigenous peoples of the North can hardly be overestimated.

Most breeders in the North breed deer for one purpose—to obtain valuable products and resources necessary for life in harsh conditions. However, we can also meet large farmers in the tundra who came to this region specifically to earn money on animal husbandry. Today, more and more people are trying to adhere to the principles of a healthy diet and refuse fatty pork and beef in their ration. In addition, do not forget about the constant reduction in the livestock, so the prices for venison will only grow. It should be noted that the Russian people who are engaged in deer breeding do this not just for observance of traditions. It is just that in a harsh climate, there are practically no alternative options (Syrovatskij and Neustroev 2007).

At the regional level, measures are being taken to support the management of reindeer, for example: in the Trans-Baikal and Krasnoyarsk Territories, the Republic of Sakha (Yakutia), the Khanty-Mansiysk and Yamalo-Nenets Autonomous Districts, laws have been adopted on state support for traditional economic activities and traditional crafts of the indigenous peoples of the North, Siberia and the Far East, which protect both the habitat and the traditional way of life of the population, including northern deer breeders (Zakon 1–7).

In the Republic of Sakha (Yakutia) in 2016, 668.2 million rubles were allocated for state support of deer breeding, including 42.5 million rubles from the federal budget and 625.7 million rubles from the republican budget. Such measures are aimed at contributing to the sustainable development of domestic management of reindeer in the country (Sever i severyane 2012; Balanov 2016).

2.2   The Velvet Antler Industry

The velvet antler industry is a branch of animal husbandry that specializes mainly in the breeding of *Siberian elk and sika deer*. Its main product, velvet, is an important medicinal raw material for the medical industry. Along with velvet also venison, skins and by-products (secondary) products—blood, tails, penises, veins, and fetus (embryos) are marketed. These products are also used in traditional medicine.

The traditional breeding zone of Siberian elk is the Altai, and of sika deer the Primorsky Krai. Over the 150 years of the existence of the velvet antler industry Russian scientists and practitioners have developed systems for Siberian elk and sika deer management including feeding conditions and harvesting of the products, methods for preserving the velvet, methods of breeding management and selection. The diseases of Siberian elk and sika deer have been studied and measures for their prevention and therapy have been developed and are being successfully implemented. Research on methods for obtaining medicinal products from the velvet antler industry products and the study of their therapeutic effectiveness are being conducted.

The second chapter of this review covers in detail the topic of the velvet antler industry in Russia.

2.3   Deer Parks and Farms

Deer parks (enclosures) appeared in Russia in the second half of the twentieth century and are settled exclusively in hunting sectors for the purpose of keeping and breeding animals for shooting or release into the wild (hunting grounds). By-products are live animals for sale, meat, etc. During periods when hunting is not carried out, the parks can be used for ecological tourism.

The center of development of these areas of deer breeding is concentrated in the European part of Russia.

Farmed deer breeding in Russia began to develop actively at the beginning of the twenty-first century and is based on the experience of New Zealand and Europe.

The main breeding species are red deer, sika deer and Siberian elk.

The main products are live animals for the creation of new farms and hunting enclosures. Secondary products are venison, velvet, and trophy animals for hunting.

Chapter "Diseases of Farmed Red Deer in New Zealand" of this review is devoted to the topic of deer farming in more detail.

3 Velvet Antler Industry

3.1 Production of Velvet Among the People of Siberia, the Far East, and China

For many centuries velvet has been the base product from which various medicines are made in the school of oriental medicine. In the "Chinese Consolidated Pharmacopoeia" of doctor and pharmacologist Li Shi-Zhen (1518–1593) there is a special section dedicated to velvet and ossified deer antlers, which were included in more than 400 recipes. It is not surprising that the antlers of velvet deer were in the greatest demand among the peoples of China (Bogachyov and Bogachev 1993). As late as the seventeenth century, hunters in the Russian territories bordering China and Mongolia did not know anything about the healing properties of velvet, since the merchants did not say what they needed them for. The local people joked, saying that even if they only cook soup from these antlers, they still pay excellent money, and they preferred Siberian elk to any other forest game. By the middle of the nineteenth century, Russian hunters could sell a pair of velvet antlers to Chinese merchants for up to 50 rubles, depending on their condition. A cow at that time cost only 4–5 rubles. In China, they were already being sold for 250 rubles, while in shops the prices sometimes reached 600 rubles. And even so, this trade was still very economically advantageous for the hunters. The velvet of manchurian wapiti cost up to 75 rubles, and of sika deer—up to 300 rubles (Lunicyn 2004).

The main centers for buying velvet in Russia were in the village of Nikolskoe (Ussuriysk Territory), the town of Kyakhta (Transbaikalia), and the city of Biysk (Altai). From Kyakhta alone, in the period from 1852 to 1882, over 4 million antlers of velvet deer were exported to China annually (Frolov 2014).

3.2 The History of Deer Breeding in Altai, Thansbaikalia and Primorye (Late Seventeenth to Nineteenth Centuries)

The settlement of the southern Altai by Russians began at the end of the seventeenth century. The Siberian elk figured prominently in the hunting industry of the settlers, but the Russians learned about the value of the Siberian elk velvet only in the 1760s, when the border with China was established along the River of Bukhtarma (now the Charyshsky District of the Altai Territory).

In China velvet has been valued for thousands of years for its special healing qualities. The high price of this raw material prompted hunters to hunt stags of Siberian elk, Manchurian wapiti and sika deer throughout the vast expanse of Asia—

from Turkestan to the Far East. The widespread increase in hunting for Siberian elk led to a rapid decrease in their number in the taiga. As a result, hunting for Siberian elk has become more difficult and less successful, and velvet raw materials are even more scarce and expensive. This contributed to the emergence of the idea of the need for the domestication of wild deer (Druri and Mityushev 1963).

Until now, the question remains, who and when was the first to tame velvet deer. The literature data on this issue are very contradictory (Frolov 2014). There is an opinion that Russian fugitive peasants began to keep Siberian elk as livestock during the reign of Empress Catherine (1762–1796). The names of Afanasy Chernov (the 1820s), of the peasants of the Oshlakovs in the Uimonsky district and of Semyon Lubyagin from the village of Yazovoy are also named. In the 30s, in the southern Altai, Savely Ushakov kept a captured Siberian elk in the village of Fykalki in the upper reaches of the Bukhtarmy river, then sold it to Avdey Sharypov. His sons Afanasiy, Roman and Sidor, continued to catch Siberian elk to keep them in captivity. In 1835 they built the first Siberian elk enclosure. To obtain the velvet products, they caught several bucks of Siberian elk and created conditions for their permanent management in their farm. Without killing the stags, they annually cut down their non-ossified antlers and obtained valuable raw materials for medicine. Such entrepreneurship allowed them at home to remove from one Siberian elk during its life up to 10–12 pairs of non-ossified antlers instead of one pair obtained when shooting while hunting (Lunicyn and Borisov 2012).

Reliably, according to N.A. Frolov (2014), are the following facts:

1. Homeland of velvet antler industry is Altai. So far, no literary sources, including those on Transbaikalia and the Far East refuting this conclusion, have been found.
2. The first experiments on breeding of Siberian elk were carried out by the Sharypov family, the village of Fykalki (the upper reaches of the Bukhtarma river). In 1917, 11 families of three generations of the Sharypovs were already engaged in Siberian elk breeding in this village.

In 1897, almost 60% of Altai Siberian elk breeding was concentrated in the Bukhtarma area (Table 4, Map).

Other regions of Altai lagged significantly behind in the development of Siberian elk breeding farms (Table 5).

In Oirotia, the Siberian elk enclosure of the Fomin brothers was built in 1866 in the western part of the Ust-Kansk region (modern territory of the Altai Republic), on the Talitsa River. A few years later, the merchant Popov built a Siberian elk enclosure in the village of Shebalino (now the regional center of the Altai Republic) (Fig. 1) (Tihonin 1932).

Subsequently, Siberian elk breeding in Gorny Altai developed successfully; the number of peasant households engaged in Siberian elk breeding grew rapidly. By the end of the nineteenth century, more than 200 Siberian elk enclosures contained about 3000 Siberian elk (Lunicyn and Borisov 2012).

Table 4 The rate of creation of Siberian elk enclosures in the Bukhtarma region in 1830–1897 (according to Frolov 2014)

Nos. of Siberian elk enclosures opening by decades						Data for 1897 (numbers)		
1830–1839	1850–1859	1860–1869	1870–1879	1880–1889	1890–1897	Siberian elk enclosures	Siberian elk breeders	Siberian elk
1	1	4	6	53	43	115	163	1854

Table 5 The state of Siberian elk breeding in Northern Altai in 1897 (according to Frolov 2014)

Altai regions	Data for 1897		
	Siberian elk enclosures	Siberian elk breeder	Siberian elk
Northwestern	25	27	417
Northeastern	15	21	168

3.2.1 Manchurian Wapiti Breeding

In the Yenisei province, primarily in the Sayan Territory, there is a zone of "transition" from the subspecies "Siberian elk" to the subspecies "Manchurian wapiti," which the Chinese have never divided and called the same—"ma-lu." Hunters from Transbaikalia hunted this eastern subspecies of red deer. Manchurian wapiti breeding in Eastern Siberia had much less development. In the 1850s, researchers noted that since the 1830s some peasants have kept Manchurian wapiti in their households to get velvet. In 1843 the Manchurian wapiti household of A.A. Neskornomny was established. After 30 years about 300 householders kept 800 heads. In the 1890s more than 1000 Manchurian wapiti were kept in a semi-industrial state, including 100 females and 100 calves. The largest number of them

Fig. 1 (**a**) Peasant-Siberian elk breeder from the village of Shebalino, 1913. (**b**) Cutting of antlers of a Siberian elk that has been cast. Photo by A. Silantyeva. (According to Frolov 2012)

were in the Verkhneudinsky and Chita districts (the modern territory of the Republic of Buryatia and the Trans-Baikal Territory), (Lunicyn and Borisov 2012).

3.2.2 Sika Deer Breeding

Sika deer breeding in the Far East began somewhat later than Manchurian wapiti breeding in Altai (Chikalev et al. 2014). The useful properties of sika deer have long been known to the inhabitants of eastern countries. The deer were hunted and mercilessly exterminated. As a result, the area of sika deer, which once occupied a vast territory from the Ussuri region to southern China, Korea, and northern Vietnam, has significantly decreased. In most of China, deer have been completely exterminated or are extremely rare. In Korea, there was a specialized clan of sika deer hunters—"kavando," who hunted in the forests of Primorye. In the early twentieth century, in Primorye, there were up to 10 thousand heads of sika deer, in the 1960s the number decreased to 500 heads.

The first deer breeders in Primorye were the peasant-old believer S.Ya.Ponosov and the horse breeder M.I. Yankovsky with his son. Ponosov had a small settlement in the Shkotovsky region, where he was the first who started to breed sika deer from 1888. Yankovsky, a political immigrant, tamed a herd of wandering deer in 1897 on a peninsula in the Amur Bay near the Sidimi Cove. Their farms served as primary breeding grounds for sika deer. The third breeding ground was the Askold Park.

In Primorye, deer breeding developed in two ways: either in the form of park keeping in large areas in a semi-wild state with animal feeding only in the winter months, or in the form of small domestic deer parks, in which hand-raised animals were artificially fed all year round (Lunicyn and Borisov 2012).

3.3 Velvet Antler Industry in the Twentieth Century (up to 1917)

The late nineteenth and early twentieth-centuries period was marked by the active development of Siberian elk breeding in Altai. For comparison with the data in Table 4, we quote the data for 1905 for three settlements of Bukhtarma (village of Berel, village of Berezovka, village of Chernovskoe) in Table 6.

In this way, in three settlements of Bukhtarma in 1905, there were as many Siberian elk breeders and Siberian elk as in 1897 in the entire southern Altai.

High prices for velvet in China were a powerful incentive for the emergence and expansion of Siberian elk breeding. From 1892 to 1907, an average of 100 poods (1 pood = 16 kg) of velvet worth 35,000 gold rubles were annually exported to China through the Ongudai customs, that is 20–25 rubles in gold for 1 kg. We recall

Table 6 The state of Siberian elk breeding in three settlements of Bukhtarma in 1905 (according to Frolov 2014)

Siberian elk breeders (numbers)	Siberian elk (numbers)		
	Stags	Female siberian elk	Total
113	783	1089	1872

that a cow at that time cost only 4–5 rubles. With good management, one Siberian elk produced 2–3 kg of canned velvet per year. This is more than 50 gold rubles. The total amount of gross income of Siberian elk farms in Altai in 1914 was about 50,000 gold rubles (Lunicyn and Borisov 2012).

The First World War in 1914 reduced the pace of development of this industry in Russia, but Siberian elk continued to be a profitable type of economic activity for peasants until 1917. In 1917, there were about 350 peasant-Siberian elk breeders in Altai, whose households had about 8000 Siberian elk. During this period, the center of Siberian elk breeding moved to the village of Berelskoye (now in the territory of the East Kazakhstan region, Kazakhstan).

In the early twentieth century, Siberian elk breeding spread from Altai to other regions of Russia. So, in Tuva (now the Republic of Tyva) in the most developed Todzhynsky region in 1915 there were 857 Siberian elks.

3.3.1 Manchurian Wapiti Breeding

In the Sayan Territory the old believers were engaged in deer breeding, and in 1915 in the district of Minusinsk (the modern territory of the Republic of Khakassia) there were about 1400 head of deer. In 1909–1915 they sold about 3150 antler pairs of Manchurian wapiti (Kuznecov 1938).

In the Verkhneudinsky district (the modern territory of the Republic of Buryatia) in 1915 there were 574 Manchurian wapiti in the home keeping, in the Chita district (the modern territory of the Trans-Baikal Territory) there were 260 heads. However, this Manchurian wapiti breeding was, so to speak "reduced," since in most households only stags and no female were kept, therefore, and it did not have young stock. To resume the population, wild Manchurian wapitis had to be caught again. Only a few households kept females and obtained young stock. Most of the sold velvet of Manchurian wapiti were hunted—3000 in 1900, and only 1000 from domesticated deer.

Significant changes in Manchurian wapiti breeding took place after 1893, when by the decision of the Governor there was a ban imposed on catching of Manchurian wapiti in pits, and this was the only way to restock these animals. This ban was so unexpected and incomprehensible to the Manchurian wapiti breeder that they understood this circular as a requirement to release all the Manchurian wapiti kept in captivity. According to the Siberian Yearbook, already in 1913, there were only 23 domestic Manchurian wapiti in Transbaikalia (Frolov 2014). More than thousands of people were deprived of the prospects for profitable trade. This testifies to the complete decline of Manchurian wapiti breeding in Transbaikalia already before the First World War in 1914.

3.3.2 Sika Deer Breeding

In the early twentieth century and up to 1928, 104 sika deer breeding households were organized in Primorye.

On the Yankovsky Peninsula by 1917 there were about 2000 heads. Annual slaughter for obtaining of velvet reached 100 heads, and about the same amount of velvet was obtained by cutting. The household of Startsev had 1500 deer, and the

Vladivostok Hunting Enthusiast Society had 4000 deer. A rather large deer breeding household was owned by S.I. Konrad on the Peschaniy Peninsula. In 1909 he purchased 60 male fawn from the household of Yankovsky and on the Askold Island. By 1920 there were 378 deer on his household (Lunicyn and Borisov 2012).

During the years of the revolution, civil war and collectivization, the population of Siberian elk, Manchurian wapiti, sika deer declined sharply, all private Manchurian wapiti- and deer enclosures were looted and destroyed. The animals were slaughtered en masse or released and the fences were cut. For example, A.S. Popov, the Altai Siberian elk breeder had in the village of Shebalino a Siberian elk enclosure of up to 500 heads, a tannery, and fattened cattle for meat. But with the establishment of Soviet rule, the herd was liquidated (Lunicyn and Borisov 2012). In the 1920s, Popov, pressured by the authorities, gave up his property and worked at «Sibgostorg» (Frolov 2012). In Altai, in the village of Berel (now located on the territory of the East Kazakhstan region, Kazakhstan) out of 7000 heads (in 1920), only 245 animals remained by 1928. The deer breeding industry during this period was badly damaged, like the whole economy of the country.

3.4   Velvet Antler Industry in the USSR

The recovery of deer breeding began in 1924. The first Siberian elk breeding state farm was established in the village of Chernovaya, Katon, Karagai region (now the village of Katonkaragai, East Kazakhstan region, Kazakhstan). The main livestock were Siberian elks, confiscated from sole wealthy peasants. In Oyrotia, in the village of Shebalino, the catch of Siberian elk was organized, more than 100 heads were got, which formed the core of the herd of the state farm of Shebalinsky (Tihonin 1932). In 1929 the state farms "Abayskiy," "Talitskiy," "Kaytanakskiy," "Nizhne-Uymonskiy," "Verkhne-Uymonskiy" were created. These farms were not large, they had a total of 2620 Siberian elk. In addition to Siberian elk, they began to breed sika deer in the 1930s in the Altai farms, which were originally brought from the Far East to the Siberian elk state farm of Shebalinskiy (222 heads). Later, sika deer were bred on several households (Chikalev and others, 2014).

In the Far East in 1924, the state farm of Sidimi was organized for breeding of sika deer. By the beginning of 1928, the state of sika deer in Primorye was as follows: domestic deer households—1657 heads, in parks—3837. By 1935, the number of velvet deer was: Siberian elk—8218 heads, sika deer—10,304 (Lunicyn and Borisov 2012).

After the state farms of deer breeding were organized, this industry followed the path of rapid growth. The first two 5-year plans were the time of reconstruction. During this period, the system of keeping the Siberian elk and sika deer, the methods of cutting and velvet treatment, were changed, and the feeding of animals has significantly improved. Thanks to this, the number of Siberian elk increased by 51%, sika deer—by 72%, velvet output, respectively, by 82% and 282%. The best Siberian elk breeding brigades had a productivity of 6.7 kg per stag, 1.2 kg for a sika deer, the mortality rate decreased to 2–3%, the young stock yield was 52–59% (Mityushov et al. 1950).

3.4.1 Scientific Study of Velvet and Velvet Antler Industry

With the organization of deer breeding state farms, problems were identified in the industry, that required scientific understanding. It was necessary, first, to prove that:

1. Velvet as the main products of households, really did have therapeutic properties and, in addition to export supplies, could be used in domestic medicine.
2. It was necessary to develop techniques for conducting the industry with the prospect of enlarging households.
3. To develop a scientific basis for animal feeding
4. To find new ways of canning of velvet and by-products, to develop standards for them
5. To study deer diseases and find ways to treat and prevent them

To solve these problems, a research laboratory for velvet antler industry was created in 1933, later renamed into the central one (CNILPO). Its founder and its first director was a prominent scientist, pathophysiologist Professor S.M. Pavlenko, who headed the department at the 1st Sechenov Moscow Medical Institute (Galkin 1983).

In the first phase of the laboratory's existence, its activities focused on actively studying the medicinal properties of velvet, their scientific justification.

The first studies of the histological structure and chemical composition of velvet were made by Mejsel and Grigoreva (1933), Timofeeva and Kolli (1936) and others.

Professor S.M. Pavlenko developed a technique to produce the pantocrine, a medical preparation made from velvet. Then began its comprehensive study (Pavlenko 1936). S.M. Pavlenko headed a large group of scientists who conducted clinical trials of the drug. It was proved that pantocrine has tonic and stimulating properties. It is a drug of general biological action and, acting on the main physiological systems, it produces complex functional rearrangements in the body. The results of these works were published in the collections "Pantocrine" for 1936 and 1969. A.S. Tevi, S.M. Pavlenko, L.N. Shchepetilnikova developed a tablet form of pantocrine, registered in 1966 as an invention. Then its mass production was established.

Until 1962, the attention of researchers was focused on the Siberian elk and Sika deer velvet. In subsequent years, the velvet of Eurasian elk (Gladilov 1962; Tevi and Zhuravleva 1969), reindeer, European red deer, and roe deer (Brekhman et al. 1969; Dobryakov and Brekhman 1972) were studied. It was found that they also have biological activity, depending on the species and ecological and physiological factors. Under the guidance of Professor I.I. Brekhman a new drug from velvet of reindeer "Rantarin" (in tablets) and its liquid analog "Velcornin" were developed in 1972. The works of A.B. Silaev (the professor at Moscow State University and the specialist in the field of natural compounds) made it possible to reveal the antitumor properties of pantocrine (Silaev 1971).

Today in Russian pharmacy there are several directions for the manufacture of treatment agents from deer velvet. The first is the processing of red deer velvet into various forms of an alcoholic extract such as the "Pantocrine" preparation, the second is the processing of reindeer velvet into "Rantarin" and "Epsorin" tablets. The third is the production of pantogematogen, it is a dietary supplement based on blood, isolated from velvet of Siberian elk. In the last decade, velvet baths have become popular—water procedures using velvet concentrate. From the ossified horns, a food supplement with an immunostimulatory effect "Cygapan" is produced (Cygankov et al. 2000).

In close cooperation between scientists and practicing deer breeders, a method of preserving the velvet of sika deer and Siberian elk was developed. Having studied the Chinese method and the experience of one who boils Siberian stag antlers in Altai and Primorye, the laboratory proposed *a combined method of preserving velvet*, the constituent phases of which were cooking in hot water (96–98 °C), heat chamber (75–80 °C) and wind drying. The number of cooking days has reduced from 5–6 to 3, and the number of immersions of antlers into water (by 40–60 s) per day of each cooking from 70 to 9. This combined method is currently used by all Russian who boils Siberian stag antlers (Galkin 1983; Zabolotskih 2010). Exploring mechanical cooking, devices (machines) were invented that allow simultaneous cooking of up to 30 pieces of Siberian elk velvet and up to 60 or more pieces of sika deer velvet. The method of preserving velvet in an atmosphere of saturated steam made it possible to mechanize the cooking process, to increase productivity by 7–10 times, and to improve the medicinal qualities of raw materials.

In 1946 the laboratory developed GOSTs[1] for velvet of Siberian elk, Manchurian wapiti and Sika deer, which operated in the USSR until 1976, and in 1977 they were significantly processed taking into account the increased productivity (GOST 4227–76—preserved velvet of Siberian elk and Manchurian wapiti, GOST 3573–76—preserved velvet of sika deer antlers). GOSTs determine the requirements for cut velvet, their distribution by varieties, regulation on acceptance, packaging and transportation, they are currently in force.

Based on scientific research and economic experience, a progressive system of velvet antler industry management was proposed. It consisted of the following main technological elements:

1. Separate keeping of stags and females with the specialization of service teams for receiving velvet and raising young stock.

---

[1]GOST—the state standard, which formulates the requirements of the state for the quality of products, works and services that are of cross-sectoral importance. GOST is established on the basis of the application of modern achievements of science, technology and practical experience, taking into account the latest editions of international standards or their projects. The GOST system was developed and launched in the USSR. Since 1992, the state standard of the Russian Federation has been designated GOST R. It confirms that the products have been tested and meet all safety requirements. On the territory of the Eurasian Economic Union, GOSTs are applied voluntarily.

2. Keeping of large groups of stags in three separate herds—young (from 6 months to 2 years), middle (from 2 to 5 years) and older (from 5 years). Differentiated feeding.

3. Application of rations with increased doses of silage and supplementary feeds, feed mixtures- and granules preparation, mechanized feed distribution.

4. Development of the basics of breeding, indicators of super valuation of stags, methods of animal's selection. Creation of nuclear stock in 15–20% of the total number of livestock, estrus in isolated paddocks with a sex ratio of 1:25, the use of the best breeder-stags.

5. Calving in small groups in isolated paddocks, obligatory registration of newborn calves.

6. Early (before the estrus) separation of young stock and its differentiated rearing.

7. Velvet preservation with the use of mechanisms (velvet cooking device, steam chamber).

8. Siberian elk breeding (deer breeding) complex of structures and buildings, located and interlocked in such a way as to organize cutting and canning of velvet in a cyclical flow method.

9. Pasture rotation based on the alternate use of the park's gardens by animals of all age and sex groups. Grass stand feeding in 3–4 cycles using each section for 6–7 days, allocating one garden for a month's rest to restore the grass stand and sanitation.

10. Prevention measure and treatment of infectious and invasive diseases[2] (Galkin 1983).

3.4.2   State and Geography of Velvet Antler Industry in the Period up to Year 1941

After the end of the civil war in Russia (USSR) in 1924, the velvet antler industry went through several stages: the restoration of velvet antler industry, its development based on the introduction of scientific developments, the increasing of the livestock and of production volumes (see Sect. 3.4).

The geography of velvet antler industry has expanded during this period. Sika deer were brought from the Far East to collective farms in Altai (Altai Territory and the Altai Republic (Map 2)), and Kazakhstan (which was then part of the USSR). Sika deer acclimatized well in new territories, but the snow cover became the

---

[2]Diseases of velvet antler deer are divided into: internal non-infectious (traumatic, shock, dental diseases, obstruction of the esophagus, congestive dystonia of the proventriculus), infectious diseases (tuberculosis, actinomycosis, anthrax, rabies, trichophytosis, leptospirosis, emphysematous carbuncle, necrobacillosis, infectious anaerobic enterotoxemia, foot and mouth disease, pasteurellosis, brucellosis), invasive diseases (theileriosis, eimeriosis), helminthiasis (seteriosis, elaphostrongylosis, varestrongylosis, dictyocaulosis, dicroceliosis, cysticercosis, echinococcosis, paramphistomatosis, ashvortiosis, oesophagostomiasis, trichocephalosis, verdicmansiosis, nematodyrosis, moniesiasis), entomoses (booponuosis, hypodermatosis, pharyngomyosis) and ectoparasitic diseases (Lunicyn 2004).

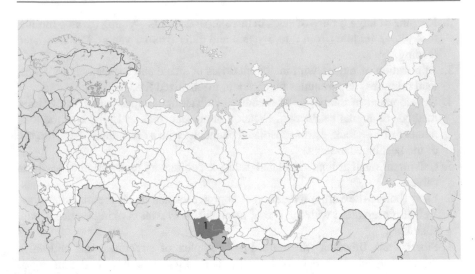

Map 2 The map of the Altai Territory (1) and the Altai Republic territories (2)

limiting factor for them. With the height of snow cover more than 30 cm, they needed additional feeding (Lunicyn and Borisov 2012).

After the organization of collective farms (a collective farm is a cooperative corporation of peasants) and state farms (a soviet farm is a state agricultural enterprise), the development of velvet antler industry as a branch of animal husbandry followed the path of rapid growth. In 1928, there were 107 large deer breeding farms in Primorye, which contained 1691 heads of sika deer, in addition, 1657 sika deer were kept in domestic deer farms and 3837 heads were kept in deer parks. As of 1 January 1935, there were 8218 Siberian elk (Altai and Kazakhstan) and 10,304 sika deer (Primorye) in the velvet antler industry of the USSR. By 1940, the total number of velvet antler deer in the USSR reached 20,000 (Lunicyn and Borisov 2012). Before the start of the Second World War, the velvet antler industry transformed into a developed branch of animal husbandry.

During the *Second World War*, the production performance of households decreased significantly. With the end of the Second World War, the supply of households increased, qualified personnel came to collective and state farms, feed production increased, and the pre-war level of velvet production and livestock population was reached by 1950 (Galkin 1964).

3.4.3   Development of the Velvet Antler Industry in the USSR from 1950 Till 1990

In the 1950s–1970s, more than 400 Siberian elk and deer households operated on the territory of the USSR. The industry continued to develop. The geography of velvet antler industry had expanded by 1992. In addition to the three traditional regions (Primorye, Altai, Eastern Kazakhstan) velvet antler deer began to be bred in the Kaliningrad region, the North Caucasus and the Baltics, it has been continued works

Table 7 The population of velvet antler deer in the USSR

Years, region	Number of households	Population	
		Siberian elk	Sika deer
1970 Primorye			40,300
1976 Primorye			45,000
1976–1980 USSR		35,700	63,100
1976–1980 Kazakhstan		7800	2600
1985 Altai Territory	15	25,000	8000
1988 Altai Mountains	17	20,000	9000

on the acclimatization of animals in the natural environment of the Urals, in the central zone and on the south of Russia, in the Stavropol Territory, the Dagestan Autonomous Soviet Socialist Republic, Kabardino-Balkaria (Matyushin 1954; Dormidontov 1968; Smirnov 1968).

Altai Siberian elk from Siberian elk state farms, most often from Shebalinsky, were widely used for settling to acclimatize in other regions. By the end of the 60s of the last century, about 500 Siberian elk were settled from the Urals to the Baltics and southern of the Ukraine (Chikalev et al. 2014). In the 50s the USSR also sold animals for export—to New Zealand, England, and China (Lunicyn and Borisov 2012).

By 1970 the population of sika deer in Primorye was 40,300, and by 1976—45,000 heads. In 1985 the 15 largest households in the Altai Territory had 25,000 Siberian elk and 8000 sika deer (Kostin 1975; Eger and Deev 1992). The available statistics for the period under review are somewhat scattered; we have summarized them in Tables 7 and 8.

3.5 Velvet Antler Industry in Russia in the 90s Years and in the Twenty-First Century

In the early 1990s new economic relations, high profitability and export destination of the industry contributed to its intensive development. By the year 1992, 38 Siberian elk farms and 64 sika deer farms operated in the Altai Territory and the Altai Republic, their population totaled 62,139 Siberian and sika deer. In 1998 the velvet herd in Russia consisted of 27,000 Siberian elk and 66,000 sika deer (Lunicyn and Borisov 2012).

In Primorye, during the same time, the population of sika deer decreased to 12,000. Compared to Siberian elk, their productivity is lower, and the costs of keeping and value of production are approximately equal. Therefore, in most deer farms sika deer are gradually replaced by Siberian elk (Table 9). The same trend is shown by the data for the years 2003–2017 in Table 10 (Dalisova et al. 2019).

By the year 2000, 112 households of various forms of ownership specialized in velvet antler industry in Russia: 38 households bred sika deer, 74—Siberian elk. The number of Siberian elk was 55,000 and sika deer—33,000 (Lunicyn and Borisov 2012).

Table 8 Productivity of velvet antler deer in the USSR

Years, region	Velvet produced (ton)				Purchased velvet on (rubles)
	Raw		Canned		
	Siberian elk	Sika deer	Siberian elk	Sika deer	Sika deer
1976–1980 RSFSR			121		
1976–1980 Kazakhstan	28,1				
1980 USSR			33		
1980 RSFSR			20,7	9,4	
1981 RSFSR			29,8		
1981 Kazakhstan			5,6		
1988 Altai Mountains	47	3,7	18,2	1,36	
1989 Altai Mountains		20			
1990 Altai Mountains		21			
1990 Far East					5,5 million
1990 Kazakhstan					1,94 million
1990 Kaliningrad region					0,323 million

Table 9 Population of velvet deer in Altai region and Altai Republic (1999–2001) (according to Lunicyn 2004)

Region	Siberian elk			Sika deer		
	1999	2000	2001	1999	2000	2001
Altai region	13,681	14,144	14,172	9090	8621	7501
Altai republic	30,476	34,407	35,207	5740	5667	5255

Table 10 Population of velvet deer in Altai Republic (2003–2017) (according to Dalisova et al. 2019)

Siberian elk				Sika deer			
2003	2008	2013	2017	2003	2008	2013	2017
50,477	56,276	51,788	49,304	4341	2795	838	615

In 2005, there were 184 enterprises of various forms of ownership in the Russian Federation in the Siberian elk- and velvet antler industry: 28 enterprises bred sika deer, 156 were engaged in Siberian elk breeding. The total population was 97,000 Siberian elk and 25,000 sika deer. In the Krasnoyarsk Territory, Khakassia, Kaluga, and Tver Regions, there were small households of Siberian elk breeding. Sika deer breeding farms are mainly located in the Kaliningrad region, Stavropol and Krasnodar regions, Kabardino-Balkaria. The leading place in the professional

Photo 1   Velvet drying (Photo by A. Mytsikov)

breeding of Siberian elk is traditionally occupied by the Altai Territory and the Altai Republic; sika deer are professionally bred in the Primorsky Territory. These three regions provide up to 98% of all products of velvet antler industry in Russia (Dalisova et al. 2019).

In 2015, on the territory of Russia, there were velvet households in the North Caucasus, in Novosibirsk, Kemerovo, Irkutsk, Kaliningrad regions; Republics of Altai and Tuva; Altai, Krasnoyarsk, Trans-Baikal and Primorsky Territories (Kambalin 2015).

The main products of these households are velvet of Siberian elk and sika deer (Photo 1), as well as velvet slices, pantohematogen, velvet candles, balsams, meat products, delicacies, and live animals.

According to data from the collection "Altajskij kraj. APK: istoriya i perspektivy" (2012), Altai Siberian elk velvet are valued 1.5–2 times more expensive than New Zealand, American, Korean, Chinese velvet and 10–12 times more expensive than reindeer velvet (Maralovodstvo i pantovoe olenevodstvo na Altae 2012).

For the Altai Republic, velvet antler industry is currently one of the most significant branches of agriculture (Photos 2 and 3). In 2012, 53,900 Siberian elk and sika deer were kept here, and 106 tons of velvet were produced. In 2005, the Republic of Altai sold velvet products for $ 10 million. This is 38% sales revenue of

a

b

Photo 2 Gorno-Altai Republik, 2021 (photo by A. Mytsikov). (**a**) Feeding of Siberian elk in enclosures, (**b**) feed distribution

a

b

c

d

Photo 3 Handling system for Siberian elk when velvet cutting, Altai Territory, 2018. (photo by N. Bystritskaya). (**a**) Side view, (**b**) bottom view, (**c**) inside view, (**d**) in working position

all agricultural products in the republic and 58% of the export revenue (Maralovodstvo i pantovoe olenevodstvo na Altae 2012).

In the southern regions of the Altai Territory is also engaged in Siberian elk breeding. In total, at the beginning of 2009, 34 Siberian elk and deer breeding farms were registered in the Altai Territory with a livestock of 22,173 Siberian elk and 3897 sika deer, from which 14,919 kg of preserved velvet were procured in 2008 (Maralovodstvo i pantovoe olenevodstvo na Altae 2012).

Table 11 Breed of Siberian elk (Cervus elaphus sibiricus) (Gosudarstvennyj reestr 2019)

Application	Name	Year	Originator/patent holder	Category
®9252515	ALTAY-SAYANS	2007	11421, 11422, 11423, 11424, 11425/6, 11421, 11422, 11423, 11424, 11425	Breed
9357654	DOMESTICATED FORM	1993	1291, 5730, 8937	Breed

It should be noted that the products of Siberian elk breeding in Russia itself are not in particular demand. Outside the Altai region, neither meat of Siberian elk, nor medicinal preparations based on their blood and horns are in particular demand. The profitability of a business is highly dependent on the buyer of the product—an importer in Korea, or a local reseller in Altai. For reference, in the Korean market, the cost of velvet, depending on the variety, varies from $ 60 to $ 400 per kilogram (Maralovodstvo 2018).

The creation of special economic zones of a tourist and recreational type on the territory of the Altai Territory and the Altai Republic gave an impetus to the development of Siberian elk breeding. The development of tourism has expanded the domestic sales market for velvet products of the Altai region. The active use of the products of velvet antler industry for medical and health purposes made it possible to build tourist hotel bases and complexes near the enclosure for Siberian elk. Here they offer medical and health procedures based on products of velvet processing, the most popular are velvet baths (Maralovodstvo i pantovoe olenevodstvo na Altae 2012). The All-Russian Research Institute of Velvet Antler Industry SO RASH (VNIIPO, earlier—the All-Russian Scientific Research Experimental Station of Velvet Antler Industry, established in 1956; Barnaul) is engaged in the study of the breeding, keeping of Siberian elk, as well as processed products in the Altai region (http://www.wniipo.ru).

In the Altai velvet-breeding households, back to the Soviet Union, work has been started on the selection of Siberian elk. In 1993, according to the State Register the domesticated form of the Siberian elk was included (Table 11) (Gosudarstvennyj reestr 2019). Based on this breed, the Altai-Sayans breed of Siberian elk® has been created, it was included in the State Register in 2007. The live weight of adult stag Siberian elk of the Altai-Sayans breed ranges from 250 to 300 kg. The antlers reach their maximum growth at the age of 12–13. The mass of velvet at the age of 10 is 8 kg, the length of beam is 63 cm, the circumference is 18 cm (Gosudarstvennyj reestr 2019).

Currently, along with velvet breeding, there is a tendency to use the Altai-Sayans breed for meat deer breeding (Maralovodstvo i pantovoe olenevodstvo na Altae 2012).

In August 2018, the VII World Deer Congress was held in Russia (Altai Territory, Belokurikha) (Minselhoz 2018). More than 300 experts from 17 countries of the world (Australia, Austria, England, Kazakhstan, Canada, New Zealand, Ukraine, Switzerland, and others) took part in it. As a result of the work of the congress, it was

Table 12 Dynamics of the Siberian elk population in the federal districts of the Russian Federation for 10 years (as of the 1st of July) (thousand heads) (Itogi 2008, 2018)

№	Region	Households of all categories, 2006	Households of all categories, 2016
1	Russian Federation	72,978	86,236
2	Central Federal District	500	2440
3	Northwestern Federal District	51	2013
4	Southern Federal District	4	–
5	North Caucasian Federal District	–	1598
6	Privolzhsky Federal District	–	1283
7	Ural Federal District	655	362
8	Siberian Federal District	69,831	78,051
9	Far Eastern Federal District	1937	489

Table 13 Dynamics of *Siberian elk* population of the Russian Federation in households of different categories (thousand heads) (Itogi 2008, 2018)

	Households category	As of July 2006	As of July 2016
		1	2
	Households of all categories Including:	72,978	86,236
	Agricultural holdings, including:	67,064	72,009
1	Large and medium-sized business	45,093	17,202
2	Small business	21,530	54,118
3	Subsidiary farming of non-agricultural organizations	441	689
4	Peasant (farm) households	4966	13,467
5	Individual entrepreneurs	640	594
6	Private households	308	166

noted that in the Russian Federation at least 20 regions are engaged in deer-, reindeer- and Siberian elk breeding. Siberian elk breeding is a cost-effective sub-industry. In Russia, there are about 100,000 Siberian elk, the leaders of Siberian elk breeding are the Republic of Altai and the Altai Territory with a total livestock of about 80,000 animals. The population of Siberian elk in Russia is planned to double by 2025—up to 200,000 heads. This should contribute to the development of velvet processing and the deer breeding industry in general. It is planned to increase the number of Siberian elk mainly of breeding animals. Now in Russia, the breeding herd of Siberian elk is a little over 11,000 heads. It is assumed that the breeding stock will be increased to 15% of the total number by 2024 (Minselhoz 2018).

The following statistical materials based on the results of the All-Russian Agricultural Census 2006 and 2016 (Itogi 2008, 2018), allow us to judge the situation in velvet antler industry in Russia at the beginning of the twenty-first century and the trends of its development (Tables 12, 13, 14, and 15).

Table 14 Dynamics of *sika deer* population in the federal districts of the Russian Federation for 10 years (as of the 1st of July) (thousand heads) (Itogi 2008, 2018)

№	Region	Households of all categories, 2006	Households of all categories, 2016
1	Russian Federation	15,061	9464
2	Central Federal District	827	3218
3	Northwestern Federal District	1675	136
4	Kaliningrad region	1674	133
5	Southern Federal District	1308	241
6	Republic of Adygeya	164	0
7	North Caucasian Federal District	Provided after 2006	48
8	Privolzhsky Federal District	254	175
9	Ural Federal District, including Khanty-Mansiisk autonomous district—Yugra	0	55
10	Siberian Federal District	7603	2676
11	Far Eastern Federal District	3394	2965
12	Primorski Krai	3394	2809

Table 15 Dynamics of *sika deer* population of the Russian Federation in households of different categories (thousand heads)

	Households category	As of July 2006 1	As of July 2016 2
	Households of all categories including:	15,061	9464
	Agricultural holdings, including:	13,247	7431
1	Large and medium-sized business	9977	5027
2	Small business	3073	1797
3	Subsidiary farming of non-agricultural organizations	197	607
4	Peasant (farm) households	1231	1340
5	Individual entrepreneurs	538	138
6	Private households of population	45	555

The livestock of Siberian elk increases in all districts, except for the South, Ural, and Far East district (Table 12). Growth is observed in small and peasant (farm) households (Table 13). The livestock of sika deer, in contrast, is decreasing (Table 14) in the above-entitled economic considerations and their active sale in deer parks of hunting area. Since large farms replace sika deer with Siberian elk, prices for live sika deer are not high. Growth of their livestock in households of population and subsidiary farming of non-agricultural organizations, are probably connected with this circumstance.

4   The Modern Deer Breeding in Russia: Deer Parks and Deer Farms

Since we have covered reindeer management in northern regions and the velvet antler industry in Chaps. "Farming Red Deer in New Zealand: Industry History, Structure and Administration" and "The Management of New Zealand Farmed Deer", here we will discuss park and farm deer breeding.

4.1   Enclosure Breeding

The first hunting deer parks appeared in the USSR in the early twentieth century, and their main task was to ensure guaranteed hunting. A prominent example of such a deer park is the history of the state natural complex "Zavidovo," located on the border of the Tver and Moscow regions, the foundations of which were laid even before the revolution in 1917.

In 1929, its lands were transferred to the Military Hunting Society (an integration of hunters-militaries). During the winter of 1941–1942, the front line was through Zavidovo, and the hunting areas suffered great damage but were restored. It was decided to create a demonstration hunting ground in Zavidovo. Since 1971, it has received the status of the state scientific-experimental Zavidovo Reserve Ministry of Defense of the USSR. During the 1960–1970s intensive measures were taken to bring and release large quantities of game ungulates: wild boars, Siberian elk, sika deer.

In these years "Zavidovo" became a government residence. The leaders of the state Nikita Khrushchev and Leonid Brezhnev often visited this place. They have not only rested and hunted in Zavidovo but also received the leaders of foreign states. At present, there has been formed the National Park "Zavidovo" on this territory, where the official country residence of the President of the Russian Federation "Rus" is located.

At the time, the hunting ground "Zavidovo" became, in fact, the first large center for the acclimatization of large ungulates in our country. The acclimatization of Siberian elk was successful. Although the first animals, brought to "Zavidovo" in 1937, were almost completely exterminated during Second World War, the follow-up releases of 338 Siberian elk, combined with comprehensive biotechnical measures and high security, marked the beginning for a stable local population. This subspecies, being stronger and more resistant to low temperatures, turned out to be better adapted to local habitat conditions than the European red deer (Egorov 2004).

Other state decisions on the direction of hunting grounds on the territory of Russia are the development of state experimental hunting grounds, currently subordinate to the Ministry of Natural Resources and Environment of the Russian Federation. These tasks include keeping and breeding of game animals in the deer park in semi-free conditions and in an artificially created habitat (terminology adopted in the legislation of the Russian Federation and related, among other things, to the keeping of ungulates in the deer park) (Federalnye gosudarstvennye byudzhetnye.. 2013).

Similar initiatives are being pursued in the regions, for example, on the territory of the reserve "Byuzinsky" in the Krasnoyarsk Territory where a deer park was created to increase the number of the Siberian elk population. Animals raised in deer park will be used for future reintroduction to the wild (Belenyuk and Belenyuk 2019).

Hunting for ungulates in natural conditions is regulated by hunting legislation, establishing objective restrictions on the number of individuals allowed to be hunted. There are no such restrictions in the deer park, therefore the park keeping and deer breeding in private hunting grounds in Russia has become quite widespread at the beginning of the twenty-first century and has mainly practical purposes: raising trophy animals for hunting or selling live animals, organizing effective hunts, taking animals for sale for meat.

As a rule, the owners of enclosure complexes restrict breeding to the following species of the deer family (Cervidae):

- European red deer (*Cervus elaphus hippelaphus)*
- Siberian elk (*C.e. sibiricus*)
- Sia deer (*Cervus nippon L.)*
- European fallow deer *(Dama dama L.)*
- Roe deer (*Capreolus capreolus L.*) (Kabelchuk et al. 2013; Ovechkina and Kozlov 2016)

Deer parks mostly consist of leases from the state forest plots and agricultural lands (areas) owned by the owner. There is a common fence on the outer perimeter, inside the territory can be divided into zones for the purpose of quarantine, separate keeping and breeding of animals by species, age, and sex (Aliskerov 2012).

The great majority of deer parks are located in the most populated regions of the European part of Russia (Map 3). According to our estimates, there are more than 300 deer parks in hunting areas for 2021.

We are seeing on our data in Table 16 that the number of deer kept in the deer parks has been constantly growing in the last decade. At the same time, if at the beginning of the reporting period there were three times more sika deer than red deer and five times more than fallow deer, then by the end of 2020 the number of red deer had already surpassed that of sika deer by 1.4 times. Such dynamics of the deer population is associated with the different availability of these species for the buyer, because sika deer have been bred in Russia for a long time to obtain velvet, and it was quite easy to purchase them in large quantities within the country. The European red deer was present in Russia in small numbers only in three places: in the Voronezh, Rostov, and Belgorod regions. To meet an emerging demand, thousands of European red deer have been imported from Europe since 2010, the breeding of which has changed the ratio between the two species – sika deer and red deer. Of particular interest was the obtainment of trophy antlers of the European red deer, hence the outstripping growth in their number.

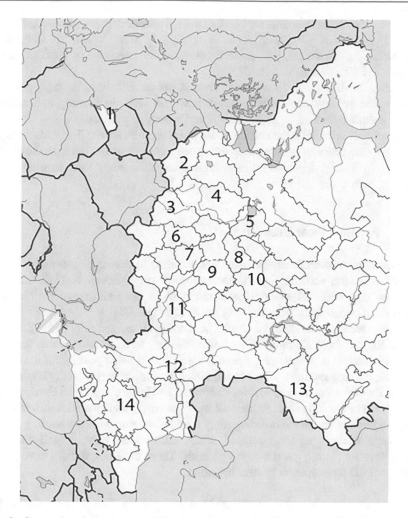

Map 3 Geography of deer parks and deer farms of the European part of Russia

Table 16 The numbers of the main species of deer held in hunting parks in the Russian Federation at the end of each reporting year (according to data from the Rosstat https://rosstat.gov.ru)

Species	2010	2011	2012	2013	2014	2015	2016	2017	2018	2019
Red deer	598	1232	1720	2149	2828	3605	4736	6020	6644	7735
Sika deer	1552	1645	2378	2583	3139	3950	4394	4667	5015	5497
Fallow deer	318	699	976	1126	1364	1705	1902	2381	2873	3239
Roe deer	26	83	124	301	343	487	496	707	841	967

There are no legal standards for keeping the number of deer per hectare in the deer parks. It is forbidden to keep together species of deer, which are capable of hybridizing. In this regard, it is worth mentioning the existence of the opinion that red deer and sika deer form hybrids.

Feeding in deer parks in winter is carried out in full: hay, haylage, grain. In summer, the amount of feed is reduced, and deer feed on existing cultivated or natural pastures, feeding on grass, shrubs and woody vegetation.

For animals drinking large and small natural reservoirs are mainly used. Rarely, if necessary, an artificial reservoir is equipped, or automatic drinking-bowls are installed. In winter, depending on the region, animals satisfy their requirement of drinking, both water and snow.

4.2   Farmed Deer Breeding

The history of modern intensive farm deer breeding in Russia begins around 2010, and the first farm according to the technological principles of New Zealand and European deer breeders was designed and built in the Kaliningrad region by an experienced expert from Poland—Bartlomiej Dmuchowski.

Today, interest in modern deer farms in Russia is localized only in the European part of Russia: from Kaliningrad to Bashkiria, and from Karelia to the Stavropol Territory. Interest in breeding European red deer was the first priority for the owners of private hunting grounds. Why? In the 90s, a stratum of very wealthy hunters was formed, mainly these were residents of Moscow and large cities in the European part of Russia. They could afford to hunt all over the world and saw a lot of interesting things in hunting and game breeding, they could acquire and keep a hunting ground of a large territory and agricultural lands, as well as publish well-illustrated magazines about hunting and hunting grounds. These three possibilities have played in favor of the development of deer breeding:

- The hunters have seen and appreciated the beauty of the European red deer and its antlers, and decided to keep them in their hunting deer parks.
- They learned that the European red deer can be easily bred on the farm.
- These hunters had their own hunting grounds and enclosures of a large area, as well as agricultural lands.
- In expensive hunting magazines they told everyone, who was interested, that it is possible to have in personal property a beautiful big game—the European red deer.

Through publications in magazines and personal communication with those interested in red deer breeding, a great contribution to the popularization of intensive deer breeding in Russia made the scientist and enthusiast Thomas Landette Castilejos (Spain) (Landete-castillejos 2012, 2013). As the founder of IDUBA (International Deer &Wild Ungulate Breeders Association), Thomas came to

Russia several times with a view to involving as many wealthy hunters as possible—owners of hunting areas in deer breeding.

These main factors marked the beginning of a steady interest in farmed deer breeding among owners of hunting areas, in agricultural lands in the zones where conventional farming can be difficult in the center and north of the European part of Russia. The development of this interest was facilitated by:

– The novelty of the first deer farms, publications and TV programs about them, which attracted the attention of a large number of hunting grounds and agricultural producers in the European part of Russia, who were looking for alternatives to traditional farming methods in market conditions
– The opportunity to communicate with foreign colleagues within the events of FEDFA, IDUBA, the International Deer Congress, via the Internet and through publications in magazines

The most active entrepreneurs and specialists establish sustainable business linkages for the development of technologies for keeping and breeding of red deer, European fallow deer, organizing the supply of deer from Europe to Russia, satisfying the growing demand for animals of these species for breeding in deer parks and on farms. The studied technologies of keeping and breeding of European red deer and fallow deer in the conditions of farms in New Zealand, Australia, European farms are actively adapting to the climatic and other features of Russia and are used for the design and construction of modern farms for breeding of European red deer, Siberian elk, sika deer.

Despite a strong interest in deer breeding, a significant impact on the growth of the number of farms is a lack of wide demand for deer breeding products, primarily meat. Currently, it is most profitable for deer breeders to sell live animals of different ages to newly created deer parks and farms.

Since 2017, the All-Russian conference "Game breeding" is being held in Russia every year, with the support of the Ministry of Natural Resources and Environment of the Russian Federation, which brings together government officials, hunters, farmers, zootechnicians, veterinarians. This conference is of great interest among agricultural land owners, agricultural companies, private and state hunting grounds. Following these conferences, the Russian Game Breeders Association was created in 2019, it unites 19 members for 2021. Their farms have in total: European red deer—3908, Siberian elk—750, sika deer—1065, European fallow deer – 1949 heads.

For 2021, 21 deer farms using intensive technology are located in 14 regions of Russia (Map 3). This small figure shows that today the economic benefit of deer breeding on the farm is low, but the strong interest in deer breeding shows perspective. Individual farms are interested and experimenting with artificial insemination technologies for breeding of European red deer and Siberian elk.

5   Conclusion

Deer breeding as a type of business is developing quite successfully in Russia and has now reached such a level that companies specializing in the construction of deer parks and deer farms from the design stage to the complete construction of all structures and fencing have begun to appear.

On the initiative of the owners of deer parks and deer farms, the Russian Game Breeders Association was established in 2019 (https://rugba.ru/), which became a member of FEDFA (Federation of European Deer Farmers Associations). The main objectives of the Russian Game Breeders Association are the:

- Creation of legal and economic conditions for the production, processing, and development of the market of game breeding products, an increase of competitiveness and improving the quality of game breeding products on the territory of the Russian Federation.
- Development and maintenance of game breeding standards, based on the use of modern technologies, the need for environmentally friendly products, the effective use of agricultural and other lands, as well as the principle of biodiversity conservation
- Representation and defence of the rights and interests of Association members in state and municipal authorities (local governments) at various levels, public and other organizations, including international ones
- Coordination of business activities of Association members
- Resolution of disputes and conflicts arising between the members of the Association

References

Aliskerov SV (2012) Zveri v volerah. Voler - dolgosrochnyj proekt. Nacionalnyj ohotnichij zhurnal OHOTA, №7, Moskva

Altajskij kraj. APK: istoriya i perspektivy (2012) Barnaul

Balanov IM (2016) Severnoe olenevodstvo Yakutii. https://arktika.sakha.gov.ru/news/front/view/id/2649084. Accessed 21 Mar 2021

Baskin LM (1970) Severnyj olen: ekologiya i povedenie. Nauka, Moskva

Belenyuk DN, Belenyuk NN (2019) Opyt sozdaniya maralovodcheskogo pitomnika s celyu vosstanovleniya chislennosti populyacii blagorodnogo olenya v Krasnoyarskom krae. Vestnik of the Krasnoyarskogo gos agranago un-ta 2:103–110

Bogachyov AS, Bogachev SA (1993) O syre narodnoj mediciny – zhelchi, pantah, zhirah i drugom. PGSHA, Ussurijsk, p 112

Borozdin EK, Zabrodin VA, Vagin AS (1990) Severnoe olenevodstvo. Agropromizdat, Leningrad

Brekhman II, Dobryakov Y, Taneeva AI (1969) Biologicheskaya aktivnost pantov pyatnistogo olenya i drugih vidov olenej. Izvestiya SO AN SSSR Seriya Biologicheskaya. 2:112–115

Chikalev A, Petruseva N, Bessonova N (2014) Pantovoe olenevodstvo: Uchebnik. Infra-M, KURS, Moskva

Cygankov VV, Ivanov AA, Shandala NK (2000) Rol mineralnoj komponenty pishchevoj dobavki "Cygapan" v radiozashchite. tezisy dokladov IV Mezhdunarodnogo simpoziuma Biologicheski aktivnye dobavki k pishche: XXI vek, Sankt-Peterburg, 22–24 maya 2000, s 90–92

Dalisova NA, Rozhkova AV, Stepanova EV (2019) Eksport produkcii maralovodstva i pantovogo olenevodstva sibirskih regionov. Socialno-ekonomicheskij i gumanitarnyj zhurnal Krasnoyarskogo GAU 1:35–45

Dobryakov YI, Brekhman II (1972) Biologicheskaya aktivnost pantov razlichnyh vidov olenej. lekarstvennye sredstva Dalnego Vostoka. DVNC AN SSSR, Vladivostok 11:197–201

Dormidontov R.V. (1968) Pantovoe olenevodstvo na Severnom Kavkaze. Krolikovodstvo i zverovodstvo. 6:32–34

Druri IV, Mityushev PV (1963) Olenevodstvo. Selhozizdat, Moskva, Leningrad

Drury IV (1955) Olenevodstvo. Gosudarstvennoye izdatel'stvo sel'skohozyaystvennoy literatury, Moskva, Leningrad

Eger VN, Deev NG (1992) Sostoyanie i perspektiva razvitiya pantovogo olenevodstva v RSFSR. sb. nauch. tr. CNILPO, Novosibirsk, pp 3–6

Egorov AN (2004) Istoriya stanovleniya nacionalnogo parka «Zavidovo». http://www.konakovo.org/goskompleks-zavidovo/istoriya-stanovleniya/. Accessed 20 Mar 2021

Frolov NA (2012) Aleksej Stepanovich Popov. Ekonomicheskoe rassledovanie sudby krupnejshego maralovoda dorevolyucionnogo Altaya. Izd-vo Altajskogo un-ta, Barnaul

Frolov NA (2014) U istokov pantovogo olenevodstva Rossii Bijsk: Izd-vo Altajskogo un-ta, Barnaul

Galkin VS (1964) Issledovanie i razrabotka racionalizacii soderzhaniya i kormleniya pantovyh olenej v sovhozah Gornogo Altaya. Dissertaciya avtoreferat Moskva

Galkin VS (1983) Pantovoe olenevodstvo: dostizheniya i perspektivy (k 50-letiyu CNILPO). Ohota i ohotniche hozyajstvo 11:1–2

Gladilov IA (1962) Issledovanie pantov losya. Ohota i ohotniche hozyajstvo 5:32–34

Gosudarstvennyj reestr selekcionnyh dostizhenij, dopushchennyh k ispol'zovaniyu. Tom 2 «Porody zhivotnyh» (oficial'noe izdanie) (2019) FGBNU «Rosinformagroteh», Moskva

Gurvich IS (1966) Etnicheskaya istoriya Severo-vostoka Sibiri, Moskva

Itogi Vserossiyskoy sel'skohozyaystvennoy perepisi 2006 goda (2008) IITS «Statistika Rossii», Moskva. ISBN 978-5-902339-66-3

Itogi Vserossiyskoy sel'skohozyaystvennoy perepisi 2016 goda (2018) IITS «Statistika Rossii», Moskva. ISBN 978-5-4269-0066-0

Kabelchuk BV, Direganov EV, Lysenko IO, Verzun TG (2013) Ekologiya, razvedenie i soderzhanie pyatnistogo i blagorodnogo olenej v poluvolnyh usloviyah v Stavropolskom krae. Argus, Stavropol

Kambalin VS (2015) Ocenka razvitiya i perspektiv maralovodstva v Sibirskom federalnom okruge. sovremennye problemy ohotnich'ego hozyajstva Kazahstana i sopredel'nyh stran. Mat mezhdunarod nauch konf, RGP "In-t zoologii", Almaty 11–12 marta 2014, pp 550–559

Kostin AI (1975) Sostoyanie pantovogo olenevodstva i perspektivy ego razvitiya v sovhozah Primorskogo kraya. sb nauch tr CNILPO. Barnaul, 20–22

Kuznecov BA (1938) Izyubrennyj promysel i razvedenie izyubrya v Zabajkalskoj oblasti. Irkutsk

Landete-castillejos T (2012) Na puti k sovmestnoj evropejskoj strategii menedzhmenta olenej. Nacional'nyj ohotnichij zhurnal «Ohota», Moskva, 11–12:24–28

Landete-castillejos T (2013) Chastnyj menedzhment v Rossii – istoriya budushchego. Nacional'nyj ohotnichij zhurnal «Ohota», Moskva, 6–8

Lunicyn VG (2004) Pantovoe olenevodstvo Rossii. RASHN Sib otd-nie VNIIPO, Barnaul

Lunicyn VG, Borisov NP (2012) Pantovoe olenevodstvo Rossii: RASHN VNIIPO. Azbuka, Barnaul

Maralovodstvo i pantovoe olenevodstvo na Altae (2012) Altajskij kraj. APK: istoriya i perspektivy. Barnaul, 116–119

Maralovodstvo: osobennosti otrasli i razvedenie zhivotnyh kak biznes (2018). http://технологи я-бизнеса.рф/biznes-plany-rukovodstva/zoobiznes/biznes-na-pantakh-otkryvaem-olenyu– fermu. Accessed 04 Jul 2021

Matyushin AI (1954) Akklimatizaciya pyatnistyh olenej pod Moskvoj V: sb nauch tr Moskovskogo pushno-mekhovogo instituta. Moskva, 5:114–119

Mejsel NM, Grigoreva GA (1933) Stroenie pantov pyatnistogo olenya. vestnik DVF AN SSSR. Moskva, 11–3, pp 115–121

Minselhoz RA (2018). http://mcx-altai.ru/novosti/29-publikatsii/1197-vii-vsemirnyj-kongress- olenevodov-2018. Accessed 07 Jul 2021

Mityushov PV, Lyubimov MP, Novikov VK (1950) Pantovoe olenevodstvo i bolezni pantovyh olenej. Moskva

Muhachyov AD (1990) Olenevodstvo. VO Agropromizdat, Moskva

Muhachyov AD, Lunicyn VG, Pluzhnikov NV (2004) Olenevodstvo. Bolshaya rossijskaya enciklopediya. https://bigenc.ru/agriculture/text/2684640. Accessed 04 Jul 2021

Novikova NI, Funk DA (eds) (2012) Sever i severyane. Sovremennoe polozhenie korennyh malochislennyh narodov Severa, Sibiri i Dalnego Vostoka Rossii. Izdanie IEA RAN, Moskva

Ovechkina N.N., Kozlov V.M. (2016) O plotnosti naseleniya kopytnyh v volerah. Agrarnaya nauka Evro-Severo-Vostoka. Moskva, 2 (51) 58-62

Pavlenko SM (1936) Pantokrin (hualukrin). Moskva-Leningrad, 2, pp 3–9

Pavlov PM (ed) (2013) Federalnye gosudarstvennye byudzhetnye uchrezhdeniya Minprirody Rossii – gosudarstvennye opytnye ohotnich'i hozyajstva i «Centrohotkontrol'» (informacionnyj obzor). Moskva. http://www.ohotcontrol.ru/documents/publication/2013. Accessed 04 Jul 2021

Pomishin SB (1990) Proiskhozhdenie olenevodstva i domestikaciya severnogo olenya. Nauka, Moskva

Rosstat. https://rosstat.gov.ru/storage/mediabank/ohota4.xls

Silaev AB (1971) Biologicheski aktivnye veshchestva pantov i perspektivy ih prakticheskogo primeneniya. pantovoe olenevodstvo. Gorno-Altajsk 3:s 103-110

Smirnov YA (1968) Akklimatizaciya pyatnistyh olenej na Altae. Avtoreferat dissertacii kand biol nauk, Moskva

Syrovatskij DI, Neustroev MP (2007) Sovremennoe sostoyanie i perspektivy razvitiya severnogo olenevodstva v Rossii: Rekomendacii. FGNU «Rosinformagrotekh», Moskva

Tihonin IY (1932) Shebalinskij gosudarstvennyj maralnik: materialy ekspedicii po izucheniyu maralovodchestva severnogo Altaya. OGIZ, Moskva, Irkutsk

Timofeeva AM, Kolli EA (1936) K izucheniyu himii pantokrina. V sbornike nauchnyh trudov CNILPO. Moskva-Leningrad 2:16–21

Tevi AS, Zhuravleva VE (1969) Materialy po izucheniyu lekarstvennoj cennosti neokostenevshih rogov losya. Pantokrin, sbornik nauchnyh rabot, vol 2. Laboratoriya pantovogo olenevodstva, Moskva, pp 132–139

Zabolotskih YS (2010) Pantovoe olenevodstvo. Uchebnoe posobie dlya studentov vysshih uchebnyh zavedenij, Vyatskaya GSHA, Kirov

Legislative Instruments

Zakon Zabajkalskogo kraya ot 29 marta 2010 g. № 354-ZZK «O gosudarstvennoj podderzhke tradicionnyh vidov hozyajstvennoj deyatelnosti i tradicionnyh promyslov korennyh malochislennyh narodov Severa, Sibiri i Dalnego Vostoka v Zabajkalskom krae» (2010). Zabajkalskij rabochij № 55 (in press)

Zakon Krasnoyarskogo kraya ot 25 noyabrya 2010 g. № 11-5343 «O zashchite iskonnoj sredy obitaniya i tradicionnogo obraza zhizni korennyh malochislennyh narodov Krasnoyarskogo kraya» (2010) Vedomosti vysshih organov gosudarstvennoj vlasti Krasnoyarskogo kraya № 62 (433). Krasnoyarsk (in press)

Zakon Respubliki Saha (Yakutiya) ot 25 iyunya 1997 goda № 179-I «O severnom domashnem olenevodstve» (s izmeneniyami na 27 maya 2020 goda) (v red. Zakonov Respubliki Saha (Yakutiya) (2020). Yakutskie vedomosti (in press)

Zakon Respubliki Saha (Yakutiya) ot 13 iyulya 2006 goda «O territoriyah tradicionnogo prirodopolzovaniya i tradicionnoj hozyajstvennoj deyatel'nosti korennyh malochislennyh narodov Severa Respubliki Saha (Yakutii) (2006). Yakutskie vedomosti № 58 (in press)

Zakon Hanty-Mansijskogo avtonomnogo okruga ot 28 dekabrya 2006 goda «O territoriyah tradicionnogo prirodopolzovaniya korennyh malochislennyh narodov Severa regionalnogo znacheniya v Hanty-Mansijskom avtonomnom okruge — Yugre» (2006). V: Sobranie zakonodatelstva Hanty-Mansijskogo avtonomnogo okruga – Yugry. Hanty-Mansijsk, № 12 s 1488

Zakon Yamalo-Neneckogo avtonomnogo okruga ot 5 maya 2010 g. № 52 (2010) ZAO «O territoriyah tradicionnogo prirodopolzovaniya regionalnogo znacheniya v Yamalo-Neneckom avtonomnom okruge». Krasnyj Sever, № 54 (specvypusk)

Part V

Management Systems and Diseases of Farmed Deer in China

The Chinese Deer Farming Industry

John Fletcher

Abstract

This short chapter provides a brief account of the history of deer enclosures in China, their current status and management systems. Deer have been exploited as a source of traditional Chinese medicine throughout history and this remains the reason for enclosing deer today. The possible reasons why deer and specifically their growing antlers are so highly valued in Chinese traditions are discussed and the ways in which deer are farmed in China for their products. The current status of musk deer kept for the production of musk is briefly discussed.

Keywords

Chinese traditional medicine · Chinese deer farming · Musk deer · Artificial insemination · Sympathetic medicine · Confucianism · Homeopathy

1 Historical Background

Deer were enclosed in China as early as the Shang dynasty in about the eleventh century BC, and we also know that in the thirteenth and fourteenth centuries AD during the Mongol rule in the Yuan dynasty deer were apparently embarked in their thousands (Yu 2004). This was almost certainly in extensive parks where hunting was the chief objective, for such hunting parks served to enhance the reputation of rulers and to provide practice for military campaigns and were widespread in Asia (Allsen 2006; Fletcher 2011).

The use of deer products as medicines also has a long history. Silk scrolls from a Han dynasty tomb in China dated to 168 BC indicate that antler was being used at

J. Fletcher (✉)
Reediehill Deer Farm, Auchtermuchty, UK
e-mail: johnfletcher@deervet.com

that time (Haigh and Hudson 1993). The traditional Chinese medical pharmacopoeia has included growing antlers as one of the two most important medicines for many centuries; the other vital ingredient being ginseng. Wild deer were hunted to provide both hard and growing (velvet) antler and other deer tissues: sinews, pizzles (male genitalia), tails, etc., as well as embryos and blood for many hundreds and probably thousands of years (see Figs. 1 and 2). Enclosed deer within parks are thought to have provided deer products for elite high-ranking officials from an early date (Yu 2004).

By the early twentieth century, some deer were being confined specifically to provide a source of traditional medicines. A fictionalised account of the capture of wild sika in Eastern Mongolia entitled *'Jen Sheng – the root of life'* by Mikhail Prishvin recounts the establishment of one of those early farms and in a foreword to the book Julian Huxley emphasises how this represented the beginnings of a new domestication (Prishvin 1936). Prishvin describes the use of a drop floor crush to restrain the stags and permit the removal of their antlers. China was thus probably the first nation in the world to adopt intensive deer management, keeping their deer in paved and walled pens usually providing access to adjacent deer houses then as they still do for the most part today (see Fig. 3a and b).

With the establishment of the People's Republic of China in 1949, deer were taken from the few private deer farms and gathered into the State's North Eastern Sika Farm (Yu 2004). The industry grew from that point, first in the north of China with sika and during the 1960s sika were also introduced to the central and south of China and a wapiti breeding station was established in the Xinjiang area (Yu 2004).

Fig. 1 Slicing dried velvet antler for market

2 Present Status and Products

The Chinese deer farming industry grew and by 2011 involved around 1 million deer (Wang 2011a, b). Most of these deer are still kept in paved and walled enclosures with forage often gathered and fed by hand as a chief part of their ration. These deer are kept almost entirely for the production of antler velvet and other by-products for the traditional Chinese medicine trade (Zheng 2011) (Fig. 2).

Experiments have been conducted in many parts of the world to analyse what beneficial effects growing antler might have on human health and there is a considerable literature on the subject. Reported pharmacological benefits, chiefly in laboratory animals, include gonadotrophic, haematopoietic, and hypotensive effects, as well as protection from stress and shock, improved recovery following liver damage and whiplash injury, and in muscles after exercise, and antlers are also reported to have stimulated growth in young animals and the retardation of ageing (see, for example, Sim et al. 2001; Suttie et al. 2004).

It is perhaps tempting for a sceptical westerner to think that the use of antler is a classic example of the application of 'sympathetic medicine' which was so historically popular in seventeenth century Europe, and which still forms the basis of beliefs in homeopathy. The astonishingly rapid growth and regeneration of an entire organ, the antler, is unique within mammals and perhaps because it is occurring only in the male stag, it is easy to comprehend how it might have been linked to

Fig. 2 Deer by-products include not only antler but almost all other organs including sinews, tails and even hearts

masculinity as well as regeneration, vigour, and strength: all qualities associated with Yang. The growth of antlers in the summer strengthens the association of growing antler with Yang in Confucianist philosophy. Traditionally in the balance of the two primordial cosmic forces of Yin and Yang, antler is supposed to provide a Yang influence and was historically associated with stimulating male potency. Of course, the antler in fact grows at a time when testosterone levels are at their annual nadir when they become almost unmeasurably low.

From the belief in antler as a universal remedy it is perhaps unsurprising that virtually all parts of deer are also associated with beneficial medicine. From the Chinese deer farms come blood, especially collected when the growing antlers are removed, as well as if a deer is killed, also the tail, the male genitalia, the sinews of the leg, embryos from pregnant female deer, as well as still births, bones, skins, etc. All these products find a place in the market as elements within traditional Chinese medicine. The development of pills and liquid medicines derived from these products is taking an increasingly large role. Many of these products are now being marketed as 'nutroceuticals' and have a growing body of users in western society.

Venison is increasingly finding a market partly in response to marketing efforts from New Zealand but continues to be only a very small part of the Chinese deer industry.

3   Musk

The traditional Chinese medicine trade still values the musk secretions from the preputial glands of musk deer very highly and continues to keep musk deer in captivity in order to supply musk which is also traded from the diminishing numbers of wild musk deer killed in Asia. According to a non-governmental organisation, TRAFFIC, which specialises in investigating the trade in wild animals and plants, the population of wild musk deer in China had declined from more than 3 million in the 1950s to only 200,000 to 300,000 by the 1990s (Parry Jones www.TRAFFIC. org). By 2017, more than 20,000 musk deer were in captivity in China being kept to provide musk (www.xinhuanet.com, 22nd November 2018) As 'musk deer' are no longer classified as cervids we will not consider them further here.

4   Velvet Antler Production

Most deer farmed in China re kept within walled enclosures and all their feed is cut and carried (see Figs. 3a and b). During the last 15 years, some deer have been farmed using New Zealand systems on grazed pasture with others being enclosed in small paddocks and the feed supplied but most of these systems also incorporate a level of housing (Wang 2011a, b). These are rather labour-intensive systems with approximately one labour unit for 80–100 sika or 60–80 wapiti (Wang 2011a, b).

Artificial insemination is used very widely and there is growing use of embryo transfer so that antler weights are likely to continue to increase.

Fig. 3 (**a, b**) Most deer are kept within walled enclosures with access to shelter

The procedures adopted for the removal of growing antlers has been fully described earlier in this volume. The assessment of the optimum time for cutting the antlers is important as grading of the product and the resulting prices paid are complex. In sika deer farmed in China the antlers are removed when either two tines have been grown at 40–45 days after casting of the 'buttons' which are the bases of the antler left after the antlers were cut off in the previous year, or three tines (65–70 days) (see Figs. 4a and 4b). Chinese annual production of velvet is around 400 tonnes fresh weight with the largest wapiti antler weighing 30 kg and the largest sika antler 17.2 kg (Wang 2011a, b). Most deer farms have their own resident veterinarian and also have adjacent processing factories.

Fig. 4 (**a, b**) Antlers are cut at two (Fig. 4a) or three (Fig. 4b) tine stage

5   Diseases

The most important causes of economic loss on Chinese deer farms are stress and distress including fear and distress during transport, vitamin and trace element deficiencies including copper and cobalt deficiency, filariasis, and also brucellosis and tuberculosis (Wang 2011a, b). High neonatal mortality levels are common within the housed deer but this is not a total loss since the carcases of young deer have value as taxidermy specimens (see Fig. 5).

Fig. 5 Young deer have value as taxidermy specimens

References

Allsen TT (2006) The Royal Hunt in Eurasian history. University of Pennsylvania Press, Penn

Fletcher TJ (2011) Gardens of earthly delight – the history of deer parks. Windgather Press, Oxford

Haigh JC, Hudson RJ (1993) Farming wapiti and red deer. Mosby-Year Book, St Louis, MO

Prishvin M (1936) Jen Sheng: the root of life. English version by George Walton and Philip Gibbons. Andrew Melrose, London

Sim JS, Sunwoo HH, Hudson RJ, Jeon BT (eds) (2001) Antler science and product technology. ASPTRC, Edmonton, AB

Suttie JM, Haines SR, Li C (2004) Advances in antler science and product technology. Gold Mountain, Deer Industry New Zealand and AgResearch, Hamilton, New Zealand

Wang Q (2011a) Deer farming in China. In: Proceedings of the Deer Branch of the NZVA, Queenstown, New Zealand, pp 33–34. www.sciquest.org.nz/deer

Wang Q (2011b) Deer diseases in China. In: Proceedings of the Deer Branch of the NZVA, Queenstown, New Zealand, pp 37–38. www.sciquest.org.nz/deer

Yu Z (2004) Development of the Chinese deer farming industry and the market for deer. In: Suttie JM, Haines SR, Li C (eds) Advances in antler science and product technology. Deer Industry New Zealand and AgResearch, Hamilton, New Zealand, pp 233–238

Zheng B (2011) Deer products, processing and use in China. In: Proceedings of the Deer Branch of the NZVA, Queenstown, New Zealand, p 35. www.sciquest.org.nz/deer

Part VI

Management Systems and Diseases of Reindeer

Husbandry and Diseases of Semi-Domesticated Eurasian Tundra Reindeer in Fennoscandia

Morten Tryland, Ingebjørg Helena Nymo, Javier Sánchez Romano, and Jan Åge Riseth

Abstract

This chapter presents the most relevant diseases of semi-domesticated Eurasian tundra reindeer (*Rangifer t. tarandus*) in a historical and contemporary perspective. Human relation to these animals is very ancient, and disease history is context dependent. Historical epizootics, such as anthrax and the "Reindeer pest," could be devastating for reindeer and man, but not always well documented in older sources. Others, such as necrobacillosis, were known to be associated with high animal density and corralling due to milking practice. Contemporary diseases, such as rumen acidosis and ruminal tympani, are usually connected to feeding, whereas diseases like infectious keratoconjunctivitis and contagious ecthyma are associated with high animal density and stress. The ongoing and rapid transition from traditional reindeer herding to increased husbandry and feeding may lead to an altered disease prevalence, possibly introducing new pathogens and altering the epidemiology and appearance of "old" pathogens. Chronic wasting disease is also a serious threat, but so far not detected in

M. Tryland (✉)
Department of Arctic and Marine Biology, UiT The Arctic University of Norway, Tromsø, Norway

Department of Forestry and Wildlife Management, Inland Norway University of Applied Sciences, Koppang, Norway
e-mail: morten.tryland@inn.no

I. H. Nymo
Research Food Safety and Animal Health, The Norwegian Veterinary Institute, Tromsø, Norway

J. Sánchez Romano
Department of Medical Biology, UiT The Arctic University of Norway, Tromsø, Norway

J. Å. Riseth
Department of Social Sciences, NORCE Norwegian Research Center, Narvik, Norway

Finnmark Land Tribunal, Tromsø, Norway

J. Fletcher (ed.), *The Management of Enclosed and Domesticated Deer*,
https://doi.org/10.1007/978-3-031-05386-3_19

413

semi-domesticated reindeer. In addition, climatic changes may contribute to altered ecosystems, including new distribution patterns for insects, which can be vectors for reindeer pathogens. Changing conditions for reindeer herding create new challenges for both reindeer health and the sustainability of the traditional reindeer herding industry.

Keywords

Climate change · Fennoscandia · Infectious diseases · Reindeer feeding · Reindeer herding · Management

1 Introduction

1.1 In the Beginning: From Reindeer Hunting to Herding

Reindeer and caribou (*Rangifer tarandus*) are distributed throughout the northern Holarctic region as a typical representative of the large mammalian fauna of this region. The species has several subspecies. The Eurasian tundra reindeer (*Rangifer tarandus tarandus*) is distributed across the tundra region and mountain areas of Eurasia, both as wild and semi-domesticated, altogether three to four million animals (Røed et al. 2018). In this chapter, we will concentrate on Fennoscandia (i.e., Norway, Sweden, Finland, and the Kola Peninsula, Russia) and the semi-domesticated variant (Fig. 1), which currently dominates this region with about 685,000 animals, against about 25,000 wild reindeer in southern Norway.

Eurasian tundra reindeer have been exploited for food and other subsistence since the last glaciation (Kofinas et al. 2013). In Fennoscandia, large herds of wild reindeer migrated between coast and inland twice a year and were of high importance for all residents. Wild reindeer were hunted by trapping pits, guiding fences and shooting with bows and arrows. This led to the development of large systems of enclosures, fences, and pitfall traps, typically in Varanger, northeastern Norway (Nieminen 2018). Archeological evidence shows that a kind of reindeer keeping started as early as 500 AD (Mulk 1988). An early example of domesticated reindeer along the coast of Troms, Northern Norway, was reported at the end of the 800s by the North-Norwegian chieftain Ottar (Ohthere) when he visited King Alfred the Great in England in 890. Ottar told that he, in addition to his herd, had a few tame reindeer, which could be decoys for catching wild reindeer or transport animals (Bately and Englert 2007; Bjørklund 2013). Reindeer have properties that eases domestication; docility, a trusting nature and allowing themselves to be handled and trained. Reindeer is the only deer species in the world that is domesticated (Nieminen 2018). Reindeer domestication seems to have been a complex transition process that lasted for centuries. The South Sámi case is illustrating:

Fig. 1 Semi-domesticated Eurasian tundra reindeer, Troms, Norway. (Photo: Morten Tryland)

As long as the wild reindeer herds were large and the resources rich, there was no point in having large, domesticated reindeer herds. It was not only more labor intensive, but also almost an impossible combination. But when wild reindeer herds shrink . . . there was a need to change strategy. Larger herds that could ensure both milk production and meat needs, was an ingenious solution . . . .

The readiness for such a change already existed in the days of the hunting community: We know that the Sámi had domestic reindeer as decoy for use in wild reindeer hunting, that they probably also had some females that were milked, and that reindeer were used for transport at least in the early Middle Ages. In addition, the Sámi hunting community had technology and methods for handling wild reindeer that have clear parallels to today's reindeer husbandry (translated from Norwegian; Fjellheim 1999, p. 58).

Across Northern Eurasia, indigenous peoples of the North continued hunting wild reindeer for meat and kept a limited number of domesticated decoy and transport reindeer for several centuries, in some regions through a millennium (Istomin and Dwyer 2010; Bjørklund 2013; Anderson 2014; Stépanoff 2017). Further, from the seventeenth and eighteenth centuries, the domestic herds of several Arctic peoples, the Sámi included, began to grow surprisingly quickly. This development coincided with the Little Ice Age, a period of cool climate conducive to improved health and fertility of reindeer (Krupnik 1993). For improved understanding of the transition process, Bjørklund (2013) calculated the household need of reindeer for subsistence to live off reindeer alone. He found that the *"tipping point to qualify as pastoralism...seems to have been around 200–250 in a winter herd"* (ibid.). Historical sources substantiate that family herd sizes up to the nineteenth century hardly exceeded 100 animals. Another fact is that 30–40 animals also is considered the maximum level a nuclear family was able to herd and manage (Bjørklund 2013). These findings have several implications.

First, reindeer herding in the seventeenth and eighteenth century needs to have been part of a complex adaptation that also involves hunting, fishing, and trading, which is also confirmed by archival data and historical analyses (Hansen 2008). Second, the work requirement might have been the background for the emergence of the Sámi herding siida, a group of several households collectively sharing the herding work (Bjørklund 2013; Paine 1972). Third, with larger herds, the social processes among the animals became determining, as reindeer like to gather in large herds in open landscape. The wider effect of the latter implication is that open tundra and mountains became the main pastoral niche for large-scale herding (Stépanoff 2017). For Fennoscandia, this implies that out of three major types of reindeer herding a *tundra type* (I) with intermediate to long seasonal migrations became dominant in Norway and Sweden and historically also in northernmost Finland. Further, a *coastal-oriented type* (II) with local seasonal migrations is also found in Norway, while a *taiga type* (III) based on small-scale mosaic in forested landscapes dominates Finland and is also found in Sweden (Riseth et al. 2018), cf. Fig. 2.

In Finland, settlers and peasants adopted reindeer herding from the Sámi and became organized as a livelihood connected to peasantry already in the eighteenth century (Nieminen 2018). Moreover, the Nordic states' internal colonization processes got momentum from the late nineteenth century on. Bilateral and unilateral legislation both closed national borders for transnational reindeer herding and restricted Sámi rights and adopted a top-down control of reindeer herding. The main motivation was promotion of non-Sámi farming settlers in Sámi areas. It was justified by Social-Darwinistic and nationalistic ideologies (Pedersen 2001; Päiviö 2007; Brännlund 2015; Riseth et al. 2016). The effects were both dramatic and extensive and included major relocations, by several hundred kilometers at most, of at least 70 families and the range of 20,000 to 25,000 reindeer, maybe more, with repercussions for several generations of Sámi herders in Northern and Middle Fennoscandia. The impacts included reducing the scope of the tundra type (I) and shortening down migration patterns while increasing the extent of the two other types, partly beyond their ecological ranges (Riseth and Oksanen 2007). In the nineteenth century, pastoral herds continued to be small scale with an intensive reindeer husbandry economy with small herds for meat and milk (Falkenberg 1985; Vorren 1998). One account of the role of milking is:

> In the intensive reindeer husbandry, the milking played a crucial role. In older times the does were milked twice during the whole summer half, just to the slaughtering time in mid-October. Once or twice a day they were driven to the turf hut, to some snowdrift, or in the autumn to some specific pen, where they were tied and milked. . . Reindeer milk was an utmost important part of Sámi food. . . even in wintertime, either prepared to cheese or conserved by freezing or drying etc. (translated from Swedish; Hultblad (1936, p. 20).

The livelihood was diverse and could also include cattle, sheep, or goats on a permanent or temporal basis (Hultblad 1936; Hansen 2008). Family networks and

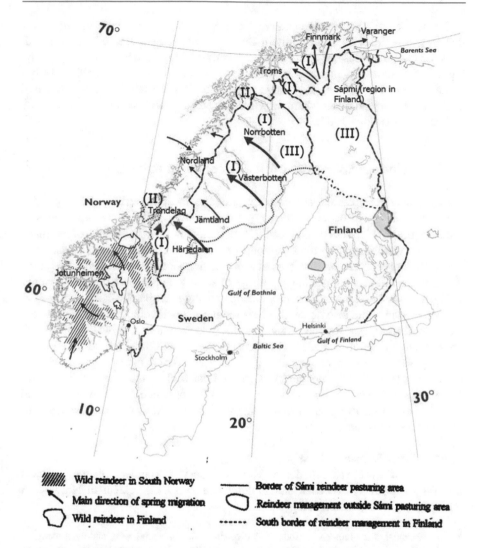

Fig. 2 A map of Norway, Sweden and Finland, depicting wild reindeer and reindeer herding. Wild reindeer populations exist in the shaded areas of southern Norway. Arrows are indicating the direction of spring migration of reindeer. The three types of reindeer herding (I, II and III) are indicated (see text for details)

trade contacts contributed to flexibility to the socio-economic adaptation (Brännlund 2015). The external pressure due to agricultural expansion, infrastructure development, mines, etc., decreased both pasture areas and mobility possibilities. Cumulative effects of these encroachments, border closures and stricter governmental control made the conditions for intensive herding increasingly difficult (Hultblad 1936; Brännlund 2015; Riseth et al. 2016).

From the mid-eighteenth century, there were large outbreaks of infectious diseases in reindeer herds in northern Sweden, with large losses of reindeer and, more or less, a collapse of entire reindeer herding communities. At the end of the eighteenth century, the population of Torne Lappmark (Sweden) was reduced from about 1300 to 800 people, and many reindeer herders left reindeer husbandry and settled in Norwegian fjords where they lived as peasants with small farms combined with fisheries (Ruong 1982).

1.2   Reindeer Epizootics in Historical Perspectives

Today, infectious diseases are normally not the most important challenges for reindeer herding. Historical sources, however, are describing severe epizootics in semi-domesticated reindeer that could kill hundreds and thousands of animals, having a crucial impact on the livelihood for the Sámi communities. Diseases like the "Reindeer pest" and anthrax are not reported in Fennoscandia today, while some diseases are still present, and some pathogens seem to appear in a different manner today, such as *Fusobacterium necrophorum* causing necrobacillosis.

The examples and disease descriptions mentioned below indicate how important reindeer herding was for these small northern communities. It was a time with limited knowledge of diseases and infectious agents, and thus limited possibilities for prophylactic measures and treatment of diseases, often with a poor prognosis as a result. From the descriptions and reports, it is also clear that the techniques for sampling and laboratory investigations limited the outcome of the diagnostic and scientific approaches to gain understanding of the causative agents and the epidemiology of these diseases.   .

1.2.1   The "Reindeer Pest"

Outbreaks of what was called the "Reindeer pest" ("Reindeer plague," "*Pestis tarandi*") appeared in Sweden and Finland and to a lesser extent in Norway. A report from Sweden stated that:

> In 1821 most of the reindeer in Sodankylä died due to a devastating pest, and even many horses and sheep died of the same disease"... In 1831, The reindeer pest appeared in Sodankylä again. During the years 1750, 1751 and 1752, the same pest ravaged among the reindeer, and only a few were left alive. (translated from Swedish; Fellman 1906).

A Gram-positive, rod-shaped bacterium was isolated and believed to be *Clostridium septicum* (Horne 1915). This bacterium causes the acute and severe disease gas gangrene in people and ruminants. The bacterium is present in the environment worldwide but is also found in the gastrointestinal tract of animals, and its spores, a robust dormant stage, may survive in the soil for years (Underwood et al. 2015).

The outbreaks usually appeared during warm summers. The disease was transmitted rapidly within a herd and calves and young reindeer were most susceptible. Animals were often found dead with no previous disease signs, or they displayed coughing and thick foul-smelling flood from their nose, regional lymphadenopathy and swellings with gas under the skin (Nordkvist 1960; Qvigstad 1941). Outbreaks of the "Reindeer pest" were recorded in Karasjok (Finnmark, Norway) and in Jokkmokk (Norrbotten, Sweden) in 1896, but no further outbreaks were recorded (Horne 1915; Nordkvist 1960).

Descriptions of the disease and the diagnostic techniques often included cultivation of the bacterium under aerobic conditions (Horne 1915), however, *C. septicum* is known to require strict anaerobic growth condition. There is therefore doubt whether the "Reindeer pest" was caused by *C. septicum*, or if these outbreaks represented a different disease, possibly caused by a different pathogen (Josefsen et al. 2019).

1.2.2 Anthrax

Anthrax is a well-known and feared disease caused by the spore-forming Gram-positive rod-shaped bacterium *Bacillus anthracis*. It may cause severe disease and disease outbreaks in livestock, wildlife and humans. Reindeer may display tremors, running nose and eyes, dark spots in the oral mucosa, anorexia and strained respiration (Qvigstad 1941). Toward the end of the eighteenth century, anthrax outbreaks in reindeer were reported from Sweden:

> The disease (anthrax) was since 12-14 years ago (before 1754) unknown in Lule Lappmark (Sweden). Now, since it is highly contagious, it has during the past five years ravaged in Torne Lappmark and killed almost all the forest reindeer." (Translated from Swedish; Hoffberg 1788).
>
> In four years, more than 100 persons from Jukkasjärvi (Sweden) alone were forced to move, and most of them travelled to Norway to live from fishing. (Translated from Swedish; Gissler 1759).

As with the "Reindeer pest," the microbiological documentation of the anthrax outbreaks was poor, and whether *B. anthracis* in fact was the causative agent of these reported epizootics in Fennoscandia during the eighteenth century has been disputed (Nordkvist 1960; Josefsen et al. 2019). In contrast, Anthrax has been well documented in reindeer in Russia, where the disease killed as many as 100,000 reindeer during a single outbreak in 1911 as well as being fatal to some of the herders (Gainer and Oksanen 2012). Vaccination of reindeer in Russia started in 1928, and it is claimed that no outbreaks appeared in areas where vaccination was conducted (Nikolaevskii 1961). A recent (2016) outbreak among reindeer in Russia is discussed in Sect. 5.2 in this chapter.

1.2.3 Digital Necrobacillosis

Digital necrobacillosis in reindeer is a wound infection caused by the bacterium *F. necrophorum*, affecting the digits and the distal parts of the feet. The bacterium may also infect the mucosa of the mouth and other parts of the gastrointestinal tract (i.e., oral necrobacillosis). Digital necrobacillosis is characterized by swellings of the distal parts of the feet and lameness, usually affecting one foot only. The disease is called "slubbo" (i.e., "club") in Sámi, describing the club-shaped swelling of the foot.

The disease has a poor prognosis, and reindeer with digital necrobacillosis often die of starvation or predation or are euthanized for animal welfare reasons. The disease could affect hundreds or even thousands of reindeer. The first written report of digital necrobacillosis in semi-domesticated reindeer is almost 300 years old (von Linné 1732), with more thorough and classic descriptions presented by Horne (1898) and Bergman (1909). Digital necrobacillosis was associated with holding reindeer in fences or in restricted areas over time, to have easy access to draft and milking animals (Skjenneberg and Slagsvold 1968). A report from the Norwegian Sámi villages in 1903 stated:

> In the Sami villages, almost every forth reindeer was sick, some affected in the forelegs, some in the hindlegs, and some in more than one leg. Reindeer of all ages were affected. Many reindeer died from septicemia and many were taken by volves. (Translated from Norwegian; Anonymous 1905).

Already from the eighteenth century, most accounts of reindeer herders' understanding of digital necrobacillosis were relatively unison, and traditional knowledge of the disease seems in many aspects to be in accordance with veterinary science (Riseth et al. 2020). Interestingly, reindeer husbandry practices seem to have changed during that period, to reduce the risk of the disease, either to a more extensive type without milking or by establishing many milking sites, preferably on dry ground and with frequent rotational use (Ruong 1982; Drake 1918). If the herd was large, the use of the same milking pen (cf. Fig. 3) for 3–4 days was the limit. Even for small herds and after heavy rainstorms, they had to shift the milking site and clean the used pen very soon and before 14 days. Considered as good reindeer herders were those people who milked the reindeer far away from the dwelling site, since the latter could be contaminated by feces (Drake 1918; Ruong 1937).

This practice was continued by herders in all South Sámi coastal districts to post WWII area, for several until 1964 (Fjellheim 1995), and in some districts further North, in Vesterålen (Riseth et al. 2020). Literature, oral tradition as well as informant interviews provide clear indications that reindeer herders still have historical experience that *keeping reindeer herds too close and too long in the same place increases the risk of disease.* This seems to have established itself as part of the Sámi traditional knowledge that forms part of the basis of reindeer husbandry practices or as an element of good reindeer husbandry practice (op. cit.) as a modus operandi (Drake 1918; Ruong 1937).

Digital necrobacillosis disappeared from the semi-domesticated reindeer herds in Fennoscandia after the 1950s (Skjenneberg and Slagsvold 1968). However, new herding practices with increased feeding in winter and calving in fences as protection from predators have recently been established. This may facilitate the re-emergence of diseases such as digital necrobacillosis, as discussed in Sect. 4.3.

Fig. 3 An adult male wild reindeer (*Rangifer tarandus tarandus*) with digital necrobacillosis. Some outbreaks of the disease have appeared in the wild reindeer populations in Norway over the past two decades (Photo: Erik M. Ydse, Norwegian Nature Inspectorate)

1.2.4 Pasteurellosis

The disease pasteurellosis was first reported in reindeer as large epizootics among semi-domesticated reindeer in Norway and Sweden in 1912–1914 (Horne 1915). Outbreaks in the Swedish municipality Arjeplog in 1912 and 1913 killed approximately 1600 and 1500 reindeer, respectively, 86–95% of them being calves (Magnusson 1913; Nordkvist 1960). The reindeer herders claimed they had not previously observed a similar disease in their herds. The outbreaks occurred mostly in July-August and were associated with dry and warm weather. Scattered and smaller outbreaks of pasteurellosis are still occurring, as described in Sect. 4.4.

1.2.5 Infectious Keratoconjunctivitis

Infectious keratoconjunctivitis (IKC) was mentioned with its Sámi name, *Calbmevikke* (or *Tjalme vikke* in some older reports), which means eye disease, in the beginning of the nineteenth century. The disease was first described in 1912 (Bergman 1912) after severe outbreaks in Sorsele municipality, Sweden, in 1909 and 1910. In these outbreaks, about 90% of the calves in the herds were affected and many calves developed permanent blindness, usually in one eye, and many died (Nordkvist 1960). The outbreaks were often associated with the transition of herds between pastures in spring, summer, and autumn. Smaller outbreaks of IKC occur in reindeer in Fennoscandia today, often associated with stress and supplementary feeding (Tryland et al. 2009), as discussed in Sect. 4.1.

2   Reindeer Herding Today

2.1   Extensive Herding

In Finland, intensive herding continued until WWII (Helle and Jaakkola 2008). For Norway and Sweden, the early twentieth century became a new transition period in which extensive reindeer husbandry with larger herds emerged (Ruong 1954). Mainly, this seems to be a cumulative effect of agricultural expansion, infrastructure and encroachment, and governmental regulations, resulting in crowding effects due to permanent moving of herder families (Brännlund 2015; Riseth et al. 2016). During the twentieth century, reindeer husbandry underwent several transformations. First, there was a shift from subsistence use and milking toward meat and market production. Second, a general modernization occurred, encompassing ordinary schooling for children and family sedentarization, i.e., families changed their dwellings from traditional *goahti* (turf huts) and *lávvo* (herder tents) to ordinary wooden houses like peasants. Generally, this transition was completed during the first half of the twentieth century. Third, a change from animal and human muscle power toward increased motorization with snowmobiles and cars began in the 1960s. In Finland, extensivation of herding also took place in the same period (Helle and Jaakkola 2008). In suitable landscapes, all-terrain vehicles (ATVs) and snowmobiles are used for person transport, while helicopters are increasingly used for gathering herds, and trucks and barges for animal transportation. Finally, from the 1970s onward, cooperation with governmental authorities increased along multiple dimensions. In short, this implied significant changes from an independent lifestyle toward a livelihood occupation, increased integration into the broader society, and increased dependency on the state (Riseth 2006; Helle and Jaakkola 2008; Ulvevadet 2012; Löf 2014).

Government regulations, international agreements, institutional barriers, and pressure from recreation use of the land and loss of animals due to predators significantly influence reindeer husbandry industries. Further, contemporary reindeer herders are, in addition to the effects of natural variability, increasingly feeling the impacts from climate change. Combined, these factors significantly contribute to the direct and indirect loss of grazing land for their animals (Kløcker-Larsen et al. 2017; Risvoll 2015; Löf 2014).

Modern reindeer herding is based on larger herds than historically used. Modern communication technology with telemetry, Global Positioning System (GPS) devices, and drones is also increasingly used. Despite all technology use, it is important to be aware that reindeer herding is still based on a comprehensive knowledge of social relationships, reindeer and their habitat, and response patterns to various conditions. This knowledge is still transmitted between generations while at the same time being renewed with the young people's own learning and experience building. Many districts therefore also have a higher degree of tameness of the animals than usual, at least for parts of the flock, in order to perform migration through demanding landscapes. It is also worth noting that many districts over the past decades have undergone changes in how they work with the herd, causing less stress on the animals. One example is calf marking without the use of lasso (*suopan*) but instead either by using binoculars in a special corral or a traditional method with collection on a snow patch.

Herders' work with reindeer can analytically be divided into two main parts; *herding* and *husbandry* (Paine 1972). Herding is the collective siida work within the open landscape most of the year, while husbandry is the family work in corrals, i.e., calf marking and family herd design, including decision on which animals to slaughter and which to remain in the herd.

2.2   Increased Feeding: Back to Corrals and Intensive Management?

Unmanaged wild reindeer populations tend to follow cyclical behavior, and semi-domesticated reindeer often do that to some extent too. The basic dynamics between pasture and herd size is influenced by both climatic and political events (Riseth et al. 2016). A historical review of pasture catastrophes in a Sámi Village in northern Sweden identified three dramatic winters from the 1930s to the 1960s where more than half of the reindeer population was lost. When meeting poor grazing conditions, reindeer naturally spread in search for food. During most of the twentieth century, the main adaptation to "locked" pastures was to let the reindeer spread. In several instances where small herds were moved to deviant locations, they often fared better than most. During the latest part of the twentieth century, losses from hardship winters were reduced due to severely improved transport opportunities due to modern infrastructure and vehicles and the development of supplementary feeding regimes (Päiviö 2006).

In boreal forests, arboreal lichens have traditionally been the emergency food for reindeer. In Finland, winter feeding developed in close concert with small-scale agriculture (Turunen et al. 2014). Around 1970, research on developing an industrial supplementary concentrate feed (pellets) started in Sweden and Norway. The use spread from the 1980s, then with Finland as a forerunner and Norway as the latest adapter. Grass silage (round-bales) started to be used in the 1990s. Finland's and Sweden's entry into EU in 1995 made supplementary feeding much more affordable due to subsidies, and it became a regular part of winter forage in most of Finland. In northernmost Finland, Sweden, and Norway, supplementary feeding has been mainly used in winters with poor grazing conditions because of hard/much snow and/or ground icing.

Over the past decades, reindeer grazing resources are heavily affected and reduced by infrastructure development (Skarin et al. 2015; Riseth and Johansen 2019) and in Sweden and Finland also by modern forest management (Kivinen et al. 2010). These anthropogenic alterations are leading to fragmentation of reindeer pastures and reduced availability of grazing resources, especially during winter.

These changes occur in parallel with climate change. The Arctic is an especially sensitive area with regards to climate change, with an Arctic annual air temperature increase of 2.7 °C in the period 1971–2017. In Fennoscandia, projected climate change can be summarized as increasingly warmer and wetter weather, and a higher frequency of extreme weather events (Box et al. 2019). Increased temperatures during summer will cause higher plant productivity and longer growing seasons, causing the replacement of the typical tundra vegetation (i.e., grasses, herbs, ferns and mosses) by shrubs and trees. Even though these changes may provide reindeer with richer pastures during summer, they are also associated with a reduction in winter food availability (Moen 2008; Tryland et al. 2019a). In winter, climate change will increase precipitation as rain in winter at the expense of snowfall and increase temperature fluctuations in the Arctic, leading to an increase in the frequency of rain-on-snow events and freeze-thaw cycles, as well as changes in vegetation (Box et al. 2019). As one of the major implications of climate change for semi-domesticated reindeer, these new conditions will negatively affect the availability of food in the winter due to ice-locked pastures (Hansen et al. 2013) and may also affect migration routes in spring and autumn since lakes and rivers may not freeze over (Riseth and Tømmervik 2017).

During the latest decades, supplementary feeding has generally increased and tends to become more common in most regions, and not only as an emergency action. In Norway, the regional pattern (see map in Fig. 2) is that there is still no supplementary feeding in Trøndelag, while two-thirds of the households in Nordland and most households in Troms practice feeding. In the western part of Finnmark, many herders still use hay, but some have changed to pellets, while in the eastern part of Finnmark, silage (bales) is the most used supplementary feed, but with increasing amounts of pellets (Lifjell 2019).

As a major reindeer husbandry region, we will explore the development in Western Finnmark (Kautokeino) more thoroughly, starting with a comparison to its neighbor region in Finland. In the late 1990s, the Sámi reindeer husbandry region in Finland, just across the national border to Kautokeino, adopted the feeding practice from more southern regions, and researchers questioned whether they were about to be dependent on practicing supplementary feeding (Kumpala 1999). At the time, one could be suspicious that the situation on the Norwegian side was moving in the same direction. Lichen pastures were heavily grazed down, 1996–1997 was a heavy snow winter with major losses and supplementary feed might aggravate pasture competition (Riseth 2000). An experienced official that has followed the development closely for four decades explains:

> Some herders tried out supplementary feed on individual basis already in the 1980s, …herders in easternmost Finnmark developed drying of water horsetail (*Equisetum fluviatile*), which was very successful, the animals loved it, … during the 1990s salt licking stones – for increasing tameness, were tried around here. …
>
> During the latest ten to fifteen years supplementary feed has gradually increased. It started with dry hay, then round-bales became more usual, but they were large and awkward to handle, and lately industrial concentrates have reached high quality and prices are also relatively reasonable, so now concentrates are the first choice. …
>
> …. In late winter supplementary feed has become increasingly usual to ensure hinds' body condition and by that promoting calf survival during the first critical period. A positive by-effect is increased tameness and reindeer that is easier to handle. West Finnmark reindeer husbandry probably is throughout dependent of supplementary feeding.

Although winter pastures are somewhat improved, internal competition is so hard that many invest very much to stay in business, as well of spouse income and own extra jobs. Cultural concerns are overruled by economics by most people, i.e. supplementary feed has arrived to stay.

Moreover, herders' increased competence in conduct of feeding sure was a precondition for the great feeding project that no doubt rescued the reindeer herds from an extensive hunger catastrophe the spring 2020 due to the unusual large snow masses covering all vegetation for several tens of miles (Hætta 2020).

Hence, until the 1970s–1980s semi-domesticated reindeer feeding was based on natural pastures. Since then, supplementary feeds have increased in significance. Nevertheless, with the increased feeding, there is a tendency that semi-domesticated reindeer is moving out of its ecological niche. This development may have ecological, as well as economic and cultural concerns, however, in this chapter, we will focus on the animal health and disease aspects of these changes. Examples of practical feeding are demonstrated in Fig. 4.

Feeding, especially over longer periods of time and in corrals, increases animal density and challenges reindeer with unfavorable hygienic conditions, quite similar to the experience and traditional knowledge referred to above with regard to *slubbo* and its association with milking practice (Josefsen et al. 2019). Several outbreaks of IKC, contagious ecthyma, and alimentary necrobacillosis have been documented during the past few years in reindeer herds on different feeding regimes in corrals, sometimes affecting several dozens of animals (Tryland et al. 2019b; Sánchez Romano et al. 2019a; Horstkotte et al. 2020). Increased feeding often includes more gathering and handling of reindeer, which contradicts the traditional practice of avoiding keeping reindeer too tight. Herders from Finland, Sweden, and Norway also express negative feelings about feeding, as it is not a free or wanted choice (Horstkotte et al. 2020).

Fig. 4 Left: Traditionally, lichens (*Cladina* spp.) was the most used transition and supplementary feed (Photo: Tom Lifjell). Middle: Ensilage (round bales) may be served on the pastureland or in corrals and has become more common as supplementary feed (Photo: Morten Tryland). Right: Industrial concentrates (pellets) designed for reindeer are used more and more (Photo: Morten Tryland)

Another part of this dilemma is formed by cultural factors, e.g., reindeer do not belong in fences and their meat may through intensive feeding lose the characteristic and valued reindeer taste. Further, feeding requires investments in pellet containers and other equipment, such as sleighs, more vehicles, throughs, etc., that may be considered a gradual change from herding toward farming (Ingold 1980). Furthermore, the old mitigation strategy mentioned by most of the sources above, i.e., to move the animals to alternative pastures, has become much more difficult to accomplish due to the fragmentation and loss of pastures.

3   Diseases Directly Related to Feeding

The artificial feeding system, with food served once or twice per day, is very different from a natural feeding situation. In addition, adaptation to a new regime and a new diet may pose a challenge to the health of semi-domesticated reindeer. Most challenges are related to dysfunction of the digestive system, high animal density or stress (Åhman et al. 2019).

3.1   Accumulation of Grass in the Rumen

Accumulation of grass in the rumen ("grass belly") may happen when reindeer, which are selective mixed feeders, are forced to eat only roughage with a high content of cellulose and lignin (Åhman et al. 2019). There are many reports of emaciation and starvation in reindeer in spite of having free access to hay or ensilage (Josefsen et al. 2019), and feeding experiments show that the condition is related to the reindeer´s inability to digest high-fiber roughages. The reindeer will continue eating, gradually increasing the amount of grass in their rumen and increasing their rumen volume while at the same time having an energy deficit and getting a poorer body condition (Åhman et al. 2019). Clinical signs and necropsy findings include a large amount of grass in the rumen and a poor body condition. A diagnosis can be made from dietary history and clinical signs. In most situations, the condition may be reversed by providing a low-fiber diet (Åhman et al. 2019).

3.2   Rumen Acidosis

Rumen acidosis ("rippling belly") occurs due to a high intake of easily digestible carbohydrates. The large carbohydrate amount will lead to a change in the microbial flora to bacteria that may produce acids, and the rumen pH may drop lower than its normal pH of 6–7. The microbial flora may be dominated by *Lactobacillus* spp., and the pH may drop as low as 4–5 (Snyder and Credille 2017). This leads to a life-threatening condition as the absorption of acid from the rumen may exceed the buffering capacity of the blood. Moreover, the high content of acid in the rumen leads to water entering the rumen from the general circulation, and the animal becomes dehydrated (Åhman et al. 2019). Ruminal acidosis in cattle can be present in different forms, varying from per acute life-threatening forms to chronic illness, which is more difficult to detect (Oetzel 2017), and the same is probably the case in reindeer. Necropsy findings include liquid rumen content with a low pH, soft content in the colon, sometimes diarrhea, and a rumen mucosa that does not peel off easily. Histological examination of the rumen wall will show a chemical ruminitis (Josefsen and Sundset 2014; Åhman et al. 2019). Typical clinical signs of rumen acidosis in reindeer include lack of appetite, increased thirst, lethargy, dehydration, decreased or ceased rumen contractions and diarrhea. A diagnosis can be made from the dietary history and clinical signs (Åhman et al. 2019). Treatment of ruminal acidosis in reindeer follows the same principles as in other ruminants, with correction of blood volume, assessment and treatment of rumen and systemic acid-base disturbances, restoration of a normal rumen microenvironment, and management of potential secondary complications (Snyder and Credille 2017).

3.3   Diarrhea

Diarrhea in reindeer may be a symptom of ruminal acidosis, and is mostly associated with feeding with pellets (Johansson 2006). However, diarrhea has also been observed in animals fed with high-fiber feed mixtures and ensilage (Josefsen et al. 2007). Diarrhea may also be caused by bacterial infections in the digestive tract (Josefsen et al. 2019). Treatment will vary depending on the underlying cause.

3.4   Ruminal Tympani

Ruminal tympani, or bloat, describes a condition in ruminants where the rumen is filled with gas or foam, and the animal is unable to eructate the gas. The pressure may become so high that it obstructs breathing and the blood circulation. In cattle, the condition is related to eating large amounts of lush grass or leguminous plants (e.g., clover and alfalfa) (Radostits et al. 2000). Bloat may occasionally affect reindeer in relation to feeding (Ippen and Henne 1985), but is far less common than in ruminants in more intensive production systems (Åhman et al. 2019).

3.5   Wet Belly

Wet belly is a disease with unknown etiology, but it has been speculated whether the disease may be associated with a fungal infection (Åhman et al. 2002, 2019). Necropsy findings have included emaciation and also mycotic pneumonia, abomasal and rumen ulcers, abscess formation, peritonitis, enteritis and hepatitis (Jacobsen and Skjenneberg 1979; Åhman et al. 2002). The disease manifests itself in the way that the reindeer fur becomes wet, primarily in the axilla, down the legs, along the lower part of the thorax and abdomen and sometimes on the lower part of the neck. The moisture comes from within the animals (Åhman et al. 2002, 2019). The affected reindeer are reported to appear hungry, to have a high food intake and to curl up when they lie down (Åhman et al. 2002). Wet belly is exclusively observed in relation to feeding but has not been associated with a specific feed or feeding regime. Despite this, the common recommendation for treatment is to change the feed (Åhman et al. 2019).

4   Diseases Related to High Animal Density and Stress

Reindeer herding in Sápmi is currently subjected to extensive stress factors due to increasing predator strains and nature interventions (Riseth and Johansen 2019). This occurs in parallel with climate change, which leads to unstable winter grazing and unsafe migration routes in both the spring and the autumn (Riseth and Tømmervik 2017). Since 2008, the annual loss of semi-domesticated reindeer in Norway are reported to be 10–14% of the adults and 36–45% of the calves, with some regions reporting up to 23% adult and 63% calf losses in the worst years (Landbruksdirektoratet 2020). Starvation, predation by large carnivores, extreme weather events and accidents are all common causes of reindeer loss (Tveraa et al. 2014). Density-dependent food limitations, leading to starvation, might be the most important cause of loss of animals in Norway, combined with loss to predators (Tveraa et al. 2014). Predation is also regarded as an important cause in Sweden and Finland, causing not only immediate loss of animals, but also reducing the growth rate of reindeer populations (Hobbs et al. 2012; Kumpula et al. 2017).

The lack of pasture resources during winter and predation are challenging animal welfare as well as the economy and are forcing reindeer herders to rely more heavily on corralling and supplementary feeding. The increased handling that is linked to corralling may increase the general stress load of the animals (Tryland 2012), with increased levels of glucocorticoids and immunosuppression, facilitating a higher incidence of disease (Coutinho and Chapman 2011). At the same time, corralling and supplementary feeding increase animal density, especially in the feeding areas, and create challenging hygienic conditions in the corrals, with fodder and forage leftovers and fecal material on the ground.

Reindeer, as semi-domesticated animals, may have a lower tolerance to handling than domestic livestock. When reindeer are sent to the slaughterhouse, they are not only exposed to the stress caused by the initial gathering and handling, but also stress associated with the transport itself, which is usually conducted by means of motor vehicles such as vans, trailers or special reindeer transport trucks (Laaksonen et al. 2017a). The combination of high animal densities and a high degree of stress facilitates the spread of infectious agents within the reindeer herds. This is especially important for agents that cause opportunistic or latent infections, such as CvHV2, the causative agent of IKC, Orf virus, which causes contagious ecthyma, *Pasteurella multocida*, a reindeer pathogen causing regular pasteurellosis outbreaks in reindeer, and *F. necrophorum*, the bacterium causing necrobacillosis.

4.1 Infectious Keratoconjunctivitis (IKC)

IKC is a contagious ocular infection involving both the cornea and the mucosa of the conjunctiva. It has been shown that the reindeer alphaherpesvirus, Cervid herpesvirus 2 (CvHV2), is the transmissible infectious agent and is usually present in the early stages of the disease, causing lesions in the conjunctival mucosa and corneal ulcers (Tryland et al. 2009, 2017; Sánchez Romano et al. 2020). However, bacterial agents, such as *Moraxella bovoculi*, *Chlamydia pecorum*, and others, have been isolated from affected eye lesions and may contribute to the pathogenesis of IKC (Sánchez Romano et al. 2018, 2019a). CvHV2 has been isolated from reindeer in Finland (Ek-Kommonen et al. 1986), Sweden (Rockborn et al. 1990) and Norway (Tryland et al. 2009) and the infection is enzootic in most reindeer populations (das Neves et al. 2010).

The virus enters and infects cells of the mucosal membranes and replication and production of progeny virus takes place within hours. The virus establishes a lifelong infection and may be reactivated from its latent stage if the animal is exposed to stressful events such as gathering, corralling, handling, and transport, or to other infections or immune suppression. This leads to reactivation and replication of the virus, which again may cause clinical signs and virus shedding from mucosal membranes. The virus is transmitted directly through contact and possibly by aerosols to immunologically naïve reindeer, often calves and young reindeer, which are the ones that primarily get sick during an outbreak.

Initial clinical signs include increased lacrimation, discolored hair under the eyes and periorbital and corneal edema (Fig. 5). The disease may heal spontaneously from this stage. However, it may also progress to more severe edema, corneal ulcer,

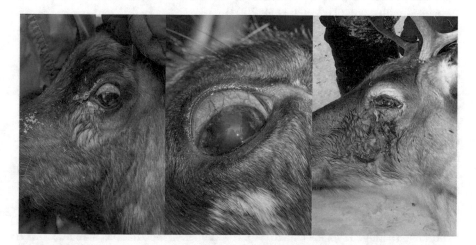

Fig. 5 Infectious keratoconjunctivitis in reindeer (Cervid herpesvirus 2, and bacterial infections), showing the initial stage with increased lacrimation and some pus formation (left), corneal oedema (middle) and a late stage (right) when the structures of the eye deteriorates (Photo: Javier Sánchez Romano)

panophthalmitis and loss of the lens, damage to the eye structure and permanent blindness (Tryland et al. 2019c). It has been suggested that CvHV2 may also cause abortions and weak born calves (das Neves et al. 2010).

Clinical signs are indicative of the diagnosis. The finding of anti-CvHV2 antibodies indicates previous exposure (das Neves et al. 2010) but may be of little value since approximately 50% of adult animals may have such antibodies and since many animals not previously exposed (i.e., calves) may be in the incubation period (7–10 days post-infection) before antibodies are present. Detection of specific viral DNA (polymerase chain reaction; PCR) may be indicative, but due to the high prevalence in presumptive healthy herds and the life-long infections as for other herpesviruses, the results may be difficult to interpret. Management should focus on minimizing stressful events, thus avoiding clinical outbreaks. Cleaning of the eyes with saline water should be conducted prior to antimicrobial treatment against secondary bacterial infections (Tryland et al. 2009).

4.2 Contagious Ecthyma

Contagious ecthyma ("scabby mouth") has been diagnosed under natural herding conditions in semi-domesticated reindeer in Sweden (Nordkvist 1973), Finland (Büttner et al. 1995) and Norway (Tryland et al. 2001). Virus characterization has shown that Orf virus (ORFV; genus *Parapoxvirus*, family *Poxviridae*) caused the outbreaks in Fennoscandia. ORFV causes contagious ecthyma in sheep and goats, and the virus is presumably transferred directly or indirectly from sheep or goats to reindeer (Tryland et al. 2001). Outbreaks in Finland have also been associated with Pseudocowpoxvirus (PCPV), which is a related parapoxvirus that is primarily associated with cattle (Tikkanen et al. 2004).

The virus enters the skin or the oral mucosa through small abrasions. Lesions in the mouth from ice crusts or coarse food or from erupting teeth may hence predispose for the disease (Tryland 2019). The incubation period is usually 3–5 days (Tryland et al. 2013). The lesions are typically located on the mucocutaneous junction around the mouth or in the oral mucosa (Fig. 6). They start as a papule, which develops into a pustule, and further into proliferative, cauliflower-like lesions (Tryland et al. 2019c). Histological examination shows a proliferative dermatitis with epidermal hyperplasia, often with elongated rete ridges, hyperkeratosis and the presence of inflammatory cells (Tryland et al. 2013).

Affected animals may be unable to chew, often having food and regurgitated matter accumulating in their mouth. There is often a foul smell from their mouth, and typical lesions can be observed upon closer inspection. Secondary bacterial infections are common and may cause complications. The animals may starve due to problems eating (Tryland et al. 2019c). Experimental inoculations in reindeer with ORFV isolated from clinically affected reindeer caused mucosal lesions that were healing about four weeks after infection (Tryland et al. 2013). The disease may affect only a few animals or appear as a regular outbreak affecting large numbers of reindeer (Tikkanen et al. 2004; Tryland et al. 2001).

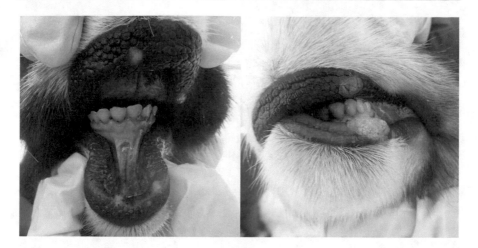

Fig. 6 Contagious ecthyma (Orf virus) with initial lesions (left) as light spots on the oral mucocutaneous junction, and a later stage (right) with the cauliflower-like proliferative lesions characteristic for the disease (Photo: Ingebjørg H. Nymo)

Diagnosis may be based on characteristic clinical signs (Tryland 2019) and histological findings (Tryland et al. 2013). A verification of the diagnosis maybe accomplished by detecting parapoxvirus-specific DNA by PCR (Klein and Tryland 2005). The humoral immune response against ORFV in sheep has been characterized as relatively poor and short lived (Damon 2007), and the presence or absence of antibodies may be of less value, both with regard to protective immunity and for diagnostic purposes (Tryland et al. 2013). The disease may be treated with supportive therapy and antibiotic treatment against secondary bacterial infections. Vaccination is not employed during natural conditions (Tryland 2019).

4.3 Necrobacillosis

Although other bacteria may contribute to the disease, digital necrobacillosis is regarded as a disease in ruminants caused by the bacterium *Fusobacterium necrophorum*, which is a part of the normal rumen microbiota (Tan et al. 1994). The bacterium has a worldwide distribution and is a major pathogen in ruminants. As described above (Sect. 1.2.3), necrobacillosis was reported as a devastating disease among semi-domesticated reindeer already in the sixteenth century, but more or less vanished from Fennoscandia after 1950 due to the termination of reindeer milking and a shift toward a more extensive reindeer husbandry form (Josefsen et al. 2019). With the increased gathering of reindeer for supplementary feeding during winter in Fennoscandia in the later years has the disease, however, re-emerged (Tryland 2019).

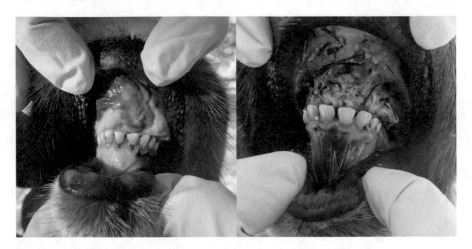

Fig. 7 Oral necrobacillosis caused by the bacterium *Fusobacterium necrophorum*. To the left is an early stage with discoloration and necrosis of the mucosa, and to the right is a later stage with severe necrosis and pus. The infection may also affect the pharynx, oesophagus and rumen (Photo: Ingebjørg H. Nymo)

Outbreaks are often associated with crowding of reindeer on wet and muddy ground contaminated with feces containing the bacterium (Josefsen et al. 2019). The bacterium is unable to penetrate whole skin or mucosal membranes, however, small abrasions on the feet or in the mouth (e.g., from teeth eruption in calves) may allow for bacterial entry. After entering, the bacterium produces toxins that cause tissue necrosis (Tan et al. 1994).

In digital necrobacillosis, the skin, subcutaneous tissues, tendon sheets, tendons, joint capsules and the articular cartilage may become necrotic. Periostitis and osteomyelitis might also occur. In chronic cases, there may be widespread fibrosis (Handeland et al. 2010). The clinical signs of digital necrobacillosis are lameness and swelling of the distal parts of often one, but sometimes several feet. After a few days, there may be ulcerations of the skin (Bergman 1909; Handeland et al. 2010). In oral necrobacillosis, clinical signs include salivation and problems with chewing and swallowing (Handeland 2012; Tryland et al. 2019c) and various parts of the oral mucosa can become necrotic (Fig. 7). Suckling calves may transfer the infection to the udder, and cases with lesions in the reproductive organs are reported during the rut (Rehbinder and Nikander 1999; Skjenneberg and Slagsvold 1968).

A tentative diagnosis may be made upon clinical history and clinical signs. A final diagnosis must be based on demonstration of the etiological agent (i.e., bacteriology, PCR, histology, in situ hybridization) (Josefsen et al. 2019). From historical reports, mortalities up to 80% were reported (Bergman 1909; Horne 1898), whereas in more recent outbreaks, a dozen animals or more may be affected, usually calves and young animals (Tryland et al. 2019b). The disease is, however, treatable with antibiotics and supportive therapy if initiated early.

4.4   Pasteurellosis

Pasteurellosisis in reindeer is caused by the bacterium *Pasteurella multocida*. Studies performed in other species indicate that some strains may be more virulent, and hence associated with clinical disease (Blackall et al. 2010), but further typing of the reindeer strains has not been performed. The bacterium causes peracute hemorrhagic septicemia or acute or subacute bronchopneumonia in reindeer (Josefsen et al. 2019). There have been multiple large outbreaks of this disease in reindeer in Fennoscandia (Josefsen et al. 2019) since the first large outbreak was described in northern Sweden and Norway in 1912–1914 (Horne 1915).

The bacterium is a commensal on mucous membranes in the upper respiratory tract in many animal species and is transmitted through direct contact with nasal secretions (Wilson and Ho 2013). This is probably also the case for reindeer, but this has not yet been confirmed (Josefsen et al. 2019). The bacterium acts as an opportunistic agent, which may cause disease when the animals' immune status is reduced by stress or other infections. Disease outbreaks may also occur if a new bacterial strain is introduced into a population, or if the virulence of a strain is altered (Wilson and Ho 2013). In reindeer, outbreaks of hemorrhagic septicemia have often been associated with dry and warm summers, and stress related to heat, insect harassment and drought likely contribute to the disease development (Nordkvist and Karlsson 1962). There have also been reported outbreaks of bronchopneumonia in calves during late winter and early spring (Skjenneberg 1957).

The animals suffering from the peracute form of the disease are often found dead with no prior signs of disease, or they may show clinical symptoms from the respiratory tract, such as coughing and nasal discharge (Josefsen et al. 2019). The pathological findings are primarily consistent with severe sepsis, such as widespread hemorrhages, edema, and hyperemia (OIE 2013). In the cases of lung disease, the most common finding is fibrinous pneumonia or bronchopneumonia (Josefsen et al. 2019).

Clinical history and pathological findings may indicate pasteurellosis, and the diagnosis can be verified by detection of the etiological agent by bacterial cultivation and PCR based on secretions from pathological lesions (Wilson and Ho 2013). Antibiotic treatment of reindeer calves with pasteurellosis has been attempted with variable success. Vaccination has also been attempted and seemed successful, however, the number of reindeer calves included in the study was too small to draw any firm conclusions (Kummeneje 1976).

5   Climate Change and Potential Impact on Pathogens and Diseases

Climate change may have direct impact on the health and diseases of reindeer. Climate change has led to an increase in the number and species diversity of blood-sucking arthropods that can act as pathogen vectors (Medlock et al. 2013; Tryland et al. 2019a) and altered the conditions of certain infectious agents and parasites (Åsbakk et al. 2014; Omazic et al. 2019; Hovelsrud et al. 2020). Some infectious diseases have therefore been classified as climate-sensitive infections (CSI) (Omazic et al. 2019).

5.1 The Reindeer Brainworm

The reindeer brainworm (*Elaphostrongylus rangiferi*) was first described in scientific literature in 1958 in Russia (Mitskevich 1958), however, the disease was mentioned much earlier in a classical book about the life of the Sámi written by Johan Turi (1910). He describes that the disease caused emaciation and hind body paresis, and sometimes, regular outbreaks with the death of several hundred animals.

The brainworm is a nematode with an indirect lifecycle with gastropods (snails and slugs) as intermediate hosts (Handeland 1994). Arctic summer temperatures are close to the developmental threshold of the parasite in the intermediate hosts (8–10 °C), and hence, under these conditions, the parasite relies on overwintering in adult gastropods, having a 2-year lifecycle. An increase in average summer temperatures in the Arctic results in faster development of brainworm larvae in the intermediate hosts, from 2–4 months to less than 2 weeks. This results in increased infection intensity in the intermediate host, possible fulfilment of the life cycle during one season and increased exposure of reindeer to the parasite (Davidson et al. 2020; Halvorsen and Skorping 1982).

Infection with the brainworm nematode occurs when the reindeer eat intermediate snail and slug hosts containing infective parasite larvae. The parasite thereafter migrates from the gastrointestinal tract via the vascular system throughout the body. Only larvae that reach the CNS develop further. The parasite is therefore often found in the spinal cord and brain (Handeland 1994, Handeland et al. 1994), where it causes the clinical signs characteristic of the disease; general weakness, poor coordination with hind limb paresis, paralysis, ataxia, knuckling of the carpus/tarsus, stumbling and falling and circling (Fig. 8) (Handeland 1994; Handeland et al. 1994; Handeland and Norberg 1992). The animal may also stand with an arched back and lowered pelvic region. Impaired tail lifting, visual impairment, somnolence and head tilt, as well as reduced growth in calves, have also been reported (Kummeneje 1974; Handeland et al. 1994). Infected reindeer may mask their clinical signs and only show them during exercise/stress (Bakken and Sparboe 1973), however, the majority of infections are believed to be asymptomatic (Davidson et al. 2020).

Clinical signs generally develop before larvae are present in feces (Handeland 1994). The diagnosis hence needs to be based on clinical signs. Ivermectin and doramectin are currently the only anthelmintics licensed for use in reindeer in Fennoscandia (Davidson et al. 2020), however, they penetrate the blood–brain barrier poorly (McKellar and Gokbulut 2012). Once the parasite has reached the CNS, anti-parasitic treatment will hence not cure the disease, as the parasite is protected by the blood–brain barrier (Davidson et al. 2020; Folstad et al. 1996). Supportive treatment is therefore the only possibility once the animals are diseased; however, clinical signs can take up to 5 months to resolve (Handeland et al. 1994).

Fig. 8 This young reindeer was found in the morning, not able to stand on its feet. The tracks in the snow clearly show the circling behavior in the attempts to rise. The animal was euthanized and brain worm infection was diagnosed (Photo: Morten Tryland)

Preventive anti-parasitic treatment can be challenging as the possible period of infection is quite prolonged, and the semi-domesticated reindeer in Fennoscandia are not readily available for treatment at all times (Davidson et al. 2020). Modeling of parasite burdens, landscape use and climate may, however, in the future, reveal trends and allow herders to develop mitigation strategies in years and seasons with high infection risk (Vineer et al. 2021).

5.2   Anthrax

Although anthrax, caused by the bacterium *Bacillus anthracis,* has not been diagnosed in reindeer in Fennoscandia during recent times, the outbreak among reindeer that took place in the Yamal-Nenets Autonomous Okrug in Siberia (Russia) in 2016 is a reminder that this may still be a relevant reindeer pathogen. Due to abnormally high temperatures during summer and the thawing of the permafrost, *B. anthracis* spores were released from the ground, leading to an outbreak of anthrax, which caused the death of over 2000 reindeer and the subsequent culling of more than 200,000 reindeer (Hueffer et al. 2020). About 100 people were hospitalized with suspicion of anthrax, which was confirmed in at least 20 cases, and one 12-year-old boy died from the disease. By the middle of August, more than 200,000 reindeer and about 12,000 people were vaccinated against anthrax (Josefsen et al. 2019). The area was known to have had large anthrax outbreaks in the past, and the hypothesis for the re-emergence of the disease was that a vast number of bacterial spores remained in the ground. When the annual vaccination in the region ceased in 2007 after several decades with no disease outbreaks, the reindeer population was again susceptible to the pathogen (Josefsen et al. 2019; Hueffer et al. 2020).

Once the bacterium infects a host, often through the digestive tract after the ingestion of spore contaminated feed, forages, and water, it produces several toxins which cause the disruption of blood vessels linings and result in leakage of blood and other fluids (Quinn et al. 2011). Clinical signs of the peracute disease in reindeer include breathing difficulties, dark blue mucous membranes and bleedings, leading to a sudden death within 2 h after the development of the initial signs. An acute form of the disease, developing over approximately 24 h, has also been described, with fever and swellings in the skin as additional clinical signs before the death of the affected animals (Nikolaevskii 1961).

Differential diagnoses may be any disease that cause sudden death or hemorrhagic septicemia (e.g., pasteurellosis, food intoxications or clostridial diseases). However, the finding of capsulated rod-like bacteria or typical bamboo canes in a stained blood smear in the microscope, and cultivation of *B. anthracis*, or positive results by PCR from blood, exudates, edematous infiltrations or organ samples are clear indications of the causative agent of the disease. Even though there are several types of antibiotics that are effective against *B. anthracis* infection, the peracute development and the lack of therapeutic options against the bacterial-induced toxemia make the treatment ineffective for sick animals (Fasanella et al. 2010).

5.3   Arthropod-Borne Diseases

Arthropod-borne diseases may have an increased relevance in the coming years. Recent studies have shown that the tick *Ixodes ricinus*, the intermediate host for several vector-borne diseases such as anaplasmosis or borreliosis that could possibly affect reindeer, has been expanding its range northward, and also from coast to inland, in recent years (Jore et al. 2014; Mysterud et al. 2017). Ticks are also becoming a common finding in reindeer herding areas (Tryland et al. 2019a). Semi-domesticated reindeer have also the potential to become tick carriers, facilitating the migration of ticks and tick-borne pathogens to inland habitats when reindeer migrate (van Oort et al. 2020).

Mosquitoes and biting midges' distribution and range have also been shown to be correlated with changes in climatic conditions (Elbers et al. 2015), but since these insects are already present in reindeer pastures, the impact on reindeer herds will be most likely associated with increased harassment. However, captive reindeer have been exposed to vector-borne pathogens that are not present in the semi-domesticated population, such as Schmallenberg virus (SBV) and bluetongue virus (BTV) (Sánchez Romano et al. 2019b). Reindeer has also been shown to be susceptible to West Nile Virus (WNV) (Palmer et al. 2004), a mosquito-borne arbovirus that is currently expanding its geographic range through Europe (Bakonyi and Haussig 2020). None of these viruses are yet detected in the Fennoscandian reindeer herds. However, if they find their way to the semi-domesticated reindeer distribution range, they may pose an additional pressure on reindeer health and reindeer herding.

The prevalence of endoparasites with life cycles partially dependent on temperature may also increase if the climate-induced alterations persist. The filaroid nematode *Setaria tundra* (Filarioidea: Onchocercidae) has mosquitoes as intermediate hosts. This parasite was the causative agent of severe outbreaks of peritonitis in semi-domesticated reindeer in Finland in 1973, 2003–2005 and 2014 (Laaksonen et al. 2007; 2017b). *Setaria tundra* infection causes peritonitis and decreased body condition, with the presence of dead and live nematodes in the peritoneal cavity of infected reindeer (Josefsen et al. 2019). The development of the parasite in the arthropod vector is temperature-dependent, with the parasites failing to develop to an infective stage at 14.1 °C, while they were completely developed in 2 weeks at 21 °C (Laaksonen et al. 2010). While the transmission period for *S. tundra* in reindeer is very short in reindeer grazing areas due to ambient temperatures, increasing temperatures may facilitate a range expansion of the parasite and cause an increased duration of that transmission period and the number of larvae being transmitted (Haider et al. 2018).

6 Emerging Health Challenges

In the past, disease outbreaks represented serious challenges to reindeer herding, and even though current health conditions of reindeer in Fennoscandia are more favorable for survival and meat production, the risk of large outbreaks and high mortalities is still there (Tryland 2012; Hueffer et al. 2020). It is also anticipated that the impacts of known diseases among semi-domesticated reindeer may become more severe with the progression of climate change, and that new and re-emerging pathogens will challenge the health status of the semi-domesticated reindeer herds (Tryland 2012).

6.1 Chronic Wasting Disease

In March 2016, chronic wasting disease (CWD) was detected in wild reindeer in Nordfjella, Norway. This was the first appearance of this disease in wild cervids outside North America since it was discovered there in the 1960s. In total (February 2021), CWD has been diagnosed in 19 reindeer in Nordfjella, and in one wild reindeer in Hardangervidda further south (Viljugrein et al. 2019). Further, an atypical and assumingly less contagious variant has been diagnosed in eight moose (*Alces alces*) and one red deer (*Cervus elaphus*), all being old animals (Vikøren et al. 2019). This atypical form has also been diagnosed in moose in Sweden and Finland (Pirisinu et al. 2018; Averhed et al. 2020). CWD has thus become one of the most serious threats for semi-domesticated reindeer and the reindeer herding industry, should the disease spread into the semi-domesticated reindeer herds (EFSA 2017).

CWD is a transmissible spongiform encephalopathy (TSE), comparable to bovine spongiform encephalopathy (BSE, "mad cow disease") in cattle and scrapie in sheep. A misfolded prion protein (PrP^{Sc}) is called a prion and is the causative and transmissible agent. The prion is shed through, e.g., urine and feces, and when the disease has been present for a period of time, the ground in that area is contaminated, and CWD prevalence starts to increase. Other crucial characteristics of CWD are the long incubation time, experimentally shown to be 17–18 months with CWD inoculation from white-tailed deer (Mithell et al. 2012), as well as the fact that the prions do not trigger the immune system, i.e., no development of herd immunity over time.

While the reindeer cases have been so far restricted to the wild population, the free-roaming nature of reindeer herds, the known contact between the affected wild reindeer herd in Nordfjella and adjacent semi-domesticated reindeer and the long incubation period for the disease make CWD entering the semi-domesticated reindeer herds and spreading to other cervid populations in Europe an immediate risk. This risk triggered the decision to cull the entire population of wild reindeer in the area affected by CWD, with the culling of more than 2000 reindeer, approximately 7% of the total population of wild Eurasian tundra reindeer (Mysterud and Rolandsen 2018). Extensive CWD testing is undergoing in Norway, and more than 125,000 cervids and about 60,000 reindeer have been examined (February 2021) (Våge et al. 2020). In September 2020, a reindeer bull shot on Hardangervidda, Norway, was diagnosed with CWD, complicating the combat against this fatal disease in Norway. How the Norwegian government will further handle the situation is currently unknown (Ytrehus et al. 2021). Sweden and Finland have developed monitoring plans in collaboration with other countries in the European Union with the aim to test 6000 animals per country in the period 2018–2020, including wild cervids, farmed deer and semi-domesticated reindeer (Averhed et al. 2020). There has been no CWD positive reindeer in Sweden or Finland (February 2021).

Reindeer herding is traditionally based on a semi-nomadic production, with animal movements and pasture rights across regions and country borders (i.e., Norway and Sweden), trade of live animals and animal products and long distances to the nearest slaughterhouse for reindeer. Preventive measures to avoid the spread of CWD if it is introduced to semi-domesticated reindeer, such as regional culling and strict regulations of reindeer movement, will undoubtedly affect Sámi culture and herders' economy (Tryland 2019). Thus, it is necessary to consider these aspects to avoid the spread of this disease while preserving the traditional reindeer herding and Sámi culture.

6.2 Altered Ecosystems: New Hosts, Reservoirs and Pathogens

New climatic conditions and landscape alterations have also contributed to the presence and altered distribution of other ungulates that may act as pathogen reservoirs for semi-domesticated reindeer. In recent years, the populations of wild boar (*Sus scrofa*) have been growing in Europe (Massei et al. 2015), and recent reports suggest that they may be able to populate currently wild boar-free areas in Northern Fennoscandia (Rosvold and Andersen 2008). Wild boars are known reservoirs of classical swine fever and African swine fever, tuberculosis, brucellosis and several other emerging and re-emerging diseases in other countries, and also reservoirs of antimicrobial resistance (Ruiz-Fons et al. 2008; Naranjo et al. 2008; Torres et al. 2020). Tuberculosis reports in reindeer are rare, but experimental inoculation showed their susceptibility (Palmer et al. 2006). However, there are no indications for the presence in Fennoscandia of mycobacteria of the *Mycobacterium tuberculosis* complex, which cause tuberculosis in several species. *Brucella suis* biovar 4 is the causative agent of brucellosis in *Rangifer* (Josefsen et al. 2019).

Roe deer (*Capreolus capreolus*) and red deer have also increased their range and increasing numbers are present in semi-domesticated reindeer pasture areas (Jaenson et al. 2012; Danilov et al. 2017). Roe deer and red deer may act as reservoir of reindeer pathogens such as *Setaria tundra* or *Anaplasma phagocytophilum* (Laaksonen et al. 2008; de la Fuente et al. 2008) and other arthropod-borne pathogens that may also infect reindeer such as Schmallenberg virus and bluetongue virus (Rossi et al. 2015; Sánchez Romano et al. 2019b).

7   Concluding Remarks

The historical descriptions of the reindeer epizootics are witnesses of a different time, with a different and more intensive reindeer herding system, where animals were kept close and corralled for easy access to draft and milking animals. Although we do not experience similar disease outbreaks in Fennoscandia today, it is important to keep in mind that the substantial changes currently happening in reindeer herding and management may severely affect the welfare and health of semi-domesticated reindeer. Especially the rapid transition from traditional reindeer herding to increased husbandry and feeding may lead to an altered disease prevalence, possibly introducing new pathogens and altering the epidemiology and appearance of "old" pathogens. In addition, climatic changes may contribute to altered ecosystems, including new distribution patterns for insects such as ticks, mosquitos, and midges, which can be vectors for reindeer pathogens. The conditions for reindeer herding in Fennoscandia are changing. With those changes follows new challenges for both reindeer health and the economic and cultural sustainability of the traditional reindeer herding industry.

References

Åhman B, Nilsson A, Eloranta E (2002) Wet belly in Reindeer (*Rangifer tarandus tarandus*) in relation to body condition, body temperature and blood constituents. Acta Vet Scand 43:85–97

Åhman B, Finstad GL, Josefsen TD (2019) Feeding and associated health problems. In: Tryland M, Kutz SJ (eds) Reindeer and Caribou. Health and disease. CRC Press, Boca Raton, FL, pp 135–156

Anderson DG (2014) Cultures of reciprocity and cultures of control in the circumpolar north. J North Stud 8:11–27

Anonymous (1905) Direktøren for det civile veterinærvesen. Beretning om veterinærvesenet og kjødkontrollen i Norge for året 1903. Kristiania 1905 (In Norwgian)

Åsbakk K, Kumpula J, Oksanen A et al (2014) Infestation by *Hypoderma tarandi* in reindeer calves from northern Finland - prevalence and risk factors. Vet Parasitol 200:172–178

Averhed G, Bröjer C, Doose N et al (2020) Wildlife disease surveillance in Sweden, 2019, SVA's report series. National Veterinary Institute, SVA, Uppsala, Sweden

Bakken G, Sparboe O (1973) Elaphostrongylosis in reindeer. Nord Vet Med 25:203–210. (in Norwegian)

Bakonyi T, Haussig JM (2020) West Nile virus keeps on moving up in Europe. Euro Surveill 25(46):2001938

Bately J, Englert A (2007) Ohthere's voyages: a late 9th century account of voyages along the coasts of Norway and Denmark and its Cultural Context. The Viking Ship Museum, Roskilde

Bergman AM (1909) Om klöfröta och andra med progressive nekros förlöpande sjukdomar hos ren. Meddelanden från Kungl. Medicinalstyrelsen 12:1–40. (In Swedish)

Bergman AM (1912) Smittosam hornhinneinflammation, dermatitis infectiosa, hos ren. Skand Vet Tidskr 2:145–177

Bjørklund I (2013) Domestication, Reindeer husbandry and the development of Sámi Pastoralism. Acta Borealia 30(2):174–189

Blackall PJ, Miflin JK, Miflin JK (2010) Identification and typing of Pasteurella multocida: a review. Avian Pathol 29:271–287

Box JE, Colgan WT, Christensen TR et al (2019) Key indicators of Arctic climate change: 1971–2017. Environ Res Lett 14(4):045010

Brännlund I (2015) Histories of reindeer husbandry resilience: land use and social networks of reindeer husbandry in Swedish Sápmi 1740-1920. PhD dissertation. Department of Historical, Philosophical and Religious Studies. Vaartoe. Centre for Sami research. Umeå University, Umeå

Büttner M, von Einem C, McInnes C et al (1995) Clinical findings and diagnosis of a severe parapoxvirus epidemic in Finnish reindeer. Tierarztl Prax 23(6):614–618

Coutinho AE, Chapman KE (2011) The anti-inflammatory and immunosuppressive effects of glucocorticoids, recent developments and mechanistic insights. Mol Cell Endocrinol 335:2–13

Damon I (2007) Poxviruses. In: Knipe DM, Howley PM (eds) Fields virology, 5th edn. Lippincott, Williams & Wilkins, New York, pp 2947–2975

Danilov PI, Panchenko DV, Tirronen KF (2017) The European roe deer (Capreolus capreolus L.) at the northern boundary of its range in Eastern Fennoscandia. Russ J Ecol 48:459–465

Das Neves CG, Roth S, Rimstad E et al (2010) Cervid herpesvirus 2 infection in reindeer: a review. Vet Microbiol 143:70–80

Davidson RK, Mørk T, Holmgren KE et al (2020) Infection with brainworm (Elaphostrongylus rangiferi) in reindeer (Rangifer tarandus ssp.) in Fennoscandia. Acta Vet Scand 62:1–5

de la Fuente J, Ruiz-Fons F, Naranjo V et al (2008) Evidence of Anaplasma infections in European roe deer (*Capreolus capreolus*) from southern Spain. Res Vet Sci 84:382–386

Drake S (1918) Västerbottenslapparna. Två Förläggare Bokförlag, Stockholm

EFSA Panel on Biological Hazards (BIOHAZ), Ricci A, Allende A, Bolton D, Chemaly M, Davies R, Fernández Escámez PS, Gironés R, Herman L, Koutsoumanis K, Lindqvist R (2017) Chronic wasting disease (CWD) in cervids. EFSA J 15(1):e04667

Ek-Kommonen C, Pelkonen S, Nettleton PF (1986) Isolation of a herpesvirus serologically related to bovine herpesvirus 1 from a reindeer (*Rangifer tarandus*). Acta Vet Scand 27:299–301

Elbers AR, Koenraadt CJ, Meiswinkel R (2015) Mosquitoes and Culicoides biting midges: vector range and the influence of climate change. Rev Sci Tech 34(1):123–137

Falkenberg J (1985) Fra nomadisme til fast bosetning blant samene i Røros-traktene (1890–1940 årene). In: Åarjel-samieh/Samer i sør–Årbok, vol 2. Saemien Sijte, Snåsa, pp 7–28

Fasanella A, Galante D, Garofolo G et al (2010) Anthrax undervalued zoonosis. Vet Microbiol 140: 318–331

Fellman J (1906) Anteckningar under min vistelse i Lappmarken. Bind I-IV, Helsingfors

Fjellheim S (1995) Det samiske kulturlandskapet. In: Fjellheim S (ed) Fragment av samisk historie, Foredrag Saemien Våhkoe Røros, vol 1994. Sør-Trøndelag og Hedmark reinsamelag, Røros, pp 58–81. (In Norwegian)

Fjellheim S (1999) Samer i Rørostraktene, Snåsa (In Norwegian)

Folstad I, Arneberg P, Karter AJ (1996) Antlers and parasites. Oecologia 105:556–558

Gaare E (1997) Naturlige betingelser for reindrift. In: Ekendahl B, Bye K (eds) Flora i reinbeiteland, Nordisk Organ for Reinforskning (NOR)/Landbruksforlaget, pp 7–18. (In Norwegian).

Gainer R, Oksanen A (2012) Anthrax and the taiga. Can Vet J 53:1123–1125

Gissler N (1759) Underrättelse om den nu i Lappmarkerna gångbara rensjukans kännande och bot. Kungl. Sven Vetenskaps Akad Handlingar 295–301

Hætta JI (2020) Interview (26.11.2020) senior advisor Johan Ingvald Hætta, Agricultural Agency, Department of reindeer husbandry, Alta

Haider N, Laaksonen S, Kjær LJ et al (2018) The annual, temporal and spatial pattern of *Setaria tundra* outbreaks in Finnish reindeer: a mechanistic transmission model approach. Parasit Vectors 11:1–13

Halvorsen O, Skorping A (1982) The infuence of temperature on growth and development of the nematode Elaphostrongylus rangiferi in the gastropods Arianta arbustorum and Euconulus fulvus. Oikos 38:285–290

Handeland K (1994) Experimental studies of *Elaphostrongylus rangiferi* in reindeer (*Rangifer tarandus tarandus*): life cycle, pathogenesis and pathology. J Vet Med B 41:351–365

Handeland K (2012) *Fusobacterium necrophorum* infection. In: Gavier-Widén D, Duff JP, Meredith A (eds) Infectious diseases of wild mammals and birds in Europe. Wiley-Blackwell, Chichester, pp 428–430

Handeland K, Norberg HS (1992) Lethal cerebrospinal elaphostrongylosis in a reindeer calf. Zentralbl Veterinarmed B 39:668–671

Handeland K, Skorping A, Stuen S et al (1994) Experimental studies of *Elaphostrongylus rangiferi* in reindeer (*Rangifer tarandus tarandus*): clinical observations. Rangifer 14:83–87

Handeland K, Boye M, Bergsjø B et al (2010) Digital necrobacillosis in Norwegian wild tundra reindeer (*Rangifer tarandus tarandus*). J Comp Pathol 143(1):29–38

Hansen LI (2008) Spesialisert reindrift eller kombinasjonsnæring? Reinholdet i Sør-Troms på 1600- og 1700-tallet. In: Andersen O (ed) Fra villreinjakt til reindrift/Gåddebivdos boatsojsujttuj. Tjálaráddjo/Skriftserie 1/2005, pp165–183

Hansen BB, Grøtan V, Aanes R et al (2013) Climate events synchronize the dynamics of a resident vertebrate community in the high Arctic. Science 339(6117):313–315

Helle TP, Jaakkola LM (2008) Transitions in herd management of semi-domesticated reindeer in Northern Finland. Annal Zool Fenn 54:81–101. https://doi.org/10.5735/086.045.021

Hobbs NT, Andren H, Persson J et al (2012) Native predators reduce harvest of reindeer by Sámi pastoralists. Ecol Appl 5:1640–1654

Hoffberg CF (1788) Cervus tarandus sub præsido D. D. Car. Linnæi propositus a carol. Frid. Hoffberg. Upsaliæ 1754. Octobr 23. (Caroli a Linne Amonenitates academicae seu dissertationes variae physicae. 1788/4.57/144)

Horne H (1898) Renens klovsyge [Reindeer foot disease]. Norsk Vet Tidskr 10:97–110. (In Norwegian)

Horne H (1915) Rensyke (Hæmorrhagisk septikæmi hos ren). Nor Vet Tidsskr 27:41–47, 73–84, 105–120 (In Norwegian)

Horstkotte T, Lépy É, Risvoll C (2020) Supplementary feeding of semi-domesticated reindeer in Fennoscandia: challenges and opportunities. Report, Umeå University

Hovelsrud G, Risvoll C, Riseth JÅ et al (2020) Reindeer herding and coastal pastures: climate change and multiple stressors. In: Nord D (ed) Nordic perspectives on the responsible development of the arctic: pathways for action. Springer, Heidelberg, pp 113–134

Hueffer K, Drown D, Romanovsky V et al (2020) Factors contributing to anthrax outbreaks in the circumpolar North. Ecohealth 17:174–180

Hultblad F (1936) Flyttlapparna i Gällivare socken. Geographica, Uppsala

Ingold T (1980) Hunters, Pastoralists and Ranchers. Reindeer economies and their transformations. Cambridge University Press, Cambridge

Ippen R, Henne D (1985) A contribution on the diseases of Cervidae. Akademie Verlag, Berlin, pp 7–16

Istomin KV, Dwyer MJ (2010) Dynamic mutual adaptation: human-animal interaction in reindeer herding pastoralism. Human Ecol 38:613–623

Jacobsen E, Skjenneberg S (1979) Forsøk med ulike fôrblandinger til rein: fôrverdi av reinfôr (RF-71). Norwegian University of Life Sciences, Ås, Norway, Vol 34/58 (In Norwegian)

Jaenson TG, Jaenson DG, Eisen L et al (2012) Changes in the geographical distribution and abundance of the tick *Ixodes ricinus* during the past 30 years in Sweden. Parasit Vectors 5(1):8

Johansson A (2006) Djurägarbehandlingar inom renskötseln. Exam thesis, Swedish University of Agricultural Sciences, Uppsala, Sweden, ISSN 1652-8697

Jore S, Vanwambeke SO, Viljugrein H et al (2014) Climate and environmental change drives *Ixodes ricinus* geographical expansion at the northern range margin. Parasit Vectors 7:11

Josefsen TD, Sundset MA (2014) Fôring og fôringsbetingede sjukdommer hos rein. Norsk Veterinærtidsskrift 2:10

Josefsen TD, Sørensen KK, Mørk T et al (2007) Fatal inanition in reindeer (*Rangifer tarandus tarandus*): pathological findings in completely emaciated carcasses. Acta Vet Scand 49:27

Josefsen TD, Mørk T, Nymo IH (2019) Bacterial infections and diseases. In: Tryland M, Kutz SJ (eds) Reindeer and Caribou. Health and disease. CRC Press, Boca Raton, FL, pp 237–271

Kivinen S, Moen J, Berg A et al (2010) Effects of modern forest management on winter grazing resources for reindeer in Sweden. AMBIO 39:269–278

Klein J, Tryland M (2005) Characterisation of parapoviruses isolated from Norwegian semi-domesticated reindeer (*Rangifer tarandus tarandus*). Virol J 2:79. https://doi.org/10.1186/1743-422X-2-79

Kløcker-Larsen R, Raitio K, Stinnerbom M et al (2017) Sami-state collaboration in the governance of cumulative effects assessment: a critical action research approach. Environ Impact Assess Rev 64:67–76

Kofinas G, Clark D, Hovelsrud GK (2013) Adaptive and transformative capacity. In: Arctic interim resilience report 2013. Stockholm Environment Institute and the Stockholm Resilience Centre, Stockholm, pp 73–93

Krupnik I (1993) Arctic adaptations: native whalers and reindeer herders of northern Eurasia. University Press of New England for Dartmouth College, Hanover

Kummeneje K (1974) Encephalomyelitis and neuritis in acute cerebrospinal nematodiasis in reindeer calves. Nord Vet Med 26:456–458

Kummeneje K (1976) Pasteurellosis in reindeer in Northern Norway: a contribution to its epidemiology. Acta Vet Scand 17:488–494

Kumpala J (1999) Developing an ecologically and economically more stable semi-domestic reindeer management - a Finnish point of view. In: Haugerud R (ed) Proceedings of the Tenth Nordic Conference on Reindeer Research. Kautokeino, Norway, 13–15 March, 1998. Rangifer Report Vol 3, Tromsø, pp 111–120

Kumpula J, Pekkarinen AJ, Tahvonen O et al (2017) Petoeläinten vaikutukset porotalouden tuottavuuteen tuloihin ja taloudelliseen kestävyyteen (Impacts of large carnivores on the productivity and economic sustainability of reindeer herding). – Luonnonvara- ja biotalouden tutkimus 12/2017, in Finnish

Laaksonen S, Kuusela J, Nikander S et al (2007) Parasitic peritonitis outbreak in reindeer (Rangifer tarandus tarandus) in Finland. Vet Rec 160:835–841

Laaksonen S, Solismaa M, Orro T et al (2008) *Setaria tundra* microfilariae in reindeer and other cervids in Finland. Parsitol Res 104(2):257–265

Laaksonen S, Pusenius J, Kumpula J et al (2010) Climate change promotes the emergence of serious disease outbreaks of filarioid nematodes. EcoHealth 7:7–13

Laaksonen S, Oksanen A, Kutz S et al (2017a) Filarioid nematodes, threat to arctic food safety and security. In: Game meat hygiene: food safety and security, vol 2. CABI International, Wallingford, pp 213–223

Laaksonen S, Jokelainen P, Pusenius J et al (2017b) Is transport distance correlated with animal welfare and carcass quality of reindeer (Rangifer tarandus tarandus)? Acta Vet Scand 59:1–8

Landbruksdirektoratet (2020) Ressursregnskap for reindriftsnæringen: for reindriftsåret 1 April 2019–31 Mars 2020. Report 43/2020. Landbruksdirektoratet, Oslo, Norway. In Norwegian

Lifjell T (2019) Miniseminar om kriseberedskap i reindriften. Unpublished pptx-presentation. NRL (Norske Reindriftssamers Landsforbund), Værnes and Karasjok

Löf A (2014) Challenging adaptability. Analysing the governance of reindeer husbandry in Sweden. PhD dissertation. Umeå University, Umeå

Magnusson H (1913) Om pasteurellos hos ren jämte ett bidrag till kännedomen om pasteurellans biologiska egenskaper. [Pasteurellosis in reindeer, with a contributionon the biological characteristics of Pasteurella]. Skand Vet Tidsskr 4(127-34):159–184

Massei G, Kindberg J, Licoppe A et al (2015) Wild boar populations up, numbers of hunters down? A review of trends and implications for Europe. Pest Manag Sci 71:492–500. https://doi.org/10.1002/ps.3965

McKellar QA, Gokbulut C (2012) Pharmacokinetic features of the antiparasitic macrocyclic lactones. Curr Pharm Biotechnol 13:888–911

Medlock JM, Hansford KM, Bormane A et al (2013) Driving forces for changes in geographical distribution of Ixodes ricinus ticks in Europe. Parasit Vectors 6(1):1

Mithell GB, Sigurdson CJ, O'Rourke et al (2012) Experimental oral transmission of chronic wasting disease to reindeer (*Rangifer tarandus tarandus*). PLoS One 7(6):e39055

Mitskevich VY (1958) Lifecycle of *Elaphostrongylus rangiferi* Miz. 1958 (in Russian). In: Boev SN (ed) Parasites of farm animals in Kazakhstan, vol 1964. Akademii Nauk SSSR Alma-Ata, Izdatel, pp 49–60

Moen J (2008) Climate change: effects on the ecological basis for reindeer husbandry in Sweden. AMBIO 37:304–311

Mulk IM (1988) Sirkas – ett fjällsamiskt fångstsamhälle i förandring 500–1500 e.Kr. Bebyggelsehistorisk tidskrift 14:61–82

Mysterud A, Rolandsen CM (2018) A reindeer cull to prevent chronic wasting disease in Europe. Nat Ecol Evol 2(9):1343–1345

Mysterud A, Jore S, Østerås O et al (2017) Emergence of tick-borne diseases at northern latitudes in Europe: a comparative approach. Sci Rep 7:16316. https://doi.org/10.1038/s41598-017-15742-6

Naranjo V, Gortazar C, Vicente J et al (2008) Evidence of the role of European wild boar as a reservoir of Mycobacterium tuberculosis complex. Vet Microbiol 127:1–9

Nieminen MKU (2018) Reindeer and man: from hunting to domestication. In: Tryland M, Kutz SJ (eds) Reindeer and Caribou. Health and disease. CRC Press, Boca Raton, FL, pp 13–23

Nikolaevskii LD (1961) Diseases in reindeer. In: Zhigunov PS (ed) Reindeer husbandry, 2nd edn. Springer, Moscow, pp 230–293. translated from Russian by Israel Program for Scientific Translations, Jerusalem, 1968

Nordkvist M (1960) Renens sjukdomar: kort oversikt. Stockholm 1960. (Lappväsendet – Renforskningen. Småskrift, 4) (In Swedish)

Nordkvist M (1973) Munnvårtsjuka – en ny rensjukdom? Rennäringsnytt 8–9:6–8. (In Swedish)

Nordkvist M, Karlsson KA (1962) Epizootisk førløpande infektion med *Pasteurella multocida* hos ren. Nordic Vet Med 14:1–15. (In Swedish)

Oetzel GR (2017) Diagnosis and management of subacute ruminal acidosis in dairy herds. Vet Clin North Am Food Anim Pract 33:463–480

OIE (2013) Haemorrhagic septicaemia. https://www.oie.int/fileadmin/Home/eng/Animal_Health_ in_the_World/docs/pdf/Disease_cards/HAEMORRHAGIC_SEPTICEMIA.pdf. Accessed 1 Mar 2021

Omazic A, Bylund H, Boqvist S et al (2019) Identifying climate-sensitive infectious diseases in animals and humans in Northern regions. Acta Vet Scand 61:53

Paine R (1972) The herd management of lapp reindeer pastoralists. J Asian Afr Stud 7:76–87

Päiviö NJ (2006) Sirkas sameby- om konsekvenser av beitekatastrofer. Ottar 1(2006):10–17

Päiviö NJ (2007) Gränsöverskridande renskötsel. Historisk belysning. In: NOU 2007:14. Lobo Media AS, Oslo, pp 499–544

Palmer MV, Stoffregen WC, Rogers et al (2004) West Nile Virus Infection in Reindeer (Rangifer Tarandus). J Vet Diagn Invest 16:219–222. https://doi.org/10.1177/104063870401600307

Palmer MV, Waters WR, Thacker TC et al (2006) Experimentally induced infection of reindeer (Rangifer tarandus) with Mycobacterium bovis. J Vete Diagn Invest 18:52–60

Pedersen S (2001) Fra bruk av naturgodene etter samiske sedvaner til forbud mot jordsalg til ikke-norsktalende. In: NOU 2001:34. Samiske sedvaner og rettsoppfatninger, Oslo, pp 289–381 (In Norwegian)

Pirisinu L, Tran L, Chiappini B (2018) Novel type of chronic wasting disease detected in moose (Alces alces), Norway. Emerg Infect Dis 24:2210

Quinn PJ, Markey BK, Leonard FC et al (2011) Bacillus species. Vet Microbiol Microbial dis. https://doi.org/10.1016/j.vetmic.2012.06.004

Qvigstad J (1941) Den tamme rens sykdommer [Diseases of domestic reindeer]. Tromsø Museums Årshefter. Naturhistorisk avdeling 59(1):1–56. (In Norwegian)

Radostits OM, Gay CC, Blood DC et al (2000) Diseases of the alimentary tract - II. In: Radostits OM, Gay CC, Blood DC, Hinchcliff KW (eds) Veterinary medicine. A textbook of the diseases of cattle, sheep, pigs, goats and horses. Elsevier, Edinburgh, pp 259–346

Rehbinder C, Nikander S (1999) Ren och rensjukdomar. Studentlitteratur, Lund, Sweden. In Swedish

Riseth JÅ (2000) Sámi Reindeer management under technological change 1960-1990: implications for common-pool resource use under various natural and institutional conditions. A comparative analysis of regional development paths in West Finnmark, North Trøndelag, and South Trøndelag/Hedmark, Norway. Dr. Scientarium Theses 2000:1. Dissertation. Department of Economics and Social Sciences, Agricultural University of Norway, Ås

Riseth JÅ (2006) Sámi reindeer herd managers: why do they stay in a low-profit business? Br Food J 108:541–559. https://doi.org/10.1108/00070700610676361

Riseth JÅ, Johansen B (2019) Inngrepsanalyse for reindrifta i Troms. Rapport 23/2018. Norut, Tromsø

Riseth JÅ, Oksanen LK (2007) Ressursøkonomiske og økologiske perspektiver på grenseoverskridende reindrift, vol 14. Senter for samiske studier, Tromsø, Norway, pp 143–163

Riseth JÅ, Tømmervik H (2017) Klimautfordringer og arealforvaltning for reindrifta i Norge. Kunnskapsstatus og forslag til tiltak – eksempler fra Troms. Rapport 6/2017. Norut, Tromsø

Riseth JÅ, Tømmervik H, Bjerke JW (2016) 175 years of adaptation: North Scandinavian Sámi reindeer herding between government policies and winter climate variability (1835-2010). J Forest Econ 24:186–204. https://doi.org/10.1016/j.jfe.2016.05.002

Riseth JÅ, Tømmervik H, Forbes BC (2018) Sustainable and resilient reindeer herding. In: Tryland M, Kutz SJ (eds) Reindeer and Caribou. Health and disease. CRC Press, Boca Raton, FL, pp 23–43

Riseth JÅ, Tømmervik H, Tryland M (2020) Spreading or gathering? Can traditional knowledge be a resource to tackle reindeer diseases associated with climate change? Int J Environ Res Public Health 17(16):6002. https://doi.org/10.3390/ijerph17166002

Risvoll C (2015) Adaptive capacity within pastoral communities in the face of environmental and societal change. PhD Thesis in Sociology, faculty of Social Sciences, University of Nordland, Bodø

Rockborn G, Rehbinder C, Klingeborn B et al (1990) The demonstration of a herpesvirus, related to bovine herpesvirus 1, in reindeer with ulcerative and necrotizing lesions of the upper alimentary tract and nose. Rangifer. Special issue No. 3, p 373

Røed K, Côté S, Yannic G (2018) Rangifer tarandus: classification and genetic variation. In: Tryland M, Kutz SJ (eds) Reindeer and Caribou. Health and disease. CRC Press, Boca Raton, FL, pp 3–13

Rossi S, Viarouge C, Faure E et al (2015) Exposure of wildlife to the Schmallenberg virus in France (2011-2014): higher, faster, stronger (than Bluetongue)! Transboundary Emerg Dis 64(2): 354–363. https://doi.org/10.1111/tbed.12371

Rosvold J, Andersen R (2008) Wild boar in Norway – is climate a limiting factor? – NTNU Vitenskapsmuseet Rapp. Zool Ser 1:1–23

Ruiz-Fons F, Segalés J, Gortázar C (2008) A review of viral diseases of the European wild boar: effects of population dynamics and reservoir role. Vet J 176:158–169

Ruong I (1937) Fjällapparna i Jukkasjärvi Socken. Geographica, Skrifter från Uppsala Universitets Geografiska institution, Uppsala

Ruong I (1954) Om renmjölkningen på sydlapsk område. In: Strömbäck D (ed) Studier tilägna Bjørn Collinder den 22. Juli 1954. Scandinavica et Fenno-Ugrica. Almqvist & Wicksell, Stockholm, pp 277–301

Ruong I (1982) Samerna i Historien och Nutiden. Aldus Akademi, Stockholm

Sánchez Romano J, Mørk T, Laaksonen S et al (2018) Infectious keratoconjunctivitis in semi-domesticated Eurasian tundra reindeer (Rangifer tarandus tarandus): microbiological study of clinically affected and unaffected animals with special reference to cervid herpesvirus 2. BMC Vet Res 14(1):15. https://doi.org/10.1186/s12917-018-1338-y

Sánchez Romano J, Leijon M, Hagström Å et al (2019a) Chlamydia pecorum associated with an outbreak of infectious keratoconjunctivitis in semi-domesticated reindeer in Sweden. Front Vet Med 6:14. https://doi.org/10.3389/fvets.2019.00014

Sánchez Romano J, Grund L, Obiegala A et al (2019b) A multi-pathogen screening of captive reindeer (Rangifer tarandus) in Germany based on serological and molecular assays. Front Vet Sci 6:461. https://doi.org/10.3389/fvets.2019.00461

Sánchez Romano J, Sørensen KK, Larsen AK (2020) Ocular histopathological findings in semi-domesticated Eurasian Tundra Reindeer (Rangifer tarandus tarandus) with infectious kerato-conjunctivitis after experimental inoculation with Cervid Herpesvirus 2. Viruses 12:1007. https://doi.org/10.3390/v12091007

Skarin A, Nellemann C, Rönnegård L et al (2015) Wind farm construction impacts reindeer migration and movement corridors. Landsc Ecol 30:1527–1540

Skjenneberg S (1957) Sykdom på reinkalver i Porsangerdistriktet våren 1956. Medlemsblad Norsk Veterinærforening 9:153–159

Skjenneberg S, Slagsvold L (1968) Reindriften og dens naturgrunnlag. Universitetsforlaget, Oslo (In Norwegian. English translation: CM Anderson and JR Luick eds. 1979. Reindeer husbandry and its ecological principles. U.S. Department of the Interior Bureau of Indian Affairs, Juneau, Alaska)

Snyder E, Credille B (2017) Diagnosis and treatment of clinical rumen acidosis. Vet Clin North Am Food Anim Pract 33:451–461

Stépanoff C (2017) The rise of reindeer pastoralism in Northern Eurasia: human and animal motivations entangled. J R Anthropol Inst 23:376–397

Tan ZL, Nagaraja TG, Chengappa MM (1994) Selective enumeration of Fusobacterium necrophorum from the bovine rumen. Appl Environ Microbiol 60:1387–1389

Tikkanen MK, McInnes CJ, Mercer AA (2004) Recent isolates of parapoxvirus of Finnish reindeer (Rangifer tarandus tarandus) are closely related to bovine pseudocowpox virus. J Gen Virol 85: 1413–1418

Torres RT, Fernandes J, Carvalho J et al (2020) Wild boar as a reservoir of antimicrobial resistance. Sci Total Environ 717:135001

Tryland M (2012) Are we facing new health challenges and diseases in reindeer in Fennoscandia? Rangifer 32:35–47

Tryland M (2019) Prions and chronic wasting disease (CWD). In: Tryland M, Kutz SJ (eds) Reindeer and Caribou – health and disease. CRC Press, Boca Raton, FL, pp 305–314

Tryland M, Josefsen TD, Oksanen A (2001) Parapoxvirus infection in Norwegian semi-domesticated reindeer (*Rangifer tarandus tarandus*). Vet Rec 149(13):394–395

Tryland M, Das Neves CG, Sunde M (2009) Cervid herpesvirus 2, the primary agent in an outbreak of infectious keratoconjunctivitis in semi-domesticated reindeer. J Clin Microbiol 47:3707–3713

Tryland M, Klein J, Berger T (2013) Experimental parapoxvirus infection (contagious ecthyma) in semi-domesticated reindeer (*Rangifer tarandus tarandus*). Vet Microbiol. 162(2–4):499–506. https://doi.org/10.1016/j.vetmic.2012.10.039

Tryland M, Sánchez Romano J, Marcin N et al (2017) Cervid herpesvirus 2 and not *Moraxella bovoculi* caused keratoconjunctivitis in experimentally inoculated semi-domesticated Eurasian tundra reindeer. Acta Vet Scand 59:23. https://doi.org/10.1186/s13028-017-0291-2

Tryland M, Ravolainen V, Pedersen ÅØ (2019a) Climate change. Potential impacts on pasture resources, health and diseases of reindeer and caribou. In: Tryland M, Kutz SJ (eds) Reindeer and Caribou – health and disease. CRC Press, Boca Raton, FL, pp 493–514

Tryland M, Nymo IH, Sánchez Romano J et al (2019b) Infectious disease outbreak associated with supplementary feeding of semi-domesticated reindeer. Front Vet Sci 6:126. https://doi.org/10.3389/fvets.2019.00126

Tryland M, Das Neves CG, Klein J, Mørk T, Hautaniemi M, Wensman J et al (2019c) Viral infections and diseases. In: Tryland M, Kutz SJ (eds) Reindeer and Caribou – health and disease. CRC Press, Boca Raton, pp 273–303

Turi J 2011 [1910] [Original title: Muitalis sámid birra] Min bok om samene (My book about the Sámi). ČálliidLágádus, Karasjok, Norway

Turunen M, Vuojala-Magga T, Giguère N (2014) Past and present winter feeding of reindeer in Finland: herders' adaptive learning of feeding practices. Arctic 67:173–188. https:/www.jstor.org/stable/24363698

Tveraa T, Stien A, Brøseth H et al (2014) The role of predation and food limitation on claims for compensation, reindeer demography and population dynamics. J Appl Ecol 51:1264–1272

Ulvevadet B (2012) The governance of Sami Reindeer Husbandry in Norway: institutional challenges of co-management. PhD dissertation. University of Tromsø, Tromsø

Underwood WJ, Blauwiekel R, Dealno ML et al (2015) Biology and diseases of ruminants (sheep, goats, and cattle). In: Fox JG, Anderson LC, Otto GM, Pritchett-Corning KR, Whary MT (eds) Laboratory animal medicine, 3rd edn. Elsevier, Academic Press, Amsterdam, pp 623–694

Våge J, Hopp P, Vikøren T et al (2020) The surveillance programme for Chronic Wasting Disease (CWD) in free-ranging and captive cervids in Norway 2019. Norwegian Veterinary Institute CWD Annual Report

van Oort BE, Hovelsrud GK, Risvoll C et al (2020) A mini-review of ixodes ticks climate sensitive infection dispersion risk in the Nordic region. Int J Environ Res Public Health 17(15):5387

Vikøren T, Våge J, Madslien KI et al (2019) First detection of chronic wasting disease in a wild red deer (Cervus elaphus) in Europe. J Wildl Dis 55:970–972

Viljugrein H, Hopp P, Benestad SL et al (2019) A method that accounts for differential detectability in mixed samples of long-term infections with applications to the case of chronic wasting disease in cervids. Methods Ecol Evol 1:134–145

Vineer HR, Mørk T, Williams DJ (2021) Modeling thermal suitability for reindeer (Rangifer tarandus ssp.) branworm (*Elaphostrongylus rangiferi*) transmission in Fennoscandia. Front Vet Sci 7:1170

Ytrehus B, Asmyhr MG, Hansen H et al (2021) Handlingsrommet etter påvisning av skrantesyke (Chronic Wasting Disease) på Hardangervidda – grunnlag for fremtidige forvaltningsstrategier. Vitenskapelig uttalelse fra Vitenskapskomiteen for Mat og Miljø. VKM Report 2021:01, ISBN: 978-82-8259-355-7. ISSN: 2535-4019. Vitenskapskomiteen for Mat og Miljø (VKM), Oslo, Norway

von Linné C (1732) Iter Laponicum Kungl. Skytteanska Samfundet

Vorren Ø (1998) Om intensiv reindrift på Helgeland. In: Åarjel-samieh /Samer i sør–Årbok, vol 6. Saemien Sijte, Snåsa, pp 63–72

Wilson BA, Ho M (2013). *Pasteurella multocida*: from zoonosis to cellular microbiology. Clin Microbiol Rev 26:631–655

Printed in the United States
by Baker & Taylor Publisher Services